India, Japan and Beyond

Rajib Shaw · Srabani Roy Choudhury
Editors

India, Japan and Beyond

Human Security, Environment, Development, Innovation and Resilience

Springer

Editors
Rajib Shaw
Graduate School of Media and Governance
Keio University
Fujisawa, Japan

Srabani Roy Choudhury
School of International Studies
Centre for East Asian Studies
Jawaharlal Nehru University
New Delhi, India

ISBN 978-981-97-3284-5 ISBN 978-981-97-3282-1 (eBook)
https://doi.org/10.1007/978-981-97-3282-1

This Springer imprint is published by the registered company Springer Nature Singapore Pte Ltd.
The registered company address is: 152 Beach Road, #21-01/04 Gateway East, Singapore 189721, Singapore

Preface

India and Japan's civilizational linkages have been the bedrock on which this relationship has been built. At the turn of the century, the changing world order reoriented this relationship from distantly friendly to a strategic partnership. The partnership works at various levels, including diplomacy, economics, trade, industry, science, technology, innovation, higher education, cultural issues, and people-to-people connectivity. All these have bolstered and invigorated this relationship.

India–Japan relations have a broad spectrum of cooperation. Diversity is often lost when it is discussed under umbrella themes. This book attempts to provide a snapshot of this diverse relationship and is divided into three significant parts. After the introductory chapter, which covers development partnership, the first part focuses on development and environment, particularly marine resources and its implications for QUAD partnership, human security issues in northeast India focusing on connectivity, disaster risk reduction, and critical rare earth elements. All these issues are strongly related to India–Japan bilateral collaboration and are part of the first pillar of environment and development.

The second part focuses on industry and innovation. Here, there are seven chapters on diverse issues of industry and innovation. The part starts with the MSMEs (micro, small, and medium enterprises), which emphasize on the immense potential of collaboration between the two countries. The following two chapters draw attention to the collaboration in energy sectors, the first on generic aspects of energy collaboration and the next on Hydrogen energy collaboration. Vaccine diplomacy and innovation are emerging areas of significance. The following three chapters focus on IT (information technology), social innovation, and agriculture technology.

The third part addresses yet another important pillar of this relationship, culture and people. Starting with a chapter on Yoga, the section has chapters on edu-tech, focusing on education and XR (extended reality, like virtual reality and augmented reality) as future education and learning tools and collaboration between the two countries. An analysis of youth voices is also provided for future potential collaboration. Finally, the potential of higher education, which can pave the way to people-to-people solid connectivity, is discussed.

As mentioned before, there is immense potential for India–Japan collaboration. The current book only provides a limited sample of the diverse issues under three parts. However, we believe that it would enhance the future scope of research and development in this field. We will be glad if the readers find the book useful.

Fujisawa, Japan — Rajib Shaw
New Delhi, India — Srabani Roy Choudhury

About This Book

The relationship between India and Japan had advanced at a faster speed than ever before. The Significant development of India–Japan strategic partnership, convergence of bilateral strategies, infrastructure development, human resource needs, broader economic relations and cultural dimensions signify that bilateral relations have entered into a "new era in Japan–India relations." With the global geo-political context, the relation has become more tighter, closer and strategic. While diplomatic relation is getting stronger, it is also important to strengthen the human security dimensions of the two countries, which include environment, development, disaster risk reduction, climate change issues as well as innovation and resilience building. Keeping this in mind, this book covers the broader aspects of human security dimensions of India–Japan collaboration. The book includes in-depth reviews as well as new data, based on case studies from India and Japan, involving multi and trans-disciplinary research. Finally, the book suggests specific policy and action measure to enhance human security through bilateral cooperation of India and Japan, which has a global impact.

Contents

Editors and Contributors

About the Editors

Rajib Shaw is a Professor in the Graduate School of Media and Governance of Keio University, Japan. He is the co-founder of a Delhi (India)-based social entrepreneur startup Resilience Innovation Knowledge Academy (RIKA) and RIKA Institute. He is the Co-chair of the United Nations Asia Pacific Science Technology Advisory Group (AP-STAG) and was CLA for IPCC's 6th Assessment Report. Professor Shaw is the recipient of the prestigious "Pravasi Bharatiya Samman Award (PBSA)" from the President of India in 2021 for his contribution to the education sector. He also received United Nations Sasakawa Award in 2022 for his lifetime contributions to the field of disaster risk reduction. He published 71 books and over 450 research papers. More about his work can be found in: www.rajibshaw.org.

Dr. Srabani Roy Choudhury is a Professor of Japanese Studies Centre for East Asian Studies, School of International Studies, Jawaharlal Nehru University, New Delhi. Her association with Japan began with the Japan Foundation Fellowship 1996–1997. She has been on visiting scholar programme at Keizai Koho Centre (2011), Ministry of Economics and Industry, Japan, Policy Research Institute, Ministry of Finance (2011), Kobe University (2014), Nagoya University (2019) and Keio University (2022). She has been part of many Indian delegations to Japan. Her recent publications are centred on economic diplomacy between Japan and India. She has to her credit four edited volumes *Japan-SAARC Partnership: A Way Ahead* (Pentagon, 2014); *India–Japan Relations @70: Building Beyond Bilateral* (Knowledge World, 2022). *The Indo-Pacific Theatre: Strategic Visions and Frameworks* (Routledge, 2023); *Japan and Its Partners in the Indo-Pacific: Engagements and Alignment* (Routledge, 2023) *India–Japan Partnership: Abe the Game Changer* (2024).

Contributors

Ariyaningsih Graduate School of Media and Governance, Keio University, Fujisawa, Japan

Titli Basu School of International Studies, Jawaharlal Nehru University, New Delhi, India

Ranit Chatterjee India Japan Laboratory, Keio University, Fujisawa, Japan

Pradeep Chauhan National Maritime Foundation, New Delhi, India

Rohit Chaurasia HCL Computers Ltd., Hyderabad, India

Mahati Chittem Liberal Arts Department, Indian Institute of Technology Hyderabad, Hyderabad, India

Abhishek Choudhury Graduate School of Media and Governance, Keio University, Fujisawa, Japan

Srabani Roy Choudhury School of International Studies, Jawaharlal Nehru University, New Delhi, India

Yuta Hikichi Embassy of India, New Delhi, India

Yoshiro Kaku NEDO, New Delhi, India

Kotaro Kataoka Department of Computer Science and Engineering, Indian Institute of Technology Hyderabad, Sangareddy, Telangana, India

Yuka Kataoka Graduate School of Media and Governance, Keio University, Fujisawa, Japan

Akiko Kato Graduate School of Media and Governance, Keio University, Fujisawa, Japan

Tomo Kawane Graduate School of Media and Governance, Keio University, Fujisawa, Kanagawa, Japan

Shashank Khandwe School of International Studies, Jawaharlal Nehru University, New Delhi, India

Devesh Lahiri National Maritime Foundation, New Delhi, India

Nabeel Mancheri The Global Rare Earth Industry Association (REIA), Brussels, Belgium

Neha Mishra Society of Indian Defence Manufacturers, Confederation of Indian Industry, New Delhi, India

Arunabh Mitra Graduate School of Media and Governance, Keio University, Fujisawa, Japan

Shraddha Namjoshi Liberal Arts Department, Indian Institute of Technology Hyderabad, Hyderabad, India

Kazuaki Ohhashi Yoga Organization of Japan, Tokyo, Japan

Takuma Otaki Graduate School of Media and Governance, Keio University, Fujisawa, Japan

Chandrali Sarkar Jawaharlal Nehru University, New Delhi, India

Megha Sharma School of International Studies, Jawaharlal Nehru University, New Delhi, India

Rajib Shaw Graduate School of Media and Governance, Keio University, Fujisawa, Kanagawa, Japan

Divya Suresh Graduate School of Media and Governance, Keio University, Fujisawa, Japan

Sachiko Takane C-Sagyoba Corporation, Tokyo, Japan

Yukio Takeyari India Japan Laboratory, Keio Research Institute at SFC, Keio University, Fujisawa, Japan

Nupur Tyagi SPHERE India, New Delhi, India

Chime Youdon National Maritime Foundation, New Delhi, India

India-Japan: Developmental Partnership for Bridging Divides

Titli Basu

Abstract While India-Japan Special Strategic and Global Partnership is positioning itself as a stabilizing force in the Indo-Pacific amid contested values, ideologies and rules- based order, this bilateral partnership has succeeded in incubating innovative ideas and translating them into strategic solutions in the post-Covid world. As India is increasingly framed in the narrative of G3 anchored on its demographic dividend, technological and economic prowess, Japan is playing a pivotal role in economic modernization of India. This chapter especially analyses the contribution of this strategic partnership on 3Ds or three verticals including (a) developmental aid (b) digitalization and (c) decarbonization not just at the bilateral level but also at the regional and global level.

Keywords India-Japan · FOIP · Developmental aid · Digitalization · Decarbonization

1 Introduction

In pursuit of a rules-based order, India-Japan "Special Strategic and Global Partnership" has positioned itself as a stabilizing force in the Free and Open Indo-Pacific (FOIP) amid fiercely contested norms and narratives, values and ideologies. Rise of China has engineered a power disequilibrium between Beijing and Washington at the global level, and Beijing and its Asian neighbours at the regional level. Amid a fragmented international system and power structures, Tokyo and Delhi alongside like-minded partners including the US, Australia and Europe especially France and the UK has succeeded in incubating innovative ideas and translating them into policy solutions in the Indo-Pacific. As Prime Minister Shinzo Abe's strategic rationale of

T. Basu (✉)
School of International Studies, Jawaharlal Nehru University, New Delhi, India
e-mail: jnu.basu@gmail.com

R. Shaw and S. R. Choudhury (eds.), *India, Japan and Beyond*,
https://doi.org/10.1007/978-981-97-3282-1_1

the 'Confluence of the two Seas' (as argued in the Indian Parliament in 2007)[1] influenced the conceptualization of a FOIP, and further operationalization of his vision of a 'Strategic Security Diamond'[2] into today's Quad, Japan has taken the lead in advancing universal value-based constructs as an important pillar of its grand strategy. India as the biggest democracy, aligned well with Tokyo's value-based diplomacy.

Adapting and aligning with the shifting geopolitics and geo-economics of the Indo-Pacific with increasing intensification of US-China strategic competition, Russian invasion of Ukraine, Israel-Hamas conflict in Gaza in addition to recovering from the seismic shocks of a global pandemic has presented monumental challenges to policy elites across the Indo-Pacific. Delhi and Tokyo have not been bystanders and instead demonstrated political resolve, both individually and bilaterally, in not only envisaging but delivering tangible policy solutions within the frame of the India-Japan Indo-Pacific Vision 2025. The strategic maturity and mutual trust in the India-Japan relations has laid the groundwork to elevate and extend this bilateral into various trilaterals including India-Japan-US, India-Japan-Australia, India-Japan-France and so on in addition to the Quad.

While there are diverging policy positions on a few critical issues- whether it is Tokyo and Delhi's respective strategic thinking on traditional security alliances or response to the Ukraine crisis or the geoeconomics of mega-free trade agreements including the Regional Comprehensive Economic Partnership (RCEP) or the Osaka Track on Data Free Flow with Trust (DFFT)- however, mutuality of strategic interests in the larger Indo-Pacific stratagem has outweighed these divergences between India and Japan. This is demonstrated in various frameworks beyond bilateralism including the India-Japan-Australia Resilience Supply Chain Initiatives (RSCI), Quad's Indo-Pacific Partnership for Maritime Domain Awareness (IPMDA) or Semiconductor Initiative or Quad Investors Network (QUIN), co-opting with India's Indo-Pacific Ocean's Initiative (IPOI) or the International Solar Alliance.

2 Strategic Rationale

While Japan's post-war treaty alliance with the US constitutes the core of Tokyo's grand strategy, China's ascendency, erosion of US regional primacy and complex challenges in the alliance management has led policy makers in Tokyo opt for internal and external balancing.[3] In this matrix, India is accorded space in Japan's National Security Strategies of 2013 and its recent revision in 2022. The India debate amongst

[1] Shinzo Abe, "Confluence of the Two Seas", Ministry of Foreign Affairs, August 22, 2007 https://www.mofa.go.jp/region/asia-paci/pmv0708/speech-2.html accessed on December 2, 2023.

[2] Shinzo Abe, "Asia's Democratic Security Diamond", Project Syndicate, December 27, 2012 https://www.project-syndicate.org/magazine/a-strategic-alliance-for-japan-and-india-by-shinzo-abe, accessed on December 2, 2023.

[3] Titli Basu, "Japan's Strategic Calculations: Constraints and Responses, IDSA Issue Brief, January 17, 2019 https://idsa.in/issuebrief/japan-strategic-calculations-titlibasu-18119, accessed on December 2, 2023.

Japanese strategic community and the academics often decodes the rising significance of Delhi amid shifting geopolitical dynamics. There is a school of thought which analyze India through the prism of rise of China. India is often weighed as an important leverage vis-à-vis Beijing. Moreover, there is a strand of scholarship which argued Japan's inability to comprehend the diplomatic weight of India without the "China factor".[4] Others have articulated that for Japan, "close cooperation with India is desirable against the backdrop of declining U.S. power in Asia vis-à-vis China".[5]

Beyond the institutionalisation of the annual Leaders' Summit, Foreign Minister's Strategic Dialogue, Defence Ministerial Dialogue, 2 + 2 Ministerial Meetings complemented by a series of bilateral and multilateral exercises across three services including the JIMEX, Malabar and La Perouse Exercises, Coast Guard Exercises and Dharma Guardian Exercises, India-Japan strategic partnership has invested in deepening the latitude of interoperability. Moreover, India has featured in their discussion concerning "Strategy on Defence Production and Technological Bases" alongside the US, the UK, France, Australia and Southeast Asia. As Tokyo enunciated "Three Principles on Transfer of Defence Equipment and Technology" in 2014, Delhi has explored the prospects of cooperation under its Make in India initiative and invited Japanese companies to participate in the defence corridors and ecosystem. However, interplay of complex variables like cost-competitiveness, technology transfer and domestic politics often presents colossal challenges for the policymakers. While technology collaboration in terms of robotics and Unmanned Ground Vehicle (UAVs) have paved the way for future defence technology cooperation and export of naval ship communication antennas to India is in the offing, taking the long view, stakeholders will have to invest greater energy in deepening understanding about each other's defence sector, and most importantly grasp the nuanced cultural distinctions. While dealing with Tokyo, it is important to understand that for Tokyo defence equipment and technology cooperation is beyond simple arms trade, and it features as a subset of Japan's vision and ambition to support 'Proactive Contribution to Peace'.[6]

While one school of thought argues that the variable of centuries old spiritual rapport and civilizational linkages facilitates India-Japan relation, international relations however dictate that one of the key reasons for Delhi to feature as a rational choice for Tokyo since the early 2000s is owing to Washington's shifting interest and attitude towards Delhi as a stabilising power in the Asia–Pacific. As Washington's net assessment started factoring in the vitality of India, US-Japan 2+2 Security Consultative Committee debates and discussions since the middle of 2000s has forcefully argued the case of cultivating 'India as a strong and enduring Asia–Pacific

---

[4] Toru Ito, "China Threat Theory in Indo-Japan Relations", in "India-Japan Relations in Emerging Asia" eds. Takenori Horimoto and Lalima Varma, Manohar 2013 pg-116.

[5] Takenori Horimoto, "Japan-India Rapprochement and Its Future Issues", JIIA Japan's Diplomacy Series, Japan Digital Library 2016 http://www2.jiia.or.jp/en/pdf/digital_library/japan_s_diplomacy/160411_Takenori_Horimoto.pdf.

[6] Titli Basu, "A Review of India-Japan Defence Technology Cooperation", IDSA Comment, August 31, 2018. https://idsa.in/idsacomments/india-japan-defence-technology-cooperation-tbasu-310818, accessed on December 2, 2023.

partner'[7] and supported India's increasing involvement in regional architectures. India, anchored at the heart of the Indian Ocean constitutes a key strategic and economic theatre, playing a pivotal role in hosting critical maritime chokepoints and highways transporting trillions of dollars' worth of trade and energy. The potency of Indian naval power and the geostrategic importance of the Andaman and Nicobar Joint Services Command in gaining an edge in the maritime great game is not lost on Japanese defence planner. India has a credible reputation in securing critical Sea Lanes of Communication (SLOC) as a "public good". Moreover, India has raised its trustworthiness as a reliable partner following the 1999 M/V Alondra Rainbow piracy incident despite hostile geopolitical pressures following the nuclear tests. As a traditional US security treaty ally, Japan adjusted its India policy since the early 2000s once the US Defence of Department recognised India as an trusted partner. Since 2009, Washington started analysing Delhi as a "net security provider" as articulated by Defence Secretary during the Shangri-La dialogue. As an extension of this strategic mutuality, US-Japan has urged advancing trilateral frameworks of engagement with India, which gradually matured from the director- generals level to the high-powered US-Japan-India ministerial trilateral in 2015.

As Delhi is increasingly framed in the Japanese narrative of G3[8] (US, China and India), as opposed to the Chinese narrative of G2, anchored on India's strategic maritime geography, demographic dividend, technological and economic prowess, Japan is performing a consequential role in the economic modernization of India as well as pushing UN's Sustainable Development Goals. Mapping India's economic prowess and emerging technological clout positions it at the heart of international politics. India aspires to be the third largest economy with a target of $5 trillion by 2027 and a $7 trillion by 2030. Culling out the trends in the IMF, World Bank and OECD projections, growth is estimated to remain robust, essentially driven by investment and services. IMF World Economic Outlook projects India to grow at 6.5 percent in 2024 and 2025 even as the global growth is likely to remain at 3.1 percent in 2024 and 3.2 percent in 2025. Meanwhile OECD projects India to grow at 6.2% in 2024.[9] Furthermore, World Bank's Global Economic Prospects has indicated that India is likely to maintain one of the fastest growth rate among global economies, edging up to 6.4 percent in 2024–25 and 6.5 percent in 2025–26.[10] Meanwhile,

---

[7] "Toward a Deeper and Broader US-Japan Alliance: Building on 50 Years of Partnership", Joint Statement of the Security Consultative Committee, by Secretary of State Clinton Secretary of Defense Gates Minister for Foreign Affairs Matsumoto Minister of Defense Itazawa, 21 June 2011, p. 5, available at http://www.mofa.go.jp/region/n-america/US/security/pdfs/ joint1106_01.pdf, accessed on 02 January 2024.

[8] Masafumi Ishii, "The lessons of the Ukraine war for the Indo-Pacific Region: a preview of, rather than a change in, what we will face in 10-15 years", JIIA, August 31, 2022, https://www.jiia.or.jp/en/ajiss_commentary/the-lessons-of-the-ukraine-war-for-the-indo-pacific-region.html accessed on December 2, 2023.

[9] "OECD Economic Outlook", Interim Report February 2024, OECD, https://www.oecd.org/economic-outlook/february-2024/ accessed on March 19, 2022.

[10] "Global Economic Prospects" World Bank, January 2024. https://openknowledge.worldbank.org/server/api/core/bitstreams/7fe97e0a-52c5-4655-9207-c176eb9fb66a/content accessed on January 21, 2024.

according to UN data, India has outpaced China in population with a huge youth demographic dividend constituting around 66 percent of the population below the age of 35. India's potential to consolidate its position as the global digital technology hub and become a skill-intensive digital talent nation is real. India has the third largest start-up ecosystem globally. It has emerged as one of the largest exporters of IT services globally. India has recently caught global attention with its successful lunar missions, quantum mission, integration of AI across various sectors and boasting one of the fastest 5G rollout programmes. In 2023, India featured 40th in the Global Innovation Index as well as the Global Competitiveness Index.

As India's rise in the global power structure is inevitable, Japan has emerged as one of the most trusted partners in the nation-building process. As India aspires to become a leading power in a multi-polar world, Japan emerged as an invaluable source for high-technology, investments and capital for infrastructure. This chapter critically analyses the intersection of policies and contribution of India-Japan Special Strategic and Global Partnership on 3Ds or three verticals including (a) developmental aid (b) digitalization and (c) decarbonization not just at the bilateral level but also at the regional and global geometries.

3 Decoding Developmental Aid

Driving India's journey through the 'Amrit Kaal' with the aim of realising an 'Atmanirbhar Bharat' is contingent on massive investments, infrastructure and developmental aid. On these three verticals Japan is proving to be an indispensable partner in the India story. India has not only emerged as the biggest beneficiary of Japan's ODA in the last decade, the aid is employed strategically, keeping it aligned with not only Japan's FOIP Vision and Development Cooperation Charter as revised in 2023 but also the 2030 Agenda for Sustainable Development. Sectoral analysis reflects that Japanese ODA towards India in 2021–2022 focussed mostly on economic infrastructure, with transportation sector accounting for 68.1 percent, water and sanitation accounting for 11.9 percent and healthcare accounting for 14.6 percent. Moreover, between 2011–2020, transport sector received around 66.2 percent of the ODA.[11] Interestingly, this aligns with the priorities of disbursements by ADB. In addition to compliance with OECD-DAC aid strategy, Japan has consistently argued the case of advancing quality growth and employing aid strategically not just in the revised Development Cooperation Charter but also in the 2022 National Security Strategy. Developmental aid, often constituting of loan, grant aid and technical assistance, serves as an important geo-economic instrument in leveraging strategic competition and expanding geopolitical influence. However, the politics of aid strategy is anchored as much on advancing trade and investment priorities of donors as on their

[11] Saito Mitsunori, "JICA's Policy In East India Development Opportunities And Challenges", JIIA, September 28, 2021. https://www.jica.go.jp/Resource/india/english/office/others/c8h0vm00009ylo4c-att/presentations_31.pdf accessed on January 21, 2024.

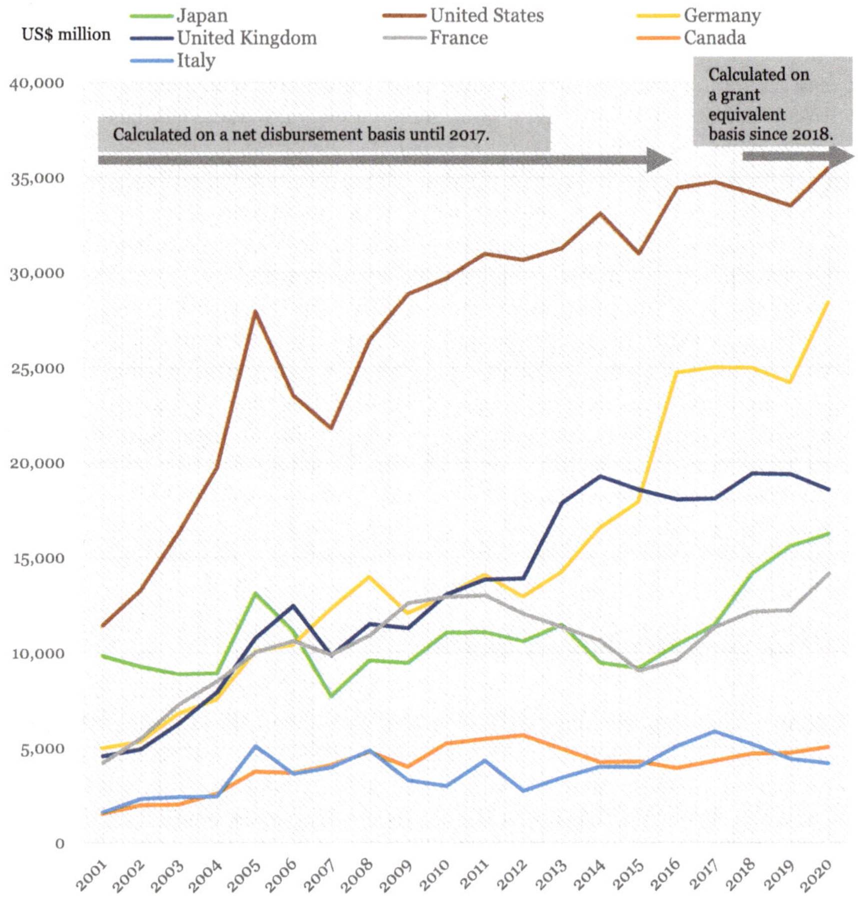

Fig. 1 ODA of DAC countries. White Paper on Development Cooperation 2022, Ministry of Foreign Affairs Japan

strategic ambitions. While Japanese ODA to Asia went southward over the decades, Asia still received 59.1 percent of the ODA in 2021 (Figs. 1 and 2).

Japan's development aid has performed a substantive role in advancing Prime Minister Modi's flagship programmes including the 'Make in India' initiative, 'Skill India', 'Digital India', 'Start-up India' and 'Smart Cities'. In this regard, Japan, being an international norm-setter in high quality infrastructure, has accorded priority to it in its development aid. Asian Development Bank (ADB) has projected that developing Asia requires US $26 trillion from 2016–2030.[12] This estimate has is

[12] "Meeting Asia's Infrastructure Needs", ADB, February 2017. https://www.adb.org/publications/asia-infrastructure-needs#:~:text=investment%20and%20development.-,Developing%20Asia%20will%20need%20to%20invest%20%241.7%20trillion%20per%20year,to%20%24250%20billion%20by%202020 accessed on March 19, 2022.

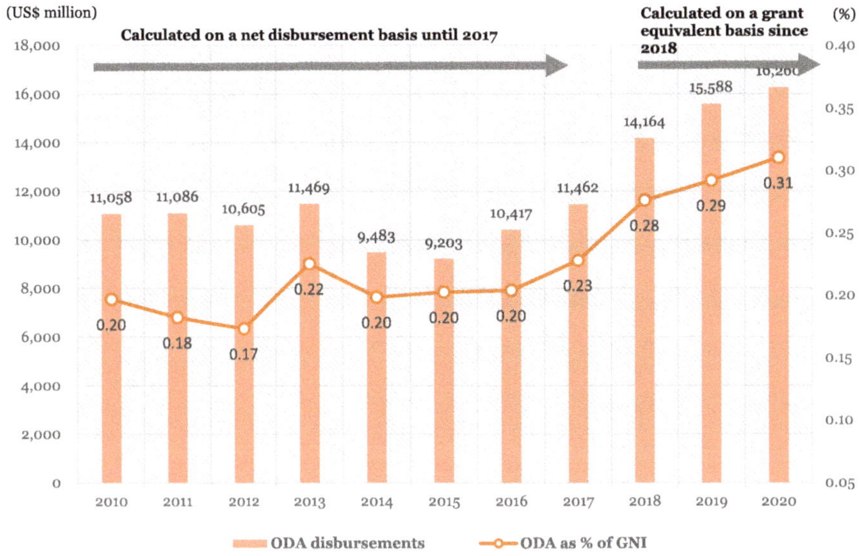

Fig. 2 Japanese ODA and percentage of GNI. *Source* White Paper on Development Cooperation 2022, Ministry of Foreign Affairs Japan

more than double since the 2009 assessment projecting US$ 750 billion annual investment.[13] India's NITI Aayog projects India to be a US$ 7 trillion economy by the next decade and in order to propel this economic story, there is imperative to build world class infrastructure that can foster manufacturing hubs and plug India into the global value chains (GVC). India's participation in the GVC, according to the Asian Infrastructure Investment Bank statistics, remains 7.4 percent below the average rate for emerging economies.[14] India's effective participation in regional and global value chains will be determined by improved infrastructure, connectivity, productivity and competitiveness (Figs. 3 and 4).

India's Economic Survey estimates that in infrastructure, we need to expend approximately US$ 1.4 trillion. Even though governments in their national budgets accord importance to infrastructure, for instance, India allocates 3.3 percent of GDP to infrastructure sector in 2024[15] especially focussing on transport and logistics, there remains a considerable gulf between the actual and the desired investment.

[13] "Meeting Asia's Infrastructure Needs", Asia Development Bank, February 2017 at https://www.adb.org/publications/asia-infrastructure-needs accessed on March 19, 2022.

[14] "Asian Infrastructure Finance 2021: Sustaining Global Value Chains", Asian Infrastructure Investment Bank, 2021, https://www.aiib.org/en/news-events/asian-infrastructure-finance/2021/introduction/index.html accessed on March 19, 2022.

[15] "Interim Union Budget 2024: What are expectations of players in infrastructure sector?" Economic Times, January 11, 2024. https://government.economictimes.indiatimes.com/news/smart-infra/interim-union-budget-2024-what-are-the-expectations-of-the-players-in-infrastructure-sector/106729939# accessed on March 19, 2022.

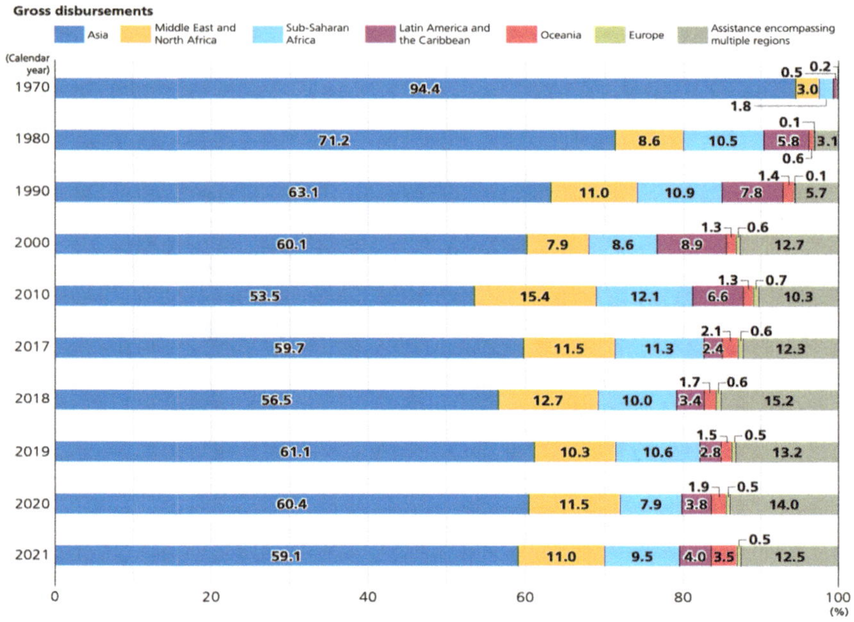

Fig. 3 Japan's ODA by region. *Source* OECD database, December 2022; White Paper on Development Cooperation 2022, Ministry of Foreign Affairs Japan

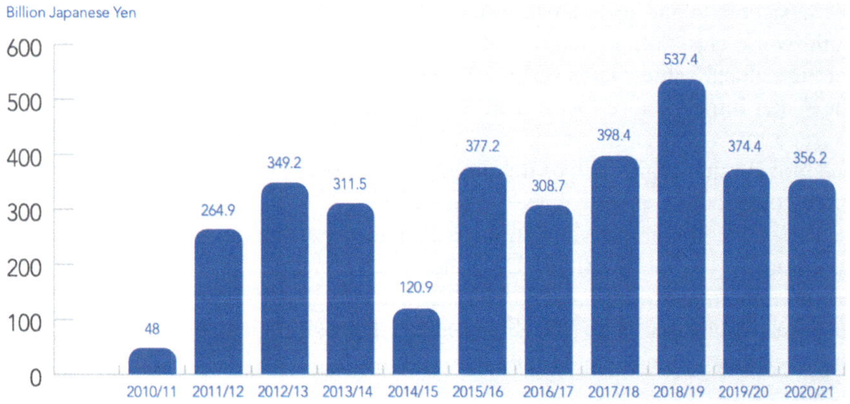

Fig. 4 ODA loan commitment towards India in the last decade. (2009/10–2021/22). *Source*: Japan International Cooperation Agency[16]

[16] "JICA Operations and Activities in India", JICA, 2020–2021. https://www.jica.go.jp/Resource/india/english/office/others/c8h0vm00004cesxi-att/brochure_24.pdf accessed on March 19, 2022.

Here, multilateral development banks and bilateral aid doners remain as an important source to bridge the infrastructure financing gap. India features as the front runner in borrowing from multilateral development banks. With regard to bilateral development aid, Tokyo is positioned as the leading donor.

As Ministry of foreign Affairs underscores, Tokyo's branding of quality infrastructure is anchored on five principles including: mobilising resources via public–private partnerships; synergising with development priorities of developing economies; advancing global standards of environmental principles; supporting quality infrastructure through low life-cycle cost, resilience and inclusiveness; and complimenting and supporting the local economy.[17] Japan's "Infrastructure System Overseas Promotion Strategy (2021–2025)" is also aligned with the policy objectives of FOIP, SDGs and advances carbon neutrality and digital transformation.

In case of Japan, exporting infrastructure enables the dual goals of generating new growth engines geared towards revitalising the domestic economy as well as reinforcing strategic relations with the Indo-Pacific stakeholders and advancing regional influence. Tokyo aligns it with their pitch for "Partnership for Quality Infrastructure" (PQI) whose latitude subsequently expanded to become the "Expanded Partnership for Quality Infrastructure" (EPQI). The larger aim is to complement the bilateral support through Japan International Cooperation Agency (JICA) and Japan Bank for International Cooperation (JBIC) with multilateral pledges through the ADB, as a smart policy for building high-quality infrastructure. The "ADB Strategy 2030" also underlines that "ADB will play an important role in supporting the global agenda of infrastructure development as a source of global growth … ADB will promote quality infrastructure investments that are green, sustainable, resilient, and inclusive."[18]

Role of Japanese developmental assistance in India's mega-infrastructure projects, for instance the Mumbai–Ahmedabad High-Speed Rail, Industrial Corridors between Delhi and Mumbai, and Chennai and Bengaluru, Dedicated Freight Corridor and urban mass rapid transport systems, especially Delhi Metro has proved to be a force multiplier in India's growth story. But what has made Japan different from other doners is its commitment to prioritise the strategic peripheries of India including the Northeast and Andaman and Nicobar Islands in its development assistance matrix. Additionally, Japan has co-opted in Delhi's regional and global infrastructure and connectivity endeavours, including the Indo-Pacific Oceans Initiative (IPOI) and the Coalition for Disaster-Resilient Infrastructure (CDRI).

In the Northeast, which is often considered as a strategic connect between India and Southeast Asian economies over Bay of Bengal, Tokyo's development assistance is productively contributing in reimagining connectivity by pursuing a "triple

---

[17] "'Quality Infrastructure Investment' Casebook", Ministry of Foreign Affairs, https://www.mofa.go.jp/files/000095681.pdf accessed on March 19, 2022.

[18] "ADB Strategy 2030: Achieving a Prosperous, Inclusive, Resilient, and Sustainable Asia and the Pacific", Asia Development Bank, July 2018 at https://www.adb.org/sites/default/files/institutional-document/435391/strategy-2030-main-document.pdf, accessed on March 19, 2022.

I" strategy advancing infrastructure, investment, and industrial value chains.[19] The primary objective is to build economic corridors, drive competitiveness, increase productivity and achieve inclusive growth. As India is imbibing global best practices, Tokyo remains a trusted ally as is demonstrated by the fact that the existing ODA projects in the Northeast stands at ¥ 246 billion. In this context, the role of the Act East Forum as an institutional anchor remains effective. Japanese ODA's contribution in the Northeast is not just confined to infrastructure projects but equal attention is accorded to strengthening people-to-people connect, promoting skill development, capacity building and bio-diversity projects.

As India envisions new economic corridors and aspires to develop robust economic integration between its Northeast region and Southeast and East Asia, "connecting the connectivities", that is leveraging the existing and ongoing cross-border infrastructure corridors is vitally important. For example, connecting Japan's "Big-B project" in Bangladesh with India's "Northeast Road Connectivity Improvement" projects has led to the conception of the latest "Bay of Bengal-Northeast India industrial value chain" project, making full use of the Matarbari deep sea port. Moreover, creating further synergy with existing regional connectivity initiatives such as the BBIN, SASEC, Asian Highways, Trilateral Highway, ASEAN Connectivity, and Japan's own Economic Corridors in Southeast Asia will prove to be a force-multiplier.[20] Japan's ODA has played a pivotal role in driving Indian's productivity, competitiveness, economic growth and employment (Map 1).

4 Digitalisation as a Transformative Force

In the age of fourth industrial revolution and digital economy, it is vitally important to be at the fore front of high-technology which has the transformative power in empowering knowledge-based economies. Since high-technology today is as much a part of geopolitics as it is about geo-economics and determines national power, countries are doubling down on fostering secure supply chains bilaterally and also within trilateral and Quad frames. Japan as a global leader in innovation has conceptualised Society 5.0 as a policy goal prioritising digital technologies, AI, Block Chain and Internet of Things. In this regard, India is leveraging its relation with Japan in fostering high-tech cooperation prioritising ICT products, AI, and cyber. In 2018, the bilateral Digital Partnership was launched prioritising digital information and communication technologies. The Digital Partnership is further buttressed by the Start-Up ecosystem, and NASSCOM and JETRO are deepening cooperation and connecting the start-ups and technology firms with venture capitalists in Japan.

[19] Titli Basu, "India and Japan: Connecting the Connectivities in the Bay of Bengal", The Diplomat, August 16, 2023. https://thediplomat.com/2023/08/india-and-japan-connecting-the-connectivities-in-the-bay-of-bengal/ accessed on March 19, 2022.

[20] Ibid.

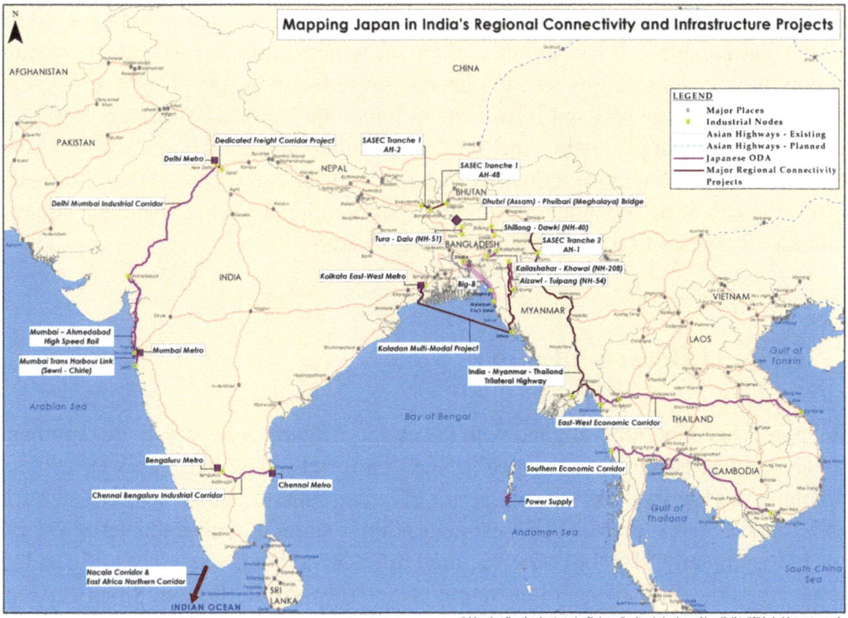

Map 1 Japan in India's Northeast: The Indo-Pacific Connect, IDSA Comment, April 19, 2022
Source Titli Basu[21]

The "Memorandum of Understanding in ICT" has buttressed partnership particularly in 5G, telecom security, and building submarine optical fibre cable. Tokyo's international competitiveness in submarine cable infrastructure gives Japanese companies an edge in supporting the "Digital India" initiative. Japanese company NEC has played a pivotal part in linking the Andaman and Nicobar with Chennai as well as Kochi and the Lakshadweep Islands. Japan is also in the lead in advancing communication networks across Asia especially in Southeast Asia. 5G has emerged as a fierce battleground amongst great powers. For instance, Japanese Guidelines on IT products and services procurement has underscored the importance of remaining vigilant on cyber-security and supply-chain risks without alluding to threats coming from China. Tokyo has created tax-incentives to stimulate investment in building safe 5G supply-chain. NTT and NEC are taking the lead in building 5G alliance. There is an urgent need to cooperate, tapping into the strengths and weaknesses in various facets of innovation and designing robust telecom supply chain. Here, India and Japan have opportunities to join forces in Open Radio Access Network (O-RAN) technology. For instance, Japanese company Rakuten's 5G network has cooperated with Sterlite Technologies and HCL, Wipro, and Tech Mahindra for hardware and software respectively.

[21] "Japan in India's Northeast: The Indo-Pacific Connect", IDSA Comment, April 19, 2022. https://idsa.in/idsacomments/japan-in-indias-northeast-the-indo-pacific-tbasu-190422, accessed on March 10, 2024.

The 2023 Memorandum of Cooperation (MoC) on the Semiconductor Supply Chain Partnership[22] is yet another pillar on which India-Japan strategic partnership will hinge since digital economies rely on semiconductors. The MoC is geared towards deepening G2G and B2B cooperation on the verticals of design, R&D, and building secure supply chain. Anchored on the bilateral Digital Partnership, complemented by the India-Japan Industrial Competitiveness Partnership (IJICP) formed in 2019, the recent Semiconductor Supply Chain Partnership will expand the strategic latitude of cooperation in the electronics ecosystem.

5 Driving Decarbonisation

Responsible economies, aligned with the targets of Paris Agreement, are employing policies towards realising the goal of net zero emissions. India and Japan being amongst the top five emitters of carbon dioxide, aims to realise net-zero status by 2070 and 2050 respectively. IEA estimates that in order to realise the net zero target by 2070, annually $160 billion is required till 2030 and thus accessing low cost long term capital is vital. To steer innovation geared towards uplifting millions of people out of poverty, India has launched 'National Hydrogen Mission', 'Green Hydrogen' and 'Green Ammonia Policy'. As India pushed for LIFE (Lifestyle for Environment) vision at G20, 2023–2024 budget has prioritised Green Growth. As decarbonisation is contingent on steady energy transition, India can be a beneficiary of Japan's Asia Energy Transition Initiative (AETI) which aims to realise sustainable development and carbon neutrality.

India's economic rise implies that energy demand will soar and low-carbon technologies have a market potential of $80 billion in by 2030.[23] Moving ahead, prioritising renewable batteries and green hydrogen will be important. Japan's low carbon technologies will be a force multiplier in India journey to develop and industrialize. Since hydrogen will define the future and has the potential to decarbonize both the power and non-power sectors, India-Japan cooperation on green hydrogen value chain from production to supply holds enormous potential. For instance, India's ACME Group dealing with renewable energy and Japan's IHI Corporation have signed an offtake term sheet enabling import of green hydrogen-based ammonia annually from Orissa to Japan from 2028. Private sector's participation in hydrogen production and hydrogen transport is the key. Japanese have a wealth of R&D experience over the last four decades on hydrogen, especially hydrogen stations and fuel cells and it has been one of the early movers into commercializing fuel cell co-generation system and fuel cell vehicles. Another sector where India-Japan

[22] "Cabinet approves Memorandum of Cooperation between India and Japan on Japan-India Semiconductor Supply Chain Partnership", Ministry of Electronics & IT, 25 Octoboer 2023, https://pib.gov.in/PressReleaseIframePage.aspx?PRID=1970784 accessed on March 19, 2022.

[23] Fatih Birol and Amitabh Kant, "India's clean energy transition is rapidly underway, benefiting the entire world", IEA, January 2022. https://www.iea.org/commentaries/india-s-clean-energy-transition-is-rapidly-underway-benefiting-the-entire-world accessed on March 19, 2022.

low-carbon cooperation is taking shape is in steel. As the 2nd and 3rd largest steel producers, India and Japan is reportedly considering employing specific technologies, possibly through Japanese project. In this context, exploring the possibility of mutual cooperation under "COURSE 50" will be useful.

"India-Japan Energy Dialogue" has culminated into "India-Japan Clean Energy Partnership" with the goal of fostering innovation and building resilient and trustworthy supply chains, especially focusing on EVs, batteries, charging infrastructure; carbon capture, utilization and storage; solar PV cells; clean coal technology, biofuels and strategic petroleum reserves. However, tapping into these opportunities and moving innovative technologies from research labs to factories to consumers will be contingent on the private sectors participation. The priority should be in R&D, tech-transfer, training, capacity building and low-cost financing.[24]

6 Road Ahead

As India-Japan further consolidates their strategic partnership in navigating the fluid balance of power and interests in the Indo-Pacific, Tokyo will continue to be an important stakeholder in building a strong India. As Prime Minister Modi aims to secure capital and technology to fuel the 'Atmanirbhar Bharat' vision, partnership with Japan, anchored on 3Ds of developmental aid, digitalisation, and decarbonisation, will be indispensable to making India a developed economy. While Japan has been a substantive contributor to key initiatives such as 'Make in India', 'Skill India', 'Digital India', 'Start-up India' and 'Smart cities', one issue area which needs greater focus in making this strategic arc most effective is deepening the people-to-people connect. Despite bipartisan support and the time tested political stability between these two Asian democracies as well as the 2nd and 3rd biggest economies, there is a huge deficit in the people-to-people connect when compared to Japan's relations with China.

Despite no historical baggage in the India-Japan relations, unlike the rocky China-Japan relations, the comparative figures, as reflected in Table 1, leaves a lot to be desired. Japan's people-to-people connect with India and China, both having comparable scale in demographics as well as market size, shows a contrasting frame. Japanese outreach in China, despite having major historical rifts and sovereignty issues often echoing intense nationalism within the domestic landscape of both countries and shaping the regional power balance, succeeded in fostering a very robust people-to-people relation. While successive annual JBIC survey features both India and China as top "promising countries for overseas business", but the above data reflects a huge gap between projection and reality. For instance, there has been a southward movement in the number of Japanese companies registered in India both

[24] "India-Japan Clean Energy Partnership", MEA, March 19, 2022. https://www.mea.gov.in/bilateral-documents.htm?dtl/34992/IndiaJapan_Clean_Energy_Partnership accessed on March 10, 2024.

Table 1 Japan's people-to-people connect with China and India

	Number of residents in Japan	Number of Japanese companies	Number of tourists visiting Japan	Bilateral trade with Japan
China	745,411 as in June 2021	31,324 as in October 2022	2,425,000 in 2023	$335.4 billion in 2022
India	43,886 as of December 2022	1400 in 2022	54,314 in 2022	$21.96 billion in 2022–23

Source Ministry of Foreign Affairs Japan and Indian Embassy in Japan[26]

in 2021 and 2022.[25] Going ahead, bridging this divide by employing enabling policies will be crucial in adding substantive depth and making India-Japan relation an enduring partnership.

[25] Japanese Business Establishments in India—2022, Embassy of Japan in India, https://www.in.emb-japan.go.jp/files/100353089.pdf accessed on March 10, 2024.

[26] "Japan–China Economic Relationship and China's Economy", MOFA, January 24, 2024. https://www.mofa.go.jp/files/100540401.pdf; "Japan–China Relations (Basic Data)", MOFA, July 6, 2023 https://www.mofa.go.jp/region/asia-paci/china/data.html; "Japan-India Relations (Basic Data)", MOFA, March 14, 2024 https://www.mofa.go.jp/region/asia-paci/india/data.html#:~:text=Japan%20and%20India%20signed%20a,countries%20have%20enjoyed%20cordial%20relations; Japanese Business Establishments in India—2022, Embassy of Japan in India, https://www.in.emb-japan.go.jp/files/100353089.pdf, accessed on March 10, 2024.

Development and Environment

Sustainable and Equitable Use of Mineral Resources: A QUAD Implication

Ariyaningsih, Chime Youdon, Pradeep Chauhan, Devesh Lahiri, Rajib Shaw, and Ranit Chatterjee

Abstract In the Indo-Pacific region, around two-thirds of the ocean lies in Areas Beyond National Jurisdiction (ABNJs), home to unique species and habitats critical to marine biodiversity. The biodiversity in ABNJ is in peril due to a patchwork of legislative frameworks. The loss of biodiversity in ABNJ impacts the ocean's ability to withstand climate change and offer resources vital for human life. It is hence essential to negotiate a secure mechanism to protect and further ABNJ governance and safeguard the marine environment and species while analyzing the impacts of human activities, creating capacity, transferring technology, and sharing the benefits of marine genetic resources equitably. This chapter will focus on sustainable and equitable use of mineral resources in QUAD region.

Keywords Indo Pacific · Resilience · Seabed mining · Mineral resources · QUAD countries

1 Introduction

The deep seabed is among one of the most unexplored regions on this planet, and is home to many species, and is the largest habitat on Earth (FAO Fisheries & Aquaculture 2021). With the discovery of deep-sea minerals, major countries have focused on promoting research and acquiring significant stakes in deep sea beds. The Indo-Pacific is one of the crucial zones that have been discussed internationally as the

Ariyaningsih (✉) · R. Shaw
Graduate School of Media and Governance, Keio University, Fujisawa, Japan
e-mail: ayya19@keio.jp; ariyaningsih@lecturer.itk.ac.id

R. Shaw
e-mail: shaw@sfc.keio.ac.jp

C. Youdon · P. Chauhan · D. Lahiri
National Maritime Foundation, New Delhi, India

R. Chatterjee
India Japan Laboratory, Keio University, Fujisawa, Japan

R. Shaw and S. R. Choudhury (eds.), *India, Japan and Beyond*,
https://doi.org/10.1007/978-981-97-3282-1_2

region, holds 40% of the total population and 60% of total GDP, has major trade routes, and technology supply chains (USA 2022). Various world leaders have identified the critical importance of the zone, such as Japanese Prime Minister Shinzo Abe, and have formulated the vision of Indian Ocean and the Pacific Ocean converging "as seas of freedom and prosperity" (Abe 2007).

Australia, India, Japan, and the United States make up the QUAD, which is another manifestation of the Indo-Pacific partnership of democratic nations that first emerged as the "Tsunami Core Group" in response to the 2004 Indian Ocean Tsunami and established a precedent for nations cooperating in a quadrilateral format to address issues in the Indo-Pacific region (Buchan and Rimland 2020). The conversations grew out of the Indo-Pacific region's humanitarian assistance and disaster relief (HADR) activities. The United States and the other QUAD partners, India, Australia, and Japan are heavily committed to ensuring practical laws regulate the Indo-Pacific and Indian Ocean as the world community's interest in the region grows.

Many bilateral meets in the region laid the ground for establishing the Quadrilateral Security Dialogue (QUAD), aimed at working cooperatively on development-related activities and securing the concept of the common good and security. With the challenges of Chinese dominance in deep seas affecting neighboring QUAD nations (Pant 2021), the QUAD has formulated novel pathways for ensuring novel pathways to generate a more resilient system. However, very little research has been documented in understanding common pathways concerning QUAD and deep-sea mining, considered reserve resources. The mineral wealth in the ocean can potentially boost the economy of various countries. Therefore, a lot of countries are looking forward to the development of techniques that would help in the mapping and exploitation of these unexplored reserves. However, REEs are "rare" because of their staggering global distribution. These elements are not present in the deposits in big mines like the other precious elements.

In the backdrop of increasing demand for marine resources, there has been a rapid increase in seabed mining and associated exploration activities. However, depending on these activities' nature, magnitude, location, intensity, etc., marine biodiversity is projected to be seriously impacted (Van Dover et al. 2017). For a long, little thought has been given to the environmental consequences of commercially exploiting marine resources. To bridge this research gap, this chapter tries to identify QUAD implication for sustainable and equitable use of mineral resources. This objective has been done by analysis the global reserves and supply chains of seabed mining and identifying its challenges and opportunities.

2 Marine Non-living Resources in the Indo-Pacific

2.1 *Role of Quadrilateral Security Dialogue*

The Quadrilateral Security Dialogue (QUAD) was established due to numerous bilateral and multilateral meetings between the partner governments. Japan was one of the first nations to incorporate the Indo-Pacific into its official foreign policy statement. It shared the vision of a "Free and Open Indo-Pacific" (FOIP) to promote peace, stability, and economic growth in the region (Ministry of Foreign Affairs of Japan 2017). Following that, the US included the idea of the Indo-Pacific in its National Security Strategy of 2017, National Defense Strategy of 2018, Indo-Pacific Strategy Report 2019, and Indo-Pacific Strategy of the United States 2022 (Trump 2017; Mattis 2018; The Department of Defense 2019; The White House 2022). This demonstrated the US's shifting emphasis from the "Asia–Pacific" to the "Indo-Pacific," setting the way for a strong foreign and security strategy to contain a rising China's aggression. In addition to its Maritime Silk Road (MSR) connectivity plan, the US's view of the Indo-Pacific region was scaled from "Africa, Europe, and the Middle East" (Pejsova 2018).

In the Shangri-La Dialogue in 2018, India's current Prime Minister, Narendra Modi, provided an overview of India's Indo-Pacific Policy and defined his country's goals for the area (Modi 2018). He emphasised the need of preserving an open, free, and inclusive Indo-Pacific and the crucial role that the Association of Southeast Asian Nations (ASEAN) governments play in boosting connectivity in the area. India has emerged as a critical pillar in the projected Indo-Pacific security architecture. It envisions playing the role of a preferred security partner and net provider for its immediate maritime neighbourhood (Agnihotri 2022). Australia described its Indo-Pacific alliances and visions in its 2017 Foreign Policy White Paper, which suggested that the Indo-Pacific was becoming the new theatre of international relations (Australian Government 2017).

The leaders of the four partner nations' first joint statement emphasized working together to produce the COVID-19 vaccine, promote collaboration on cutting-edge technology, and combat climate change (Kutty and Basrur 2021). On March 12, 2021, the leaders of the QUAD nations convened for the first QUAD Leaders' Virtual Summit to address "current challenges such as resilient supply chains, emerging and critical technologies, maritime security, and climate change" (Ministry of External Affairs 2021). The summit's efforts in the conventional security area resulted in joint military drills like the Malabar naval exercises and strengthening the strategic alliances between the four nations. The Supply Chain Resilience Initiative (SCRI) was launched virtually on April 27, 2021, by the trade ministers of the four QUAD countries. By addressing their weaknesses, employing digital technologies, and diversifying trade and investment patterns, this project aimed to protect the global supply-chain system. It also sought to ensure the region saw inclusive and balanced prosperity by directing investments away from an over-reliance on Chinese natural resources, such as rare earth elements (Krishnan 2021).

To determine the alliance's future direction, the QUAD states' leaders gathered for the first in-person summit in September 2021 in Washington, D.C. They had agreed to clarify the objectives for the mutually identified joint projects in the Indo-Pacific, such as the Covid-19 vaccine partnership, "climate change, decarbonization efforts in shipping and port operations, deployment of clean hydrogen technology, and the need for responsible and resilient clean energy supply chains." A QUAD Infrastructure Coordination Group and a working group on space cooperation were referenced in the QUAD Leaders' Joint Statement, which embodied this ambition. This group was established to talk about disaster preparedness, regional infrastructure, technical assistance, capacity-building initiatives, and satellite data sharing for observing and addressing climate change (Ministry of External Affairs 2021).

2.2 *Importance of Non-living Seabed Resources*

Rare Earth Elements (REEs) are critical hardware used in the production of renewable energy. The magnets of REEs, such as the dysprosium and neodymium, are used in offshore and onshore wind turbines (Rollet 2019). The magnets produced by these mineral resources are also used in the equipment of other renewable energy sectors like wave energy and tidal energy (Stegen 2014). Wind turbines that drive permanent magnets are highly efficient at low wind speed and are comparatively cheaper and lighter to maintain. The production of solar panels also requires terbium, praseodymium, neodymium, and dysprosium. In the production of fuel cells, yttrium is used along with other metals such as platinum and palladium (Chakarvarty 2018).

Rare Earth Elements and their associated compounds are a crucial constituent of modern-day technology, including wireless phones, cutting-edge systems, LEDs, televisions, etc. They are also used in producing phosphors to produce luminescence in various flat panel displays used in televisions, smartphones, etc. (Sharp 2019). The mineral resources are also of great significance for the defence industry. The neodymium and samarium magnets are used in guided missile systems, unmanned aerial vehicles, munitions, propulsion systems, and other defense equipment. The magnetic strength of these metals is apt for military technologies such as smart bombs and other weapon systems (Grasso 2013).

The REEs produce permanent magnet materials, i.e., samarium cobalt (SmCo) and neodymium iron boron (NdFeB), considered the world's strongest permanent magnets. The magnetic properties of these minerals provide the strength for using lighter and smaller magnets used in the defence weapon systems. These magnets retain their power even at higher temperatures and are ideal for military technologies. These minerals are used in the manufacturing and developing defense-related equipment such as lasers for detecting mines, sonar on submarines, satellite communications, missile guidance systems, motors in tanks, aircraft, missile systems, and optical equipment (Ibid). The importance of the metals can be assessed by the fact that each F-35 fighter aircraft uses 417 kg of REEs in its various equipment, which include electric motors, electronic warfare systems, and radars (Grier 2017).

2.3 Global Reserves of Critical Mineral Resources

As estimated by US Geological Survey, Nickel's estimated global reserves are close to around 94,000,000 metric tons (Ibid). Almost half of the nickel reserves are found in Indonesia and Australia, with 23% and 21% worldwide. Other countries with nickel reserves are Canada, Russia, Cuba, and the Philippines, which also hold significant reserves. According to the estimates, the REEs reserves worldwide are approximately 120 million metric tonnes. China has the largest reserves of REEs, with 44,000 metric tonnes, which is around 37% of the total world reserve. Brazil and Vietnam follow China, holding around 18% of the world's share. Russia and India also hold significant reserves with a share of 10% and 6%, respectively. Australia and the United States (US) also hold a decent amount of the reserves with 3% and 1.25%, respectively ("Rare Earths Statistics," USGS).

China sits on over one-third of the reserves and accounts for the largest share of global production. The regions that account for the highest reserves are in the provinces of Inner Mongolia, Jiangxi, Hunan, Guangxi, Fujian, Guangdong, and Sichuan. These areas account for almost 98% of China's total REE production (Tse 2011). India holds the world's fifth-largest reserve of REEs, with an estimated reserve of around 6.9 million metric tonnes (Ibid). According to the estimates, the domestic supply chain of REEs in the country has the potential to produce an annual turnover of US$ 12 billion. In addition, the REEs industry has the potential to generate net capital employment worth US$ 16 billion (Deka 2020). Further, Australia, with the sixth-largest reserve of REEs in the world, has the potential to become a significant player in the global supply. The critical minerals' reserves are spread on the country's east and west coasts. Currently, the reserves are largely untapped and only two mines are producing the critical minerals (Page and Coyne 2021).

2.4 Methodology

This study employed a method known as desk-review research or secondary research, which involves using existing data. This secondary research includes materials already published in reports, articles, or other similar documents to analyze supply chains and identify the challenges and opportunities of seabed mining. The results of its challenges and opportunities were used to formulate the approaches and recommendations for QUAD countries. The database used is research-gate, sciencedires, and google scholar with a publication date between 2015–2021. The total of documents is 34 documents.

3 Global Reserves and Supply Chains

The production of REEs is not a new phenomenon. In the last century, almost all the production of REEs was done at the Mountain Pass mine in California, U.S. With the increasing environmental regulations and the gradual shift of mining to the developing countries from the developed ones, the supply chain has been entirely shifted (Green 2019). In addition, with the availability of the labour market and low environmental protection standards, mining activity became economically viable for China (Mazumdar and Khurana 2020). The other countries could not compete economically and were eventually pushed out of the global market. China has also developed the capacity in the downstream processes of the rare-earth industry. The mining and production of crucial metals also played an essential role in the emergence of an electronics manufacturing nation (Schmid 2019). China was placed in a position where it could rule the global supply chain of the critical elements. Eventually, the whole world, including the countries possessing the elements, started sending the ores for processing to China. For instance, the US ships the rare earth mined in the country to China for downstream processing and later buys the refined elements (Yu and Sevastopulo 2021). Deng Xiaoping, the former Chinese leader, said in 1987 that while "the Middle East has oil, China has rare earth" (Hearty and Alam 2019). There came a turning point in the trade flow of the REEs, and the Chinese monopoly could be seen at the forefront. In 2010, a Chinese fishing vessel was caught close to Japan-administered Senkaku islands over which China claims its sovereignty. The commander of the Japanese vessel arrested the captain of the Chinese fishing boat, and, in retaliation, China stopped the supply of REEs to Japan. With the trade disruption, the price of the critical metals jumped by nine times. China used its capacity as a deterrent against other countries to fulfil its strategic objectives (Wagner 2019). It was only after this period that countries such as Australia entered the market of critical minerals. Furthermore, deep-sea mining is a new marine activity that, in contrast to current ocean uses, enables the precautionary approach to be incorporated into the regulatory framework before the start of commercial operations (Cuvyers et al. 2018).

4 Challenges for Sustainable and Equitable Utilisation of Marine Resources

Seabed mining and related exploration activities have increased rapidly. When deciding whether to use exploitation strategies, environmental and ethical considerations must be regarded (Banet 2020). Marine biodiversity is expected to be severely impacted depending on the nature, magnitude, location, and intensity of activities, among other factors (Van Dover et al. 2017). Furthermore, the mining industry requires energy and emits a potent greenhouse gas. Carbon emissions from the mining sector can vary greatly depending on the type of material used and how it is mined (Rüttinger et al. 2016). Furthermore, it is well known that mining deposits are becoming deeper and contain fewer valuable ore, which is likely to result in increased water needs and waste, increased energy use, and increased carbon emissions from the mining industry (Mudd et al. 2012). Deep-sea mining has the potential to disrupt some of the world's most important carbon sinks, exacerbating climate change.

Ericsson et al. (2020) highlight China's initiatives to strengthen its global mineral resource market position. China's geopolitical power can be seen in a variety of ways, but one of the most important is understanding how much of its mining activity occurs outside of China. This is due to the importance of mineral supply security to the national economy. Chinese corporations are increasing China's geopolitical strength and economic leverage in the global market by acquiring the mines and output of the next largest producers (Green and Liu 2005). Prices for rare earth elements (REE), a grouping of 17 nonferrous metals, increased by up to 600 percent in the third quarter of 2011. Territorial disputes between China and Japan, the world's second-largest REE market, exacerbated the situation (Kingsnorth 2021). It also demonstrates that China, which has emerged as the largest environmental market for REE, is producing an increasing proportion (68% in 2011) for domestic consumption. Natural resource dominance in China and its desire to use it are well-known examples (Sun 2007).

An international framework of laws, procedures, and regulations has been established to protect the marine ecosystem from deep sea mining. Still, no safeguards have been put in place to protect the ecosystem from deep-sea mining or to understand how deep-sea mining affects the ecosystem (Bosselmann 2005). Furthermore, the geopolitics of seabed mining is a tangled web of multinational corporations, the state, local and global civil society groups, and non-human elements such as the deep ocean itself. Proponents of seabed mining argue that it can contribute to economic growth. Nonetheless, they see it as the start of a new blue economy that could help lift people out of poverty and aid in the transition to green technologies (Hallgren and Hansson 2021; Kim 2017).

5 Opportunities for the QUAD Countries

Quad has a long history of geopolitics since its conceptual inception following the 2004 Tsunami. While the first Quadrilateral Security Dialogue, or QUAD 1.0, in 2021 was marred by leadership changes, misaligned interests, and divergent views on the Indo-Pacific, QUAD 2.0 was viewed as a more effective engine for addressing current and future challenges, with China's aggressive international policies being a major risk factor. The QUAD Framework 2.0 is a significant step toward a resilient, transparent Indo-Pacific region, which is consistent with the vision of the Indo-Pacific Economic Framework (IPEF) led by the United States (USA 2022). This is significant for a country like India, which is battling China's dominance in the Bay of Bengal, and so on, in terms of R&D vessels, and so on (Pant 2021). These can be accomplished in several ways. Utilizing current opportunities for collaboration with ASEAN, the EU, and other institutions against Chinese dominance would be a critical breakthrough.

Deep sea minerals are exclusive, with each mineral posing significant extraction challenges. Because of their location, Seafloor Massive Sulphides (SMS) require significant extraction force and pose operational challenges for Remotely Operated Vehicles (ROVs). This contrasts with polymetallic nodules, which require a suction mechanism to be extracted and are found on the surfaces (ibid). As a result, mining approaches differ. ROVs on the seafloor collect Massive Sulphides, which are then piped up to the surface and shipped for further processing. While readily available manganese nodules are collected using a ROV that functions similarly to a vacuum cleaner, manganese crusts are obtained using large ROVs that grind through the hard crust, creating a mixture containing the valuable minerals that is piped to the surface. Aside from these challenges, there are ongoing technical issues such as mining site delineation, system development, and so on (Sharma 2017).

Mitigating deep-sea mining and restoring ecosystems after mining will be difficult and impossible (Niner et al. 2018). Landscape changes in post-mining areas frequently differ from surrounding landscapes (Wirth 2020). Understanding the potential biodiversity loss caused by deep-sea mining will require much more boundary understanding than is currently available and understanding the technology used and its direct and indirect impacts (Clark et al. 2020; Miller et al. 2021). However, the seas remain poorly understood, particularly vulnerable deep-sea species such as shrimps, crabs, and even cold-water corals (Kim 2017; Van Dover 2014). The relationship between deep-sea habitats and ecosystem function is not well understood. Developing a comprehensive sea mechanism to protect, create, and sustainably use the oceans is a critical opportunity for QUAD countries. New directions aligned with environmental protection, ocean development, and use are also important. This is full recognition that environmental protection is essential for long-term sustainable use and development of the oceans by sound marine industries and for socio-economic stability that includes marine industries (Leal Filho et al. 2021).

6 Discussions

The supply chain forms an essential aspect of Indo Pacific region, which holds around 60% of the world's entire economy (USA 2022). With the novel challenges from COVID-19, it has become normal that a concentrated supply chain hinders the sustainability of the nations. Further, achieving all of the United Nations' sustainable development goals requires a shift toward sustainable consumption and production practices (Bengtsson et al. 2018). An initial practical step in assessing the potential need for mineral deposits required to transition to a sustainable mechanism is to converse among all stakeholders in deep-sea mining.

However, the additive impacts of marine exploration and mining are still almost entirely undetermined. They can only be delivered by improving knowledge of deep-sea ecosystems' basic and system biology and a better understanding of the complex biological interactions that allow them to function efficiently. In addition, environmental impact assessments or strategic environmental assessments at a regional level are mandatory to maintain sustainable mechanisms. It can help develop and implement techniques for establishing whether projected exploration activities in the region would significantly negatively impact vulnerable marine ecosystems or communities (Jaeckel 2020).

Nobody denies that community can be problematic because it has the potential to confuse or ignore already existing differences, differences, hierarchies, and power relations (Roche and Bice 2013b; Ryan et al. 2020). At this point, deep sea mining will not have the same direct impact on nearby groups as terrestrial mining. During this early phase, all parties must consider the impact of the project and devise processes that include local communities in determining whether the balance between benefit and impact is acceptable (De Vita 2007).

Due to the minimal skill constraints, lack of regulations, and basic technology, seabed mining is several communities' primary income source. However, such communities face environmental consequences as well as threats to their health and well-being (Cuya et al. 2021). In addition, economic exploitation for resource extraction may harm local communities' reliance on resources for food and revenue, ecotourism, and other alternative livelihoods (Quevedo et al. 2021; Rahadiati et al. 2019). Regarding resource extraction, the only significant international component relates to the ISA (International Seabed Authority), which States Parties govern the Convention and are in charge of the entire system. Land-locked countries may lose income due to seabed-based manufacture, and the International Seabed Authority ensures that a portion of profits is repaid to the international community as compensation. However, as mentioned in the previous section, calculating the mining impacts on the community is a complex task that will vary from site to site and differ on various considerations. Communities are most concerned about the impact of seabed mining because there is so little data and experience to go on in this area. In addition, there is a common lack of community knowledge about the deep seabed and its environments (Boughen et al. 2010; Mason et al. 2010).

Community capacity can be activated if a change in risk or impact perception is combined with an institutional framework that encourages networking (Adger 2006). To enhance community awareness and capacity due to mining in QUAD countries, it is suggested that the community's perception of the risks associated with a particular sector or valuing community views of risk operations is as important in their impact on a community as the real risks reinforced by scientific data (Haines et al. 2011; Marlowe et al. 2022; Todd et al. 2014). The community concerned about these risks has legitimate reasons to react in a particular way to business activity and practices. Management, scientific, and clamming community collaborations are critical to enhancing knowledge and building community capacity (Bogdan et al. 2021; Roche and Bice 2013b). A new approach to community engagement can be pioneered by this industry, which emphasises local understanding, values two-way interaction, and delegates some policymaking and responsibility to the community at large. These ideas can help raise a community's awareness of its responsibilities and the dangers it faces.

7 Potential for the Future

Based on the discussion, the authors suggest the following recommendations to improve the resilience of the QUAD Framework, which is provided at Fig. 1. They are as follows:

A. **Linking the deep sea mining to the larger QUAD economic partnership framework**: The Quad Framework 2.0 is a significant step toward resilient, transparent Indo-Pacific region, which is in lines with the vision of US-led Indo-Pacific Economic Framework (IPEF) (USA 2022). While the joint statement of Quad 2.0 (Ministry of Foreign Affairs of Japan 2022) focuses on key critical areas that are imperative to generating coastal, and ocean resource management such as commitments to support the climate change adaptation, infrastructure management, critical technologies, Maritime Domain Awareness and HADR, and space related applications (ibid), it is imperative that all countries utilizes the key opportunities. This is significant for developing countries such as India, which is battling the dominance of Chinese counterparts in Bay of Bengal, and so on, in relation to R and D vessels, and so on (Pant 2021) (GC Newsdesk 2022). This can be achieved by generating platforms for maritime diplomacy, and technology transfer, whereby deep engagements with friendly powers such

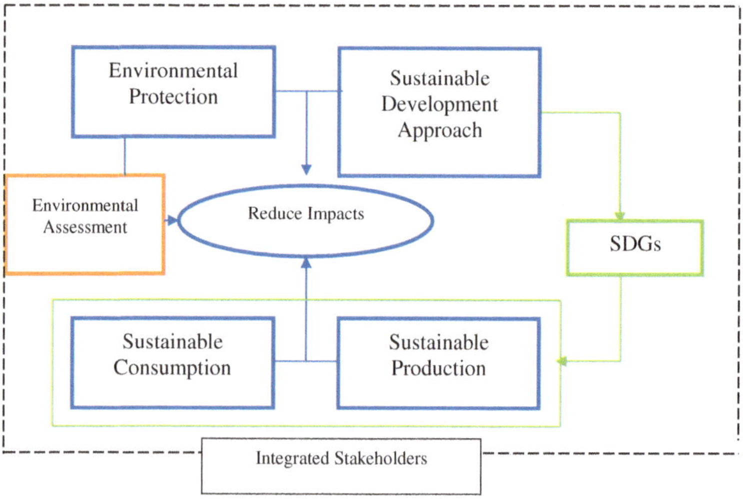

Fig. 1 Sustainable mechanism framework (*source* Authors)

as France, Japan on interoperability, and critical strategic technology transfer (Pant 2021). Further, diplomatic stances on climate action planning and carbon offsetting, which involves key consideration on the growing Indian markets, must be promoted. Further, these are essential as they can be crucial for mineral resources such as Cobalt reserves in the Pacific Ocean and the Central Indian Ocean Basin (Gateway House 2021) which are critical for EV battery productions and are a part of future reserves. Besides, operationalizing the data portals for common space database, and debt management systems (Ministry of Foreign Affairs of Japan 2022) may prompt countries from falling into the debt traps of Chinese government, which can affect the marine policies.

B. **Developing QUAD framework for ensuring supply chain management**: Supply chain forms a key aspect for Indo Pacific region, which holds around 60% of the world's total economy (USA 2022). With the advent of the Electronic Vehicle revolution to address the climate change, the demand for certain mineral resources such as Cobalt and other rare earth minerals. There are significant entry pathways for enhancing the supply chain management for India (Pant 2021; De et al. 2021). As per the authors, a critical pathway is by promoting India as an epicenter for manufacturing, and related technological innovations, where QUAD backed foreign direct investments are focused in India, in major fields such as electronics, manufacturing and so on. Further, India should engage in rare earth minerals manufacturing, and supply, and thereby reduce the Chinese dominance in the supply chain (De et al. 2021). With India being promoted as a potential zone for QUAD led investment (Pant 2021), it can address the concentrated market, and support the QUAD nations from a future conflict with China. Another key focus on enhancing the resilience of supply chain is by focusing on improving the capacity of renewable rare earth minerals, which is a nascent field (Hart 2022).

C. **Transferring technologies and best practices in the mapping, exploration, extraction, and processing of deep-sea minerals**: Mapping, extraction and processing of deep-sea minerals have been undertaken since its initial discovery in the nineteenth century during the expedition of H.M.S. Challenger (Sharma 2017). The unraveling of the economic potential concerning these minerals in the second half of twentieth century furthered investments in the field, resulting in technologies, to support the process (ibid). Various technologies have been in use ever since (STOA 2015). With key challenges such as technical ready of instruments, nascent R&D on deep-sea minerals in countries such as India and Australia (Hart 2022), it is imperative to frame strong pathways for future. Some of the key recommendations for the countries in improving its competencies include promoting research and development around the deep-sea mining ecosystem such are mining platform, and ore handling, or transfer to the sea (Sharma 2017). Further, strategic partnerships with QUAD members such as Japan, and USA for joint ventures in scientific research and design concerning the deep-sea mining technologies, mining activities, and so on could prove to be effective for India, and its mission. These could involve knowledge sharing, training and support, and technology co-creation, and licensing and so on. With

India having a long coastline, and great geopolitical advantage in the Indian Ocean, it is imperative to generate a well-defined strategy that is based on the strong foundations of investment in human resources, research, development and technology.

D. **Enhancing sustainable mechanism to reduce post-mining impacts in QUAD and partner countries**: One of the most important strategies suggested for QUAD countries is developing a comprehensive sea mechanism to protect, develop, and sustainably use the oceans. New directions that are aligned with environmental protection and ocean development and use are also important, as is full recognition that environmental protection is essential for long-term sustainable use and development of the oceans by sound marine industries, as well as for socioeconomic stability that includes marine industries (Leal Filho et al. 2021). For example, developing a win–win relationship between environmental protection and sustainable development is essential. The countries can take the case study of Australia, which has developed a leading-practice mining industry sustainability program covering indigenous community engagement; biodiversity; water and tailings management; hazardous materials management; and product stewardship (Groves 2012).

E. **Enhancing communities' awareness, capacities, engagement for sustainable coastal environment**: Understanding this heterogeneity in the local community and other social sciences disciplines is essential for effective coastal management. Roche and Bice (2013a, b) said that community interactions can occur across time, space, and scale and require various parties who may disagree with one another. They fear that national or regional discourses on ownership, authority, and cultural rights may conflict with local community discourses (Roche and Bice 2013a, b). Due to the minimal skill constraints, lack of regulations, and also basic technology, seabed mining is several communities' primary income source. Such communities face environmental consequences as well as threats to their health and well-being (Cuya et al. 2021). Economic exploitation for resource extraction may harm local. Therefore, it is essential that a new approach to community engagement can be pioneered by this industry, which emphasizes local understanding, values two-way interaction, and delegates some policymaking and responsibility to the community at large. These ideas can help raise a community's awareness of its responsibilities and the dangers it faces.

8 Conclusion

Due to the rampant industrialization of the world in the last few centuries, the global average temperature has reached a level where it has become a threat to the planet's ecological balance. The world community is working towards limiting the global average temperature rise to 1.5–2 °C above the pre-industrial level through the Paris Agreement. The commitments of the Paris Agreements emphasises the sustainable transformation of the global economy. The transformation towards the low carbon sustainable economy from the fossil fuel-based economy largely depends on the shift toward renewable energy sources such as wind turbines, solar panels, tidal energy, wave energy, and rechargeable batteries. In addition, the demand for the mineral resources that are a part of the supply chain of this renewable energy equipment is set to increase in the future, and there is a possibility that the vulnerabilities associated with the supply chain of these mineral resources will make the availability of these resources difficult and a shift towards the marine mineral resources is expected in future.

The near-shore regions are in a nation's 'national jurisdiction' while the deep seabed is in the 'area beyond national jurisdiction' (ABNJ). Presently, the rules and regulations are only applicable to the national jurisdictions, and the ABNJ is guided by the International Seabed Authority (ISA). The 'rules of exploitation of the seabed are still in the draft stages by the ISA for the better mapping and exploitation of the mineral resources. The existing 'knowledge gap' of the coastal nations, specifically the small island nations, must be addressed, and the technical knowledge must be shared. Furthermore, the environmental sustainability of deep seabed mining must be analysed before undertaking any seabed mining projects. The commercial viability of seabed mining and environmental sustainability must be considered while drafting the national and international rules and regulations related to deep seabed mining. The process of seabed mining must be inclusive, resilient, and economically sustainable in nature.

References

Abe S (2007) Confluence of the two seas. Speech by Prime Minister of Japan at the Parliament of the Republic of India, Ministry of Foreign Affairs of Japan, 22 August. https://www.mofa.go.jp/region/asia-paci/pmv0708/speech-2.html#:~:text=%22The%20different%20streams%2C%20having%20their,their%20water%20in%20the%20sea.%22&text=We%20are%20now%20at%20a,of%20freedom%20and%20of%20prosperity

Adger WN (2006) Vulnerability. Glob Environ Change 16(3):268–281. https://doi.org/10.1016/j.gloenvcha.2006.02.006

Agnihotri KK (2022) Holistic maritime security challenges facing India: mitigation imperatives and options, 31 March. https://maritimeindia.org/holistic-maritime-security-challenges-facing-india-mitigation-imperatives-and-options/

Australian Government (2017) Foreign policy white paper, chapter three: a stable and prosperous indo–pacific, indo–pacific partnerships. https://www.dfat.gov.au/sites/default/files/minisite/static/4ca0813c-585e-4fe1-86eb-de665e65001a/fpwhitepaper/foreign-policy-white-paper/chapter-three-stable-and-prosperous-indo-pacific/indo-pacific.html

Aquaculture FF (2021) Fishery and Aquaculture Statistics—Yearbook 2021. https://doi.org/10.4060/cc9523en

Banet C (2020) The law of the seabed. 1–18. https://doi.org/10.1163/9789004391567

Bosselmann K (2005) Ethical implications. Law Energy Sustain Dev 1:74–92. https://doi.org/10.1017/CBO9780511511387.008

Buchan PG, Rimland B (2020) Defining the diamond: the past, present, and future of the quadrilateral security dialogue, CSIS briefs, 16 March. https://www.csis.org/analysis/defining-diamond-past-present-and-future-quadrilateral-security-dialogue

Bengtsson M, Alfredsson E, Cohen M, Lorek S, Schroeder P (2018) Transforming systems of consumption and production for achieving the sustainable development goals: moving beyond efficiency. Sustain Sci 13(6):1533–1547. https://doi.org/10.1007/s11625-018-0582-1

Bogdan EEA, Roszko AM, Beckie MA, Conway A (2021) We're ready! Effectiveness of community disaster preparedness workshops across different community groups in Alberta, Canada. Int J Disaster Risk Reduct 55:102060. https://doi.org/10.1016/j.ijdrr.2021.102060

Boughen N, Mason C, Paxton G, Parsons R, Johns S, Parr J, Moffat K (2010) Seafloor exploration and mining in Australia: stakeholder reactions, expectations and desired level of engagement-Synthesis Report. December. https://doi.org/10.4225/08/5851894b995df

Chakarvarty U (2018) Renewable energy materials supply implications. IAEE Energy Forum. https://studylib.net/doc/25709179/renewable-energy

Christiansen S, Jaeckel A, Houghton K (2019) Ecological safeguards for deep seabed mining. http://www.umweltbundesamt.de/publikationen

Clark MR, Durden JM, Christiansen S (2020) Environmental impact assessments for deep-sea mining: can we improve their future effectiveness? Mar Policy 114:103363. https://doi.org/10.1016/j.marpol.2018.11.026

Cuya A, Glikman JA, Groenendijk J, Macdonald DW, Swaisgood RR, Barocas A (2021) Socio-environmental perceptions and barriers to conservation engagement among artisanal small-scale gold mining communities in Southeastern Peru. Glob Ecol Conserv 31:e01816. https://doi.org/10.1016/j.gecco.2021.e01816

Cuyvers L, Berry W, Gjerde K, Thiele T, Wilhem C (2018) Deep seabed mining: a rising environmental challenge. IUCN and Gallifrey Foundation, Gland, Switzerland. https://portals.iucn.org/library/node/47761

Deka C (2020) Rare earths: an ephemeral golden goose with a perpetual burden. Nature Speaks, 3 September. https://naturespeaks.in/rare-earths-an-ephemeral-golden-goose-with-a-perpetual-burden/4928/

De A, Rana V, Patel P (2021) Quad : countering Chinese dominance by building new supply chains. London School of Economics. https://blogs.lse.ac.uk/businessreview/2021/07/01/quad-countering-chinese-dominance-by-building-new-supply-chains/

De Vita CJ (2007) After Katrina: shared challenges for rebuilding communities. Urb Inst 1–7

Ericsson M, Löf O, Löf A (2020) Chinese control over African and global mining—past, present and future. Mineral Econ 33(1–2):153–181. https://doi.org/10.1007/s13563-020-00233-4

Gateway House (2021) QUAD economy & technology task force: a time for concerted action

Garside M (2022) Major countries in worldwide cobalt mine production in 2021. Statista, 24 March. https://www.statista.com/statistics/264928/cobalt-mine-production-by-country/

GC Newsdesk (2022) India's deep-sea mission push geopolitics vs environmental sustainability. GoaChronicles 1–9. https://goachronicle.com/indias-deep-sea-mission-push-geopolitics-vs-environmental-sustainability/

Grasso VB (2013) Rare earth elements in national defence: background, oversight issues, and options for congress. Congressional Research Service, 23 December, https://fas.org/sgp/crs/natsec/R41744.pdf

Green JA (2019) The collapse of American rare earth mining—and lessons learned. Defense News, 12 November. https://www.defensenews.com/opinion/commentary/2019/11/12/the-collapse-of-american-rare-earth-mining-and-lessons-learned/

Grier P (2017) Rare-earth uncertainty. Air Force Magazine, 21 December, https://www.airforcemag.com/article/rare-earth-uncertainty

Groves S (2012) The US can mine the deep seabed without joining the UN convention on the law of the sea. Backgrounder, The Heritage Foundation, no. 2746

Green SP, Liu GS (2005) Exit the dragon? Privatization and state control in China

Hallgren A, Hansson A (2021) Conflicting narratives of deep sea mining. Sustainability (Switzerland) 13(9):1–20. https://doi.org/10.3390/su13095261

Haines VA, Godley J, Hawe P (2011) Understanding interdisciplinary collaborations as social networks, pp 1–11. https://doi.org/10.1007/s10464-010-9374-1

Hart A (2022) Critical and rare: the minerals Australia can supply to the world. COSMOS 1–11. https://cosmosmagazine.com/science/engineering/rare-earth-metals-key-to-renewable-energy-econom/
Hearty G, Alam M (2019) Rare earths: next element in the trade war?. Center for Strategic & International Studies, 20 August. https://www.csis.org/analysis/rare-earths-next-element-trade-war
Indo-Pacific Strategy of the United States, February, https://www.whitehouse.gov/wp-content/uploads/2022/02/U.S.-Indo-Pacific-Strategy.pdf
IUCN (2022) Governing areas beyond national jurisdiction. Issue Briefs, March 2022, https://www.iucn.org/resources/issues-briefs/governing-areas-beyond-national-jurisdiction
Jaeckel A (2020) Strategic environmental planning for deep seabed mining in the area. Mar Policy 114:103423. https://doi.org/10.1016/j.marpol.2019.01.012
Kang Y, Liu S (2021) The development history and latest progress of deep-sea polymetallic nodule mining technology. Minerals 11(10):1–17. https://doi.org/10.3390/min11101132
Kim RE (2017) Should deep seabed mining be allowed? Mar Policy 82:134–137. https://doi.org/10.1016/j.marpol.2017.05.010
Khurana, Gurpreet S (2017) The 'Indo-Pacific' concept: retrospect and prospect. National Maritime Foundation, 2 February, https://maritimeindia.org/View%20Profile/636215922419657386.pdf
Krishnan A (2021) China wary as India, Australia, Japan push supply chain resilience. The Hindu, 28 April. https://www.thehindu.com/news/national/china-wary-as-india-australia-japan-push-supply-chain-resilience/article34431964.ece
Kutty SN, Basrur R (2021) The quad: what it is–and what it is not, 24 March. https://thediplomat.com/2021/03/the-quad-what-it-is-and-what-it-is-not/
Leal Filho W, Abubakar IR, Nunes C, Platje J, Ozuyar PG, Will M, Nagy GJ, Al-Amin AQ, Hunt JD, Li C (2021) Deep seabed mining: a note on some potentials and risks to the sustainable mineral extraction from the oceans. J Mar Sci Eng 9(5):1–16. https://doi.org/10.3390/jmse9050521
Magnuson S (2018) Some essential reading on strategic minerals. National Defense, 29 January. https://www.nationaldefensemagazine.org/articles/2018/1/29/some-essential-reading-on-strategic-minerals
Marlowe J, Jehan F, Udahemuka M, Mizero A, Jaduram R, Rotstein J, Alam ZZ, Nkessah Z, Osman M, Aung S, Jwaied SA (2022) Disaster communications with resettled refugees: six principles of engagement. Int J Disas Risk Reduct 67:102672. https://doi.org/10.1016/j.ijdrr.2021.102672
Mason C, Paxton G, Parr J, Boughen N (2010) Charting the territory: exploring stakeholder reactions to the prospect of seafloor exploration and mining in Australia. Mar Policy 34(6):1374–1380. https://doi.org/10.1016/j.marpol.2010.06.012
Mattis J (2018) National defense strategy of the United States of America. Department of Defense Washington United States, 1 January. https://dod.defense.gov/Portals/1/Documents/pubs/2018-National-Defense-Strategy-Summary.pdf
Mazumdar R, Khurana M (2020) India securing rare earth elements. India Exim Bank, September. https://www.eximbankindia.in/Assets/Dynamic/PDF/Publication-Resources/ResearchPapers/OP/132file.pdf
Miller KA, Brigden K, Santillo D, Currie D, Johnston P, Thompson KF (2021) Challenging the need for deep seabed mining from the perspective of metal demand, biodiversity, ecosystems services, and benefit sharing. In Frontiers in marine science, vol 8. https://doi.org/10.3389/fmars.2021.706161
Miller KA, Thompson KF, Johnston P, Santillo D (2018) An overview of seabed mining including the current state of development, environmental impacts, and knowledge gaps. Front Mar Sci 4. https://doi.org/10.3389/fmars.2017.00418
Ministry of Foreign Affairs of Japan (2022) QUAD joint leaders statement. https://www.mofa.go.jp/fp/nsp/page1e_000401.html

Ministry of External Affairs (2022) Quad joint leaders' statement, media centre, bilateral/multilateral documents, 24 May. https://www.mea.gov.in/bilateral-documents.htm?dtl/35357/Quad_Joint_Leaders_Statement
Ministry of External Affairs (2021) First quad leaders' virtual summit. Press Releases, 9 March. https://mea.gov.in/press-releases.htm?dtl/33601/First+Quad+Leaders+Virtual+Summit
Ministry of Foreign Affairs of Japan (2017) Diplomatic bluebook 2017, chapter 1, international situation and Japan's diplomacy in 2016. https://www.mofa.go.jp/policy/other/bluebook/2017/html/chapter1/c0102.html#sf03
Modi N (2018) Prime Minister's keynote address at Shangri La dialogue. Ministry of External Affairs of India, 1 June 2018. https://www.mea.gov.in/Speeches-Statements.htm?dtl/29943/Prime+Ministers+Keynote+Address+at+Shangri+La+Dialogue+June+01+2018
Mineral Commodity Summaries (2021) US geological survey. https://pubs.usgs.gov/periodicals/mcs2021/mcs2021-nickel.pdf
Munjal D (2022) All you need to know about the quad, the current summit, and initiatives. The Hindu, 24 May 2022. https://www.thehindu.com/news/international/all-you-need-to-know-about-the-quad-the-current-summit-and-initiatives/article65453483.ece
Mudd GM, Memary R, Northey SA, Giurco D, Mohr S, Mason L (2012) Future greenhouse gas emissions from copper mining: assessing clean energy scenarios. Cluster Res Report 1.12(1):1–32. https://www.uts.edu.au/isf
National Minerals Information Center Rare earths statistics and information. US Geological Survey. https://www.usgs.gov/centers/nmic/rare-earths-statistics-and-information
NS. Energy Staff Writer (2021) Profiling the six largest cobalt reserves in the world by country. NS. Energy, 07 June. https://www.nsenergybusiness.com/features/largest-cobalt-reserves-country/#:~:text=According%20to%20the%20US%20Geological,tonnes%20produced%20globally%20in%202020
Niner HJ, Ardron JA, Escobar EG, Gianni M, Jaeckel A, Jones DOB, Levin LA, Smith CR, Thiele T, Turner PJ, Van Dover CL, Watling L, Gjerde KM (2018) Deep-sea mining with no net loss of biodiversity—an impossible aim. Front Mar Sci 5:53. https://doi.org/10.3389/fmars.2018.00053
Okamoto N, Shiokawa S, Kawano S, Sakurai H, Yamaji N, Kurihara M (2018) Current status of Japan's activities for deep-sea commercial mining campaign. 2018 OCEANS-MTS/IEEE Kobe Techno-Oceans, OCEANS - Kobe 2018, 2013, pp 1–7. https://doi.org/10.1109/OCEANSKOBE.2018.8559373
Pant HV, MS (2021) The rise and rise of the 'quad': setting an agenda for India. ORF Brief, September
Page M, Coyne J (2021) Australia has a key role to play in reducing china's Rare-Earths Dominance. The Strategist, 25 February. https://www.aspistrategist.org.au/australia-has-a-key-role-to-play-in-reducing-chinas-rare-earths-dominance/
Pejsova E (2018) The Indo-Pacific. A passage to Europe. European Union Institute for Security Studies (EUISS), Brief Issue. March 2018. https://www.iss.europa.eu/sites/default/files/EUISSFiles/Brief%203%20The%20Indo-Pacific_0.pdf
Quevedo JMD, Uchiyama Y, Kohsaka R (2021) A blue carbon ecosystems qualitative assessment applying the DPSIR framework: local perspective of global benefits and contributions. Mar Policy 128(February):104462. https://doi.org/10.1016/j.marpol.2021.104462
Rademaekers K, Widerberg O, Svatikova K, Roel van der Veen, Consulting TE, Panella E, & Ltd M (2015) Technology options for deep-seabed exploitation: tackling economic, environmental and societal challenges. In European parliamentary research service.
Rahadiati A, Prihanto Y, Suryanegara E, Rudiastuti AW, Nahib I, Nursugi (2019) Assessment of socioeconomic vulnerability of coastal community in management of floods in Mataram. IOP conference series: earth and environmental science, vol 399, no 1. https://doi.org/10.1088/1755-1315/399/1/012098

Ringbom H, Henriksen T (2017) Governance challenges, gaps and management opportunities in areas beyond national jurisdiction. A STAP Information Paper. http://www.thegef.org/sites/default/files/council-meeting-documents/EN_GEF.STAP_.C.51.Inf_.02_Governance_Report.pdf
Roche C, Bice S (2013a) Anticipating social and community impacts of deep sea mining. Deep Sea Miner Green Econ Secr Pac Commun Suva. December 2013
Roche C, Bice S (2013b) Anticipating social and community impacts of deep sea mining. Deep Sea Miner Green Econ Secr Pac Commun Suva, Suva, 59–80
Rüttinger L, Sharma V, Arpacıoğlu CB, Er C, National Pollutant Inventory, Eurostat, Fugiel A, Burchart-Korol D, Czaplicka-Kolarz K, Smoliński A, Ghose MK, Majee, SR, Centre TH E, Excellence F. O. R., Innovation, I. N. M., Environmental Law Alliance Worldwide, Requirements, A. R., Generation, P. M. (2016). Climate change and mining. A foreign policy perspective legal notice. Guidebook for Evaluating Mining Project EIAs 63(2):45. http://www.stopajaxmine.ca/files/documents/Bellavistaopen-pitgoldmine.pdf%0Awww.adelphi.de%0Awww.adelphi.de
Ryan B, Johnston KA, Taylor M, McAndrew R (2020) Community engagement for disaster preparedness: a systematic literature review. Int J Disas Risk Reduc 49(May):101655. https://doi.org/10.1016/j.ijdrr.2020.101655
Rollet C (2019) Are rare earths used in solar panels? PV Magazine International, 28 November. https://www.pv-magazine.com/2019/11/28/are-rare-earths-used-in-solar-panels/
Sanchez-Cupido L, Pringle JM, Siriwardana AI, Hilder M, Forsyth M, Pozo-Gonzalo C (2020) Correlating electrochemical behavior and speciation in neodymium ionic liquid electrolyte mixtures in the presence of water. ACS Sustain Chem Eng 8(37):14047–14057. https://doi.org/10.1021/acssuschemeng.0c04288
Sharma R (2017) Deep-sea mining: resource potential, technical and environmental considerations. In Deep-sea mining: resource potential, technical and environmental considerations. https://doi.org/10.1007/978-3-319-52557-0
Schmid M (2019) Rare earths in the trade dispute between the US and China: a Déjà Vu. Intereconomics 54(6):378–384. https://doi.org/10.1007/S10272-019-0856-6
Sharp N (2019) Why are rare earth elements so crucial for electronics manufacturing? JJS Manufacturing, 03 October. https://www.jjsmanufacturing.com/blog/rare-earth-elements-electronics-manufacturing
Shen H (2017) International deep seabed mining and China's legislative commitment to marine environmental protection. J East Asia Int Law 10(2):489–509. https://doi.org/10.14330/JEAIL.2017.10.2.08
Smith SA (2021) The quad in the Indo-Pacific: what to know. Council on Foreign Relations, 27 May. https://www.cfr.org/in-brief/quad-indo-pacific-what-know
Stegen K (2015) Heavy rare earths, permanent magnets, and renewable energies: an imminent crisis. Energy Policy Elsevier 79(C):1–8. https://doi.org/10.1016/J.ENPOL.2014.12.015
STOA (2015) Deep-seabed exploitation-tackling economic, environmental and societal challenges. https://doi.org/10.2861/310632
Sun L (2007) Ownership reform in China's township and village enterprises. Exit the dragon? Privatization and State Control in China, December. http://books.google.com/books?id=OqGuQgAACAAJ
The Department of Defense (2019) Indo-Pacific strategy report preparedness, partnerships, and promoting a networked region, 1 June. https://media.defense.gov/2019/Jul/01/2002152311/-1/-1/1/DEPARTMENT-OF-DEFENSE-INDO-PACIFIC-STRATEGY-REPORT-2019.PDF
The White House (2022)
Trump DJ (2017) National security strategy of the United States of America. Executive Office of The President Washington DC Washington United States. https://trumpwhitehouse.archives.gov/wp-content/uploads/2017/12/NSS-Final-12-18-2017-0905.pdf
Tse P (2011) China's rare-earth industry. US Geological Survey. https://pubs.usgs.gov/of/2011/1042/of2011-1042.pdf

Todd VLG, Todd IB, Gardiner JC, Morrin ECN, MacPherson NA, DiMarzio NA, Thomsen F (2014) A review of impacts of marine dredging activities on marine mammals. ICES J Mar Sci 72(2):328–340. https://doi.org/10.1093/icesjms/fsu187

USA (2022) Indo- Pacific strategy of the United States. The White House. https://www.whitehouse.gov/wp-content/uploads/2022/02/U.S.-Indo-Pacific-Strategy.pdf

Van Dover CL (2014) Impacts of anthropogenic disturbances at deep-sea hydrothermal vent ecosystems: a review. Mar Environ Res 102:59–72. https://doi.org/10.1016/j.marenvres.2014.03.008

Van Dover CL, Ardron JA, Escobar E, Gianni M, Gjerde KM, Jaeckel A, Jones DOB, Levin LA, Niner HJ, Pendleton L, Smith CR, Thiele T, Turner PJ, Watling L, Weaver PPE (2017) Biodiversity loss from deep-sea mining. Nature Geosci 10(7):464–465. https://doi.org/10.1038/ngeo2983

Vats N (2021) India's strategy in deep seabed mining. CESCUBE. https://www.cescube.com/vp-india-s-strategy-in-deep-seabed-mining

Wagner D (2019) China, rare earth minerals and nuclear options. The Sunday Guardian, 1 June. https://www.sundayguardianlive.com/news/china-rare-earth-minerals-nuclear-options

Wirth P (2020) Post-mining regions problems, potentials, possibilities (Issue January 2012)

Yang J, Liu L, Lyu H, Lin Z (2020) Deep-Sea mining equipment in China: current status and prospect. Chin J Eng Sci 22(6):1. https://doi.org/10.15302/j-sscae-2020.06.001

Yu S, Sevastopulo D (2021) China targets rare earth export curbs to hobble US defence industry. Financial Times, 16 February. https://www.ft.com/content/d3ed83f4-19bc-4d16-b510-415749c032c1

Japan's Contribution to Human Security Through Connectivity a Case Study Investment in India's Northeast

Chandrali Sarkar

Abstract With time the concept of national security has expanded to include human security within its purview. For all available agencies and mechanisms for security, the state still plays a pivotal role, but in the Anthropocene epoch, States address and recognize a clusters of threats to human development. One such threat is spatial disparity in infrastructure and lack of infrastructure connectivity. Japan has been contributing to bridge the infrastructure connectivity gap since the end of the Second World War through the Official Development Assistance Program. It is also one of the first few countries that propelled conceptualization of human security as well. The concept has now found a dominant place in Japan's 2022 National Security Strategy and its Free and Open Indo-Pacific Vision. the Northeast Region of India is an important segment of Japan's vision where Japan's contribution to infrastructure connectivity meets with its discourse on human security. This paper analyses Japan's contribution to infrastructure development and connectivity in NER within the broader framework of promoting human security.

Keywords Infrastructure gap · Japan · India · Northeast · Human security

1 Introduction

The 1994 United Nations Human Development Report was the first document that introduced the concept of human security in the global platform. The report was released at an opportune time with emerging developmental challenges in the aftermath of the Second World War. Gradually, the focus shifted from traditional threats like armed conflicts to non-traditional threats like economic underdevelopment, resource deficiency, climate change, forced migration, ethnic conflicts, natural disasters etc. While the breach of traditional security calls for war preparation by the States, the breach of the latter leaves a profound impact on human lives and threatens human security.

C. Sarkar (✉)
Jawaharlal Nehru University, New Delhi, India
e-mail: chandralisarkar11@gmail.com

R. Shaw and S. R. Choudhury (eds.), *India, Japan and Beyond*,
https://doi.org/10.1007/978-981-97-3282-1_3

Human security requires ensuring human development. Human development requires expanding choices and inclusive growth by improving the incomes and lives of all members of society, particularly people below the poverty line. It is contingent upon generating economic growth, enhancing access to resources and basic services, and sharing benefits with the deprived. In this field, the development of infrastructure directly contributes to the lives of the people by increasing labour productivity, and accessibility and enhancing the basic quality of livelihood.

Globally an '*infrastructure turn*' is observed since 2000 (Dodson 2017). This is a term used to describe the resurgence in infrastructure investment as a strategic economic and development imperative. A rise in private and public infrastructure-driven projects is essential for filling the infrastructure deficit. Such growth projects range from the construction of new entities matching the conditions of a green economy and societal knowledge to rehabilitating archaic infrastructures.

Japan was one of the first countries to have advocated both for human security as an emerging security concept and simultaneously promoted the positive impact of infrastructure on human development. Its own experience with infrastructure had a lot to contribute to this venture. Hence, over the years, the Japanese government successfully aligned the concepts of human security with infrastructure aid. The Asian Region has been the largest recipient of Japanese infrastructural aid since the 1950s. Over the years, this region has become the centre of attraction for developed countries to invest in its growth because of its versatility in terms of resources, economic and social opportunities, and availability of cheap labour. Therefore, ensuring better connectivity with sustainable infrastructure gradually became a national security objective for Japan and like-minded countries.

The Asian Development Bank estimated in one of their studies that Asia required USD 1.7 trillion to be invested annually to counter the spatial disparity in infrastructure. Further, India is one of the fastest-growing economies in Asia (ADB 2017). It has been estimated that India's nominal GDP measured in USD terms is forecast to rise from USD 3.5 trillion in 2002 to USD 7.3 trillion by 2030 (S&P Global 2021). To facilitate this growth, India required urban infrastructure investments worth USD 840 billion by 2035 (World Bank 2022). India's requirement for sustainable and developed infrastructure is growing simultaneously with its aspiration.

Japan is playing a very positive role in India's aspiration. It is one of the strategic partners of India. The beginning of their relationship is attributed to Japan's transfer of loan aid for India's development in 1958. Since then, over 6,978 billion Japanese Yen (approximately Rs. 4,37,742 crores) in ODA loans have been committed for development across various sectors. Sector-wise transport infrastructure contributing to India's physical connectivity holds the largest position with 68.1% among other sectors (India Operations: JICA 2022). Currently, India's Northeast Region has been the emerging region for Japan's exclusive strategic investments since 2008.

Integration of this region with India has been one of the pertinent goals for India since early 2000. It was effectively prioritised after the announcement of India's Act East Policy. The strategic placement of this region makes it 'India's gateway to Southeast Asia'. Additionally, its closeness to China in the North and India's limited associability with this region makes it the most volatile region. Further, ethnic diversity, geographical factors, limited accessibility to the mainland and sharing porous borders with strategic countries generate collective concern for the human security of this region. Japan's contribution to facilitating human development in this region by connecting them with the mainland and investing in transport and developmental infrastructure, could be counted within Japan's strategic move within the prism of the human security concept. Since 2008, JICA has extended approximately INR 26,107 crore as concessional ODA loans for the development of the Northeast Region. Enhancing connectivity is one of the prioritized areas to support. Japan has extended approximately INR 10,421 crore over Phase 1 to Phase 5 of the Northeast Road Connectivity Improvement Project, and it is continuing (India Operations: JICA 2022). Currently, JICA is beginning Phase 6 of its investment.

Therefore, Japan's involvement in NER is a perfect example to justify its contribution to the concept building of human security aligning it with the strategic interests of both countries. Using infrastructure connectivity as a conduit for human development it is aligning human security with the concept of '*infrastructure turn*'. This paper thus portrays Japan's contribution to building the human security concept at the world forum while justifying the same through JICA's contribution to infrastructural connectivity for human development.

2 Japan's Contribution to Human Security Discourse

Human Security is a normative concept which suggests a shift from a State-centric to a human-centric approach in evaluating the foreign policy objectives of a nation. The International Red Cross was the first organisation that introduced the term '*human security*' followed by the United Nations Charter in 1945 and then the 1994 UNDP Human Development Report.

Theoretically, '*Security of State*' before the 21st Century was always associated with a nation's military defence policy. Military power was considered necessary for preserving the territorial integrity of the state. Surfacing of non-traditional challenges in the twenty-first century expanded the ambit of security. Developmental challenges like resource scarcity, economic underdevelopment, forced migration, climate change natural disasters, and the spread of infectious diseases now have far-reaching consequences. As a result, the integrity of a State is associated with the human development quotient.

Japan has been putting the human security concept in practice as a world aid donor since the 1960s. Japan's official Development Assistance program's commitment to developmental projects across the globe has only expanded and has become strategic with time. Infrastructure investment holds the highest share of such aid investments which reflects Japan's understanding of the positive correlation between infrastructure and human development and its priority in sustaining foreign relations. Limitations on Japan's military power post-WWII have a huge role to play in pushing

Japan towards expanding the concept of security to maintain State relevance in world politics.

A year after the release of the 1994 UNDP report then Japan's Prime Minister Tomiichi Murayama (1994–1996) was the first to endorse and offer the human security commitment. PM Murayama further stated that Japan's ODA and international cooperation emphasized the social development of "human priority" in line with the UNDP concept. The concept was very familiar to Japan's comprehensive security policy launched in the 1980s, putting greater emphasis on economic and political security over military engagements. The road to a favourable security environment must incorporate the pursuit of economic security. Thereby making infrastructure development an important quotient within expanded security concepts.

The next set of leadership also followed this path. Under Japanese Prime Minister Obuchi Keizo decisive leadership, human security as a concept was uplifted in the backdrop of the 1998 Asian Financial Crisis. His firm commitment was evident in his 1998 speech at Hanoi titled, "*Toward the Creation of a Bright Future of Asia*". He envisioned a century of peace and prosperity built on human dignity by ensuring human development and security. PM Obuchi explained the concept of human security takes a comprehensive view of all threats to human survival, life and dignity and stresses the need to respond to such threats. He also contributed 500 million yen to the establishment of the "Human Security Fund" under the United Nations even during Japan's economic crisis (JICA 2022).

Going forward in the new century, challenges around human development only aggravated. The 2000 UN Millennium Summit positioned human security as the key perspective of diplomacy. Japan took the initiative to form the International Commission on Human Security to achieve the twin goals of 'freedom from want and freedom from fear' while reporting to then-UN Secretary-General Mr Kofi Aanan. The Commission was entrusted with the responsibility of developing the concept and spreading practical awareness of human security in the era of globalisation. Representatives from Japan, Ms Sadaka Ogata, then president of the Refugee Committee (former President of JICA) along with Indian economist Dr Amartya Sen headed the Commission. It released a study report in 2003 highlighting poverty, economic insecurity, climate change, educational deprivation, and conflicts are emerging threats initiating collaboration among countries.

The same year Japan also released its revised Official Development Charter, institutionalising the concept within the Charter under '*Perspective of Human Security*', where the basic objective of ODA was to address "address direct threats to individuals such as conflicts, disasters, infectious diseases" by prioritising poverty reduction, sustainable growth, peacebuilding (2003 ODA Charter). A gradual expansion in the understanding of the concept could be seen in the 2015 ODA Charter where the basic policy was to '*Promoting Human Security*' for all vulnerable sections of society by prioritising 'quality growth and poverty eradication, sustainable and resilient international community and sharing universal values realising peaceful and secure society (MOFA 2015).

In both cases, implementing such a policy required expanding infrastructure aid. More conspicuous in the 2015 Charter than in 2003. Country-specific ODA must be allocated to resources promoting economic and social development was the angle. While the 2015 charter explicitly highlighted the strategic disbursement of aid. It further stated that aid would be dispersed taking into consideration the strategic importance of the recipient country and their developmental challenges concerning Japan. It also highlighted the importance of the development of industries and markets through improvement in infrastructure connectivity, finance, trade and investment climate imperative for quality growth and poverty eradication thus pushing a country towards a self-reliant growth trajectory (MOFA 2015).

The increase in the importance of the strategic quotient of human security gradually found its rightful place in Japan's National Security objectives. It very categorically states that ODA should be strategically used by Japan to lead international efforts for solving global issues like poverty reduction, health climate change, environment, and humanitarian issues, under the concept of human security (MOFA 2022).

3 Japan's Contribution to Human Security Through Infrastructure Connectivity

Infrastructure connectivity plays a crucial role in creating better conditions of life. Much literature firmly established the significant positive impact of infrastructure on health and education is also firmly established in the existing literature. Jeet Bahadur Sapkota (2015) stated that increased infrastructure services directly benefit business enterprises through expanded market opportunities, reduced cost of production, and increased production quality and volume of goods and services. Lall and Wahba (2021) presented global evidence on how the poor suffer the most from the lack of access to infrastructure networks. Further, Sapkota (2015) also highlighted how connectivity infrastructure is essential for promoting regional integration since many small populations across geography benefit greatly from spatial integration. Further, several kinds of literature suggest that rural community-based infrastructure, such as rural roads, rural small-scale electrification, and water supply and irrigation projects, significantly benefit small and medium-scale enterprises by increasing land and labour productivity (Sapkota 2015).

Japan's reliance on the positive impact of infrastructure connectivity on human development goes back to its colonial days when the government invested in building networked infrastructure between occupied Manchuria and mainland Japan. During the post-war period, Japan utilised aid in rebuilding physical infrastructure, like electricity, transport, and telecommunication which impacted war ridden Japanese economy positively. Similarly, as a donor country, it stresses upon infrastructure development and connectivity to make other countries self-reliant and resilient ensuring human development. Even JICA has been quite vocal about this positive correlation. Recently, it released a report on access to infrastructure and human development, where it shared a dialectic diagram of how both are connected intricately

(Sapkota 2015). The diagram is depicted below in Fig. 1. The report further defined lack of access to infrastructure as '*infrastructure poverty*' making meeting basic needs very difficult. Connectivity infrastructure such as road and transport enable accessibility, productivity, market access and community interaction, all essentially resulting in human development (Sapkota 2015).

The 1992 Annual Report compiled a narrative on 60 years' worth of Japanese ODA, highlighting that 60% of ODA was in the road and transport sector. This commitment has only taken a strategic structure in the twenty-first century (JICA Annual Report 1992). As per the latest body of literature released by JICA post-2008, it identifies investment destination based on Japan's strategic and political requirements aligning with its vision of the 'Free and Open Indo-Pacific' region. Further, these investments tend to focus on quality growth with 'inclusiveness, sustainability and resilience'. As per the literature, Japanese aid mainly engages in improving connectivity in three stages: international transportation facilitating trade and the flow of people. Second, national transportation ensures people's fair access to transportation and balanced development improving urban mobility. Lastly, rural transportation improves the living standards of rural areas by improving connectivity with urban areas through the utilization and application of information and communication technology, including intelligent transportation systems (ITS).

With regional integration in focus, Japan's infrastructure investments also have a corridor connectivity focus. Alignment with the ASEAN Master Connectivity Plan of

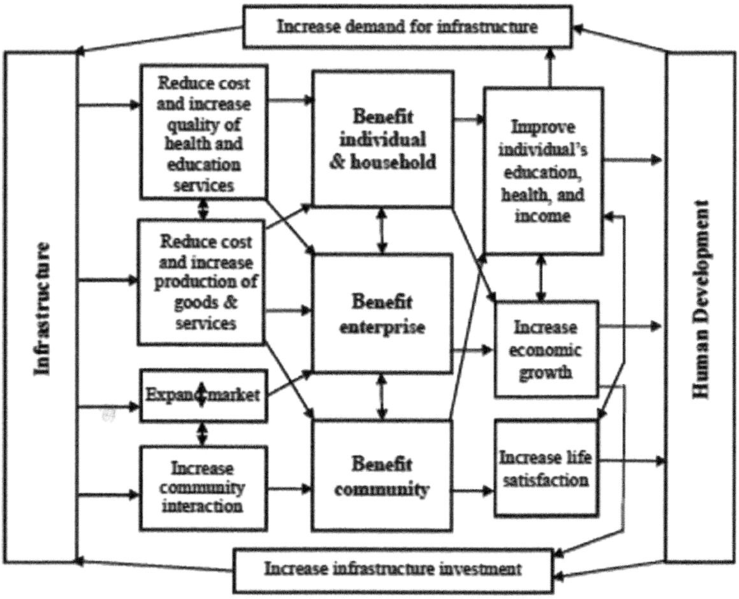

Fig. 1 Sapkota's dialectic model of infrastructure and human development. *Source* (Sapkota 2015)

creating business connectivity and strengthening three pillars of connectivity; physical, institutional and people-to-people, is a great example. Such corridor connectivity must be committed to a three-stage aid strategy for the best outcome. The East–West Corridor from Myanmar to Laos and the Southern Economic Corridor from Myanmar to Vietnam is a perfect example of the above vision. JICA's investment in ASEAN began in 1954 mainly developing public infrastructure projects. Today, it is the highest investor in infrastructure connectivity in Southeast Asia accelerating growth and addressing global issues like climate change resource scarcity, trade, improvement of business environment and intra-regional cooperation.

Similar to Southeast Asia, Japan has been instrumental in facilitating national and rural transportation development in India. Aid influx began in 1958, and currently, the largest aid donor of India is Japan. Approx. 203 billion yen cumulatively provided to India. It is indeed an important part of India-Japan political relations. With the progress in the strategic connection between both the countries aid for the transport sector took a strategic turn. This is reflected in Japan's special interest in India's North-eastern region. The location of this region serves a dual purpose of Japan's interest in India. Firstly, it fulfils the first layer of international transportation facilitating trade, supporting India Act East Policy of connecting the Indian sub-continent and Southeast Asian markets supporting India's export-led growth vision of 'Make in India'. Secondly, Japan's FOIP vision of regional integration of like-minded countries for maintaining peace, prosperity, and inclusiveness in Asia receives traction with the development NER region.

4 Japan's Contribution to Human Security in India's Northeast Region

NER occupies a strategic position in the easternmost part of India connected by a narrow stretch; the Siliguri Corridor; with the Indian mainland. Located in the sub-Himalayan region of West Bengal, between Nepal and Bangladesh. The region comprises eight states- Arunachal Pradesh, Assam, Manipur, Meghalaya, Mizoram, Nagaland, Tripura, and Sikkim. As depicted in Fig. 2 NER also shares an international border with China, Bhutan in the north, Bangladesh, and Myanmar in the South and southeast respectively, and Nepal in the Northwest.

It is one of the most untapped, energy-intensive geographical units of India. With hilly topography and a wide range of altitudes from 300–5000 m, it dons a large expanse of forest cover. The Brahmaputra also flows through this region. It has an abundance of both energy and natural resources such as reserves of petroleum, natural gas, and various minerals like high-class uranium, iron ore phosphates, kaolin China clay, dolomite, etc. Further, the Brahmaputra–Barak River system has the potential to generate 32% of the country's hydropower.

Despite its richness, this region has not been able to keep up with India's development trajectory, it contributes to only 2.8% of India's GDP, Further, this region is

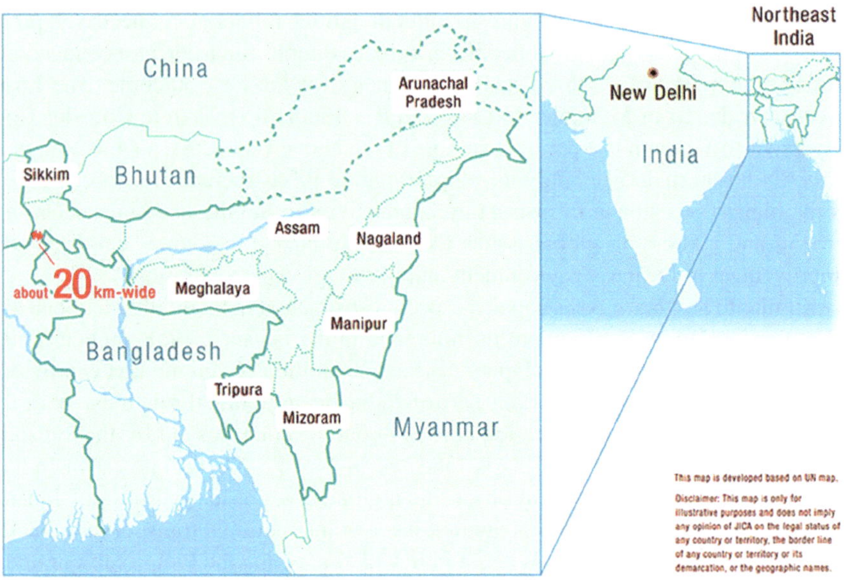

Fig. 2 Map of India's Northeast Region. *Source* JICA, Development in Northeast India: Insights from the JICA Chief Representative (2023)

ridden with human development issues cropping from lack of accessibility and ethnic conflicts, underdevelopment, etc. The Multidimensional Poverty Index released by the planning body of India; NITI Aayog, revealed an alarming picture of poverty of a population deprived of nutrition, high child and adolescent mortality rate, limited access to health care, sanitation, drinking water, electricity housing, assets, and bank account (NITI Aayog 2021). Further, the region is also prone to natural disasters like landslides, and heavy rainfall topping it, NER also lack good governance and persistent ethnic conflicts pose a huge hindrance to human development thereby threatening human security.

Transport 'infrastructure poverty' is a huge hindrance in this region. It may not be the only cause, but it is one of the central causes behind the lagging human development of this region.

A study released in 2014, estimated that NER comprises 7.98% of India's geographical area which lags in transport sector development. Road density in terms of road length per 1,000 km^2. area is very poor, especially in the hilly states of Arunachal Pradesh, Mizoram, Meghalaya, and Sikkim. Further, in terms of the quality of the road, merely 29% of the total road is surfaced. These roads mostly facilitate intra-state transportation. Poor road transport conditions in terms of surface road density affect the plying of vehicles in all seasons. It adversely affects access to health, education, and the workplace. Other means of transport like rail, air, and water have been insignificant in NER. From a qualitative point of view, 75% population of NER have had less access to 50% of surface roads. Making this place vulnerable to poverty. The United Nations Development Program (2016) highlighted that India's human development index (HDI) has a value of 0.633 which comes under the medium category of HDI. It has secured 132 positions in the world ranking, while the HDI of NER is 0.484 which is much lower than India's HDI (Marchang 2022).

Studies have suggested that states which have better road systems also have higher per capita incomes. In line with the argument, good access to roads can improve education, health, and markets for farmers, and income levels (Antonio Estache 2012). Between 1992–2004 poverty and inequality were starkly present in NER. In 2012, the urban areas of Meghalaya, Sikkim and Tripura experienced a decline in poverty and inequality. However, poverty and inequality are observed in Mizoram and Arunachal Pradesh (rural areas). The urban and rural areas of Assam, Manipur Tripura experience an increase in inequality and a decline in poverty headcounts. Sikkim is estimated to have the highest per capita income which is attributed to a sufficient supply of electricity. Hydroelectric power, tourism, pharmaceuticals, and tea have the maximum growth potential in Sikkim (Mishra and Jha 2014). Further, higher urbanisation in Assam has reduced its poverty level. This condition implies that better access to economic infrastructure has a positive effect on per capita income, which is an indicator of human development. Poverty and inequality are sharply observed in Arunachal Pradesh, Mizoram, Nagaland, and Manipur (Marchang 2022).

Northeast has longstanding ties with Japan. Recently enough, Japan started pouring a lot of aid into the NER. Such contribution has been institutionalised and aligned with the strategic interests of both India and Japan. As depicted in Fig. 3 Japan began its investment in 2008 with the Guhawati Water Supply Project in Assam, it witnessed tremendous growth since 2019 and a conspicuous shift to improving transport infrastructure improving connectivity within and without.

JICA prelim study reports on NER stated that the lack of connectivity is a huge hindrance to the region's development. Less than 30% of all the roads here are paved (the national average is about 70%), and even when it comes to national highways, the proportion comprising at least one lane on each side of the road remains at about 50% (JICA 2016). These challenging conditions hinder the stable supply of goods and access to healthcare and educational facilities for residents and serve as obstacles to economic development. As shown in Fig. 3 Japan is currently working on six road development projects facilitating both intra-state and inter-state connectivity in the region fulfilling the JICA's target of facilitating national and rural connectivity.

Further as stated in Table 1 Phase 1 of road development began in 2016 with an improvement in the North–South arterial road in Mizoram, which borders Myanmar and the north–south arterial road in the western part of Meghalaya which shares a border with Bangladesh. In 2022, Phase 6 of the national highway improvement project commenced, spanning Tripura, and connecting to a road that leads to Chattogram, the second largest city in Bangladesh, and Matarbari, where a deep-water port is being constructed with the assistance of JICA. Apart from road connectivity, JICA makes investments in other sectors like health, agriculture, sanitation, water, and electricity. Enhanced road connectivity gives accessibility to other amnesties adding to human development. Since 2008, JICA has extended approximately INR 26,107

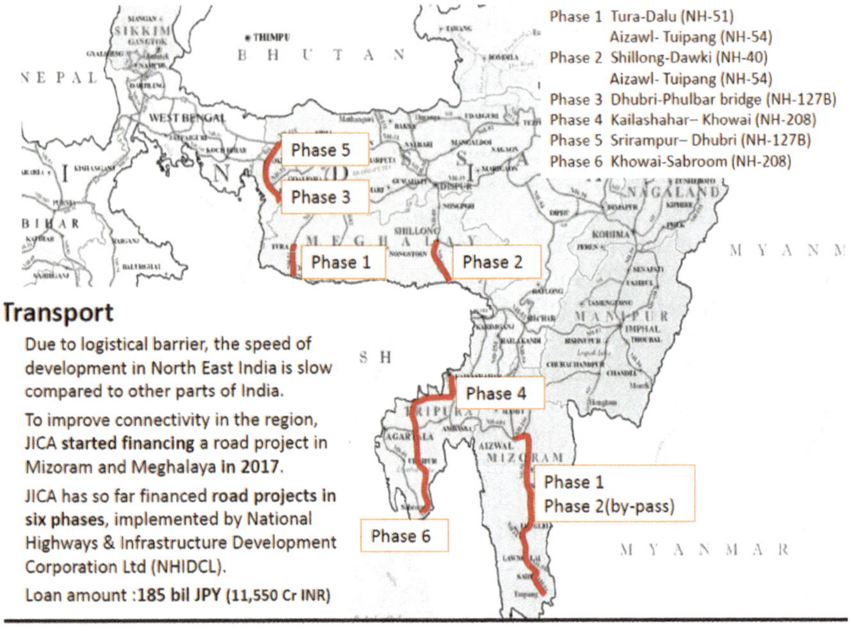

Fig. 3 JICA-assisted road connectivity projects in NER (Phase 1–6). *Source* (Taniguchi 2023)

crore as concessional ODA loans for the development of the North-east Region. Out of this Japan has extended approximately INR 11,550 crore (185 billion yen) over Phase 1 to Phase 6 of the North-east Road Connectivity Improvement Project, to help develop and improve national highways in the Northeast States, including NH-40, NH-51, NH-54, NH127B and NH-208 ((JICA 2021).

The speciality of JICA's investment in NER lies in the fact that it aligns JICA's third layer for international connectivity with India's geo-economic interests. For India, the strategic importance of NER surfaced post-1990s in the context of Look East and subsequent Act East policy. The region has been identified as the gateway to Southeast Asia. The 'North-eastern Region Vision 2020' document prepared by the Ministry of Development of the North-eastern Region (MDoNER), identified the end of the geographical border of the Northeast as the beginning of Southeast Asia (GOI 2020). Enhancing this region's connectivity to its international border adds impetus to India's geo-economic vision.

It further, enhances the credibility of existing cross-border connectivity projects launched by India. Such as the India-Myanmar-Thailand Highway Project, the road connecting Dawki and Tamabil in the Indo-Bangladesh border through Meghalaya, and the construction of the Feni Bridge connecting India and Bangladesh for transporting goods from Chittagong and Kolkata. Secondly, addressing human security issues and ensuring the development of this region allows India to have a strategic foothold in this region which shares a porous border with China in the North. The existence of human security issues juxtaposed with porous borders equally threatens

Table 1 Details of road connectivity projects in NER

Project	National connectivity	International connectivity	Year	Amount (billion yen)
Northeast connectivity phase I	Aizawl-Tuipang	Meghalaya-Bangladesh	2017	67.170
	Tura-Dalu	Mizoram-Myanmar		
Northeast connectivity phase II	Shillong-Dawki	Meghalaya-Bangladesh	2017	38.666
	Aizawl-Tuipang			
Northeast connectivity phase III	Dhubri/Phulbari	Assam-Meghalaya	2018	25.483
Northeast connectivity phase IV	Kailashahar-Khowai	Tripura-Bangladesh	2020	14.926
Northeast connectivity phase V	Srirampur-Dhubri	Assam-Meghalaya-Tripura-Bangladesh	2021	15.285
Northeast connectivity phase VI	Kowai-Tripura	Tripura	2022	23.129

Source JICA Reports 2022

a region's territorial integrity similar to so-called traditional means of territorial security threats. Hence investing in human development fends off similar threats. This is aligned with Japan's FOIP vision of transparency, peace, and prosperity, in other words limiting Chinese ascendency in the Asian region.

The effect of such road connectivity along with numerous other development projects could be observed in the current status of NER states. Most states have registered a decline in multi-dimensional poverty. Assam (32.65% to 19.35%), Tripura (16.62% to 12.11%) Nagaland (25.16% to 15.43%), Meghalaya (32.54% to 27.79%). Sikkim (3.82%) is known to have the third lowest poverty rate in India. While Arunachal Pradesh's intensity of poverty is still at 44.22%. Steps are being taken to improve accessibility to health, education, and electricity (Multi-Dimensional Poverty Index, NITI Aayog 2021).

5 Conclusion

NER is certainly a dominant example of Japan's ongoing contribution to infrastructure development which satisfies human security and national security objectives of concerned countries. Enabling cross-border connectivity to redress the landlocked nature of the region would potentially promote socio-economic activities and easy movement of goods and people. Thus, having a positive impact on issues like poverty, access to health benefits, education, electricity etc.

This region being India's window to Southeast Asia complements JICA's connectivity projects in Bangladesh including Matarbari Ports and highways and bridges from Matarbari Port to Chottogram. It has further started a new data collection survey on how to enhance cross-border services since September 2023 which focuses on target borders through NER like; Dawki (India)-Tamabil (Bangladesh), Karimganj (Assam, India)-Sylhet (Bangladesh). Interregional connectivity is the ultimate goal in Japan's contribution to building infrastructure. This would successfully support the cross-border movement of goods and services. Despite such lucrative and strategic potential of the region, JICA faces challenges in coordinating agencies, land acquisition and capacity and sustainable development of road highways in the mountainous areas.

Further, the surfacing of the COVID-19 pandemic has shown what lack of supply chain resilience does to States' economies and how crucial of an impact it has on the human security of a country. Japan, involvement in the region is quite recent compared to its engagement with Southeast Asian countries and other parts of India. It is still evaluating projects and their success rate not only in infrastructure but also in health, agriculture, forestry, sanitation, and irrigation. Further in the post-pandemic era development of NER is seen in a rather renewed light of vigour and greater interest. All in all, India's NER is a perfect example of what resolving human security could achieve and how it culminates into preserving the territorial security of the state and at the same time ensuring human development preserving the values of globalisation such as supply chain resilience, transparency, and connectivity.

References

NITI Aayog, 2 (2021) Multidimensional poverty index, 2021: Five of top six states with highest population deprived from adequate housing provision are from northeast region. Retrieved from nezine: https://www.nezine.com/info/TklienhZZ1ZEdGtVVFEvT2V3L3d2UT09/multidimensional-poverty-index,-2021:-five-of-top-six-states-with-highest-population-deprived-from-adequate-housing-provision-are-from-northeast-region.html

ADB (2017) Meeting Asia's infrastructure needs. Asian Development Bank, Philippines

Antonio Estache GG (2012) The scope for an impact of infrastructure investments on jobs in developing countries. Semantic Scholar

bank W (2022) India's Urban infrastructure needs to cross $840 billion over next 15 years: new world bank report. Retrieved from World Bank. org: https://www.worldbank.org/en/news/press-release/2022/11/14/india-s-urban-infrastructure-needs-to-cross-840-billion-over-next-15-years-new-world-bank-report

Dodson J (2017) The global infrastructure turn and Urban practice. Urban Policy Res 87–92

GOI (2020) NEC vision 2020. Retrieved from neccouncil: https://necouncil.gov.in/about-us/nec-vision-2020-0

JICA (2016) Preparatory study for North East road network connectivity improvement project. PADECO; Nippon Engineering Consultant Co., Ltd., Tokyo

JICA (2021) JICA's policy in East India development; opportunities and challenges. Retrieved from JICA: https://www.jica.go.jp/Resource/india/english/office/others/c8h0vm00009ylo4c-att/presentations_31.pdf
JICA (2022) Human security today. Japan International Cooperation Agency, Tokyo, Japan
JICA (2022) India operations. Japan International Cooperation Agency, Japan
JICA (2023) Development in Northeast India: insights from the JICA chief representative. Japan International Cooperation Agency, Tokyo
Liu JG-I (2022) The role of infrastructure in economic growth, poverty reduction, and regional integration. Retrieved from Lincoln Institute: https://www.lincolninst.edu/publications/articles/2022-08-role-infrastructure-economic-growth-poverty-reduction-regional-integration
Marchang TT (2022) Implications of infrastructure on human development in North East India: a review. ISEC Working Paper No. 535
Mishra M (2014) Prioritizing sectors for economic development in Sikkim, India. Int J Asian Bus Inf Manag 59–73
MOFA (2015) Japan's official development assistance charter. Retrieved from MOFA.GO.JP: https://www.mofa.go.jp/policy/oda/reform/charter.html
MOFA (2022) National security strategy, 2022. Retrieved from cas.go.jp: https://www.cas.go.jp/jp/siryou/221216anzenhoshou/nss-e.pdf
Lall S, Wahba S (2021) Crowded cities: new methodology in COVID-19 risk assessment. Sustainability 13:7167
Report JA (1992) Annual report. Retrieved from openjicareport: https://openjicareport.jica.go.jp/360/360/360_000_11019809.html
Sapkota JB (2015) Access to infrastructure and human development: cross-country evidence. In JICA, JICA's approach to the millennium development goals: for inclusive and dynamic development. JICA, Tokyo
Taniguchi H (2023) Enhancing connectivity in Northeast India and BBIN. Japan International Cooperation Agency, India

Towards Operationalizing Systemic Risk Management Models for Disasters: Implications to Future India Japan Cooperation

Arunabh Mitra, Chime Youdon, Nupur Tyagi, Rohit Chaurasia, and Rajib Shaw

Abstract Disasters, like floods, storms, droughts, wildfires, and landslides, are becoming increasingly common and unexpected as a result of the rapid environmental and socioeconomic changes occurring at several levels, exacerbated by the impacts of anthropogenic climate change. These disasters often result from systemic risk that outperform traditional risk management practices, posing new, unsolved policy and governance issues. While there is a growing acknowledgement of the need to look at risks holistically, current frameworks do not show the pathway to operationalizing systemic risk. Given this background, the chapter introduces a two-stage approach for understanding and operationalizing systemic risk management in the context of disasters. At Stage 1, we introduce the Systemic Risk Framework for Disasters (SRD) framework to underscore the crucial factors for evaluating and handling disaster risks. Furthermore, at the second stage, we introduce the Systemic Risk Capability Assessment (SRCA) model. This holistic methodology aims not only to assess system conditions but also to serve as a planning instrument, evaluate resilient and vulnerable systems within a given context. This can further aid to enhance the resilience specific policy and action measures between India and Japan through bilateral cooperation and it would also have a global impact.

A. Mitra (✉) · R. Shaw
Graduate School of Media and Governance, Keio University, Fujisawa, Japan
e-mail: arunabh.mitra@gmail.com

R. Shaw
e-mail: shaw@sfc.keio.ac.jp

C. Youdon
National Maritime Foundation, New Delhi, India

N. Tyagi
SPHERE India, New Delhi, India

R. Chaurasia
HCL Computers Ltd., Hyderabad, India
e-mail: rohit.chaurasia@outlook.com

R. Shaw and S. R. Choudhury (eds.), *India, Japan and Beyond*,
https://doi.org/10.1007/978-981-97-3282-1_4

Keywords Operationalizing systemic risk · Capability assessment · Port infrastructure · Disaster policy tool

1 Introduction

The increasing frequency and unpredictability of disasters can be attributed to the rapid environmental and socioeconomic changes taking place at various levels. These disasters often result from systemic risks characterized by their complexity, uncertainty, ambiguity, and the potential for cross-border consequences. Systemic risks challenge traditional risk management practices and present new and unresolved policy and governance challenges. As these risks grow, the impact of disasters on communities and entire systems becomes more severe. Various factors contribute to systemic risks, including natural disasters, climate change, ecological vulnerability, unplanned urbanization, and political or financial instability.

Systemic risks are associated with cascading and compounding effects that propagate within and between systems and sectors, such as ecosystems, health, and infrastructure. This propagation occurs through the movement of people, goods, money, and information across borders. Critical system interdependencies, exacerbated by underlying vulnerabilities, highlight the increasing need to gain a better understanding of systemic risks and potential governance solutions.

The COVID-19 pandemic serves as a prime example of how systemic risk can manifest as a catastrophe when hazards, in this case, biological ones, intersect with existing and historical vulnerabilities embedded within a system. While there is growing recognition of the need to view risks holistically, existing frameworks often lack a clear pathway to operationalize systemic risk management. The findings of this study highlight a positive shift in acknowledging systemic disaster risks. However, they also emphasize the necessity for further maturity in managing and governing these risks. To enhance the holistic management of disaster-related systemic risks and coordinate effective policy responses, this research proposes essential design principles and a conceptual framework for operationalizing Systemic Risk Management in India. This framework is intended to guide efforts to better understand, manage, and mitigate systemic risks, ultimately enhancing disaster resilience and preparedness.

This paper's main focus is on understanding systemic risk from a disaster management perspective. This understanding forms the foundation for crafting effective frameworks, policies, and strategies to assess, mitigate, and manage systemic disaster risks.

2 Understanding Systemic Risks and Its Characteristics

As defined by the Merriam Webster dictionary, a "system" is a group of items that interact regularly or depend on each other, forming a unified whole. Every system operates within specific constraints of space and time, is influenced by its surroundings, and is characterized by its structure and function. Complex systems, as described in the Complex System Journal, go beyond simple systems. They consist of numerous interconnected components and exhibit properties such as nonlinearity, emergence, spontaneous order, adaptability, and feedback loops arising from these interactions. Complex systems research is an interdisciplinary field that draws on various disciplines, including physics (for the study of self-organization and critical events), social sciences (for understanding spontaneous order), mathematics (for the study of chaos), and biology (for exploring adaptability), among others. Examples of complex systems include environments like cities, social and economic structures like ecosystems, and even the entire universe.

The roots of systemic risk research can be traced back to the field of complexity science, which gained momentum in the early 1950s and quickly advanced. What sets systemic risk research apart is its multidisciplinary nature, drawing insights from various scientific fields, including cybernetics, biology, and ecology, as well as economics.

Overall, systemic risk research has evolved as a response to the complexity and interconnectedness of modern systems and the growing recognition of the need to address and mitigate risks that transcend traditional disciplinary boundaries.

3 Attributes of Systemic Risks

Systemic risk is a concept that pertains to the possibility of an entire system breaking down, rather than just individual components or parts within that system. This term encompasses threats that possess broad, cross-sectoral, or even global repercussions, surpassing the scope of traditional risk management and national regulatory frameworks.

Systemic risks are notably intricate, characterized by numerous unknown factors, substantial ambiguities, and the capacity to affect systems beyond the original one they are associated with, novel risks are reshaping established risk management paradigms and presenting policymakers with new, unresolved challenges when it comes to decision-making in risk management emphasize that while conventional risk management has a history of effectiveness, the landscape changes significantly when dealing with interconnected, nonlinear, and globally impactful systemic risks.

Examples of such risks include climate change, the contemporary global financial system, geopolitics in the nuclear age, pandemics, and increasing wealth inequality. It is increasingly evident that modern society has given rise to hazards for which conventional risk management and governance methods appear to be insufficient, if not inadequate.

In summary, systemic risks exhibit four major attributes:

1. **Complexity**: The cause-and-effect relationships among a multitude of variables are challenging to discern and measure because they may not be readily observable.
2. **Uncertainty**: This encompasses factors like statistical variances, measurement errors, ignorance, and indeterminacy, which can erode confidence in predicting cause-and-effect chains.
3. **Ambiguity**: Ambiguity refers to the range of interpretations that different observers may derive from similar observations due to variations in the criteria or standards used to understand a situation.
4. **Ripple Effects**: Systemic risks have the potential to exert significant influence beyond their initial area of impact. When they manifest, secondary and tertiary consequences may emerge, as illustrated by the COVID-19 crisis, which has introduced new risks and opportunities across economic, social, and political domains.

In addition to these four key characteristics, systemic risks also exhibit other features that may either contribute to or result from these qualities. These include the global scope of systemic risks with transboundary or cross-sectoral implications, nonlinear cause-and-effect relationships often involving tipping points, stochastic impact structures, and their derivation from micro-level agent actions. Furthermore, differential time lags between the onset of an event and its effects, variations in signal strength, and the source of risks (exogenous versus endogenous) can also differentiate systemic risks.

4 Interconnectedness Between Disasters and Global Systemic Risk

Economic disciplines were the first to use the idea of systemic risk. Recently, it has been employed in disaster risk reduction (DRR) and resilience development too. The global disaster landscape is fast evolving. The United Nations Disaster Risk Reduction states Climate change, environmental degradation, globalization, poverty and inequality, poorly planned development, and inadequate governance are acting as significant drivers of disaster risk and contribute towards it being systemic, interconnected and cascading. The number of individuals impacted by catastrophes, including in- juries and disruptions to livelihoods, as well as the resulting economic harm, is increasing. The Interconnected Disaster Risks report published recently by United Nations University (University 2021), states that our planet is now experiencing a record number of extreme events affecting people and environment, as seen by the growing frequency of severe weather events, diseases, and man-made catastrophes.

As per a recent report by UNDRR between 2000 and 2019, EM-DAT- one of the most comprehensive international disaster databases, recorded 7348 disaster incidents connected to natural hazards (excluding bio- logical and technology disasters) around the world. In total, disasters claimed the lives of around 1.23 million people, or 60,000 every year on average, and impacted over 4 billion people. In addition, calamities cost the global economy around US$ 2.97 trillion. In comparison to the previous twenty years, these figures show a significant increase in the number of registered disaster incidents. Between 1980 and 1999, EM- DAT tracked 4212 natural-hazard-related events around the world, claiming 1.19 million deaths and affecting over 3 billion people. A total of $1.63 trillion was lost in the economy. In 2023 alone, the world has endured an alarming escalation in natural disasters, with both the frequency and intensity of catastrophic events reaching unprecedented levels. According to experts, this year has been marked as one of the most calamitous in recorded history, distinguished by a significant increase in diverse types of natural disasters. We have seen a staggering array of calamities, from earthquakes shaking the foundations of our cities to wildfires scorching vast stretches of land. Droughts have parched the earth, while floods have submerged it, landslides have reshaped landscapes, cyclones have battered coastlines, and storms have unleashed their fury across continents. These events have not only surpassed previous records but have also shown an upward trend that is deeply concerning. Data on natural disasters reveal a clear trajectory of augmentation in occurrences such as droughts, extreme temperatures, floods, mass movements, landslides, storms, and wildfires. This pattern signals an urgent call to action for global measures in disaster preparedness and climate change mitigation to prevent such events from escalating further.

Biodiversity within and between ecosystems is also declining faster than at any other point during human history. Unlike other environmental risks, Biodiversity loss and ecosystem collapse was not seen as pressing of a concern in the past. Human interventions have negatively impacted a complex and delicately balanced global natural ecosystem, triggering a chain of reactions. Over the next decade, the interplay between biodiversity loss, pollution, natural resource consumption, climate change and socioeconomic drivers will make for a dangerous mix. Given $44 trillion of economic value generation—over half the world's total GDP—is moderately or highly dependent on nature and its services and, as a result, exposed to risks from nature loss, the collapse of ecosystems will have far-reaching economic and societal consequences (Forum 2020).These include increased occurrence of zoonotic diseases, a fall in crop yields and nutritional value, growing water stress exacerbating potentially violent conflict, loss of livelihoods dependent on food systems and nature-based services like pollination, and ever more dramatic floods, sea-level rises and erosion from the degradation of natural flood protection systems like water meadows and coastal mangroves (Forum 2020).

Terrestrial and marine ecosystems are facing multiple pressure points due to their undervalued contribution to the global economy as well as overall planetary health. While not the sole drivers, at the heart of this potential catastrophe are key trade-offs and feedback mechanisms emerging from current crises. Without significant policy change or investments, the complex linkages between climate change mitigation, food insecurity and biodiversity degradation will accelerate ecosystem collapse.

The interconnectedness of nature loss and climate change is a fundamental aspect of our planet's ecological balance. A failure to address one of these challenges effectively will inevitably exacerbate the other, making it imperative that we implement mitigatory measures for both issues in our pursuit of achieving net-zero emissions. If we are unable to limit global warming to the targets of 1.5 °C or even 2 °C as outlined in international climate agreements, the consequences for biodiversity loss will be dire.

As temperatures continue to rise and precipitation patterns shift, the ongoing impact of natural disasters will emerge as the primary driver of biodiversity loss, both in terms of the composition and functionality of ecosystems. Heatwaves and droughts are already causing catastrophic mass mortality events, illustrating the vulnerability of various species to extreme climatic events. A poignant example of this vulnerability occurred in 2014 when a single exceptionally hot day resulted in the tragic death of over 45,000 flying foxes in Australia (World Wildlife Forum- Australia 2022). Furthermore, rising sea levels and intensified storms are pushing certain species to the brink of extinction, marking a troubling trend. These changes have already caused the first recorded extinctions of entire species, a somber reminder of the real and immediate consequences of climate change.

Certain ecosystems are particularly susceptible to these environmental shifts. Arctic sea-ice, warm-water coral reefs, and terrestrial ecosystems are among the most vulnerable in the near term. These environments face considerable challenges due to temperature increases, habitat loss, and altered conditions. Forests, kelp forests, and seagrass ecosystems are also at substantial risk, as climate change continues to disrupt their delicate balance.

5 Tailoring a Systemic Disaster Risks Management Framework for India's Unique Challenges

India holds a significant position when it comes to the impact of hydrometeorological disasters and various natural hazards. The country's diverse geographical and climatic conditions make it prone to a wide range of disasters, including:

1. **Floods**: India experiences both riverine and flash floods, mainly during the monsoon season. States like Assam, Bihar, and Uttar Pradesh are particularly vulnerable.
2. **Storms**: Coastal regions, especially along the Bay of Bengal and the Arabian Sea, face the threat of tropical cyclones and severe storms. Cyclones like Phailin and Amphan have had devastating impacts.
3. **Mass Movements (Wet)**: Landslides and mudslides are common in hilly and mountainous regions, particularly in the Western Ghats and the Himalayas.

4. **Extreme Temperatures**: India experiences extreme heatwaves during the summer months, with temperatures soaring to dangerous levels in some parts of the country.
5. **Droughts**: Certain regions, such as parts of Rajasthan and Maharashtra, are prone to prolonged droughts, which can have severe socio-economic implications.
6. **Wildfires**: Forest fires are a recurring issue in states with dense forests, like Uttarakhand and Himachal Pradesh.

The impacts of these disasters can be wide-ranging, affecting not only the environment but also human lives, livelihoods, and infrastructure. As such, disaster preparedness, mitigation, and response are critical aspects of India's disaster management efforts. The country has made significant progress in disaster management and response, but there are ongoing challenges, particularly in addressing systemic risks and ensuring the resilience of communities and critical infrastructure (Mitra and Shaw 2023).

Historically, India has excelled in disaster response, relief, and reconstruction efforts, but it has faced significant challenges in disaster prevention, mitigation, and preparedness. Despite its apparent strengths, India's Disaster Risk Management (DRM) framework has not comprehensively addressed systemic risks. Several critical issues need attention:

1. Inadequate Definition and Scope: Systemic risks have not been fully defined and scoped within India's DRM framework, leading to challenges in identifying and addressing these risks effectively.
2. Scale Discordance: There is a disconnect between international and national policies when it comes to systemic risks. Harmonizing these policies is crucial for a more coordinated and effective approach.
3. Lack of Established Approaches: India lacks established approaches for assessing and managing systemic risks. Developing such approaches is essential for proactive risk reduction.
4. Insufficient Database: The country lacks a comprehensive database on systemic disaster events, hindering the ability to analyze and respond to these risks adequately.
5. Institutional and Financial Gaps: Current institutional and financial mechanisms are inadequate for addressing systemic risks. New structures and resources may be needed to address these challenges effectively.
6. Integration Challenges: Integrating climate change action and Disaster Risk Reduction (DRR) measures with systemic risk management remains a challenge, despite their interconnectedness.

At a strategic level, India can benefit from adopting a systemic lens to conceptualize and interpret disaster risk comprehensively. This approach considers the interrelated nature of risks and their impacts on various sectors and systems. To bridge the identified gaps and support sustainable development, there is a pressing need to create a "Systemic Risk Capability Assessment Model" for disasters.

This model should adopt a multisectoral, multidisciplinary strategy that addresses reaction, recovery, and development simultaneously and holistically. Such an approach recognizes that no single entity possesses the mission, capability, or resources to handle all aspects of disasters, particularly in multi-hazard scenarios. Therefore, assessing the capability to manage systemic risks becomes paramount.

Currently, there is no existing model to operationalize this concept effectively. Hence, the development of the "Systemic Risk Capability Assessment Model" represents a significant contribution to India's DRM capabilities. It will provide a structured framework to identify, assess, and manage systemic risks, ultimately enhancing the country's resilience and preparedness in the face of complex disaster scenarios.

6 Framework for Systemic Disaster Risk Management

In the rapidly evolving landscape of the twenty-first century, understanding and addressing systemic threats has become paramount. The chapter underscores this urgency and lauds the establishment of the Global Risk Assessment Framework (GRAF) by the UNDRR. This significant move is pivotal in guiding decisions and behavioural shifts in response to these systemic risks. Furthermore, the GRAF's mission to enhance risk management across diverse timeframes and regions is acknowledged and appreciated.

Delving deeper into the proposed approach, the paper suggests a bifurcated approach to holistically address systemic risk management in the context of disasters. Firstly, the Systemic Risk Framework for Disasters (SRD) is introduced (as depicted in Fig. 1). Serving as an extension of the GRAF 2020–2030, the SRD places emphasis on key factors essential for risk evaluation and management. This framework is especially tailored for the Indian scenario, considering the unique challenges and opportunities the region presents. The second part of the solution introduces the Systemic Risk Capability Assessment (SRCA) Model. This innovative model seeks to bring the principles and strategies of the SRD framework down to the grassroots level. It offers a methodical way to gauge and boost disaster resilience in the face of looming systemic risks, ensuring that local communities are equipped to handle and mitigate these challenge.

Within the SRD, the 'Design' dimension stands out. It underscores the importance of establishing a broad vision for systemic resilience by integrating Climate Change Adaptation (CCA) and Disaster Risk Reduction (DRR) into strategic planning. This dimension also stresses the need to understand the dynamics of various systems associated with disaster and climate risk. Furthermore, the 'Design' dimension highlights the necessity for consistent CCA and DRR policies, risk-informed planning, and the advocacy for systems to be 'Resilient by Design'. To measure the effectiveness of these strategies, 'Key Performance Indicators' (KPIs) and 'Key Risk Indicators' (KRIs) are introduced, ensuring that shared goals are achieved and monitored. The 'Manage' dimension of the SRD calls for a thorough risk assessment, which includes a multi-hazard analysis to create a detailed risk portfolio. This dimension emphasizes strengthening interconnected systems through various actions, including the implementation of nature-based solutions. Additionally, the

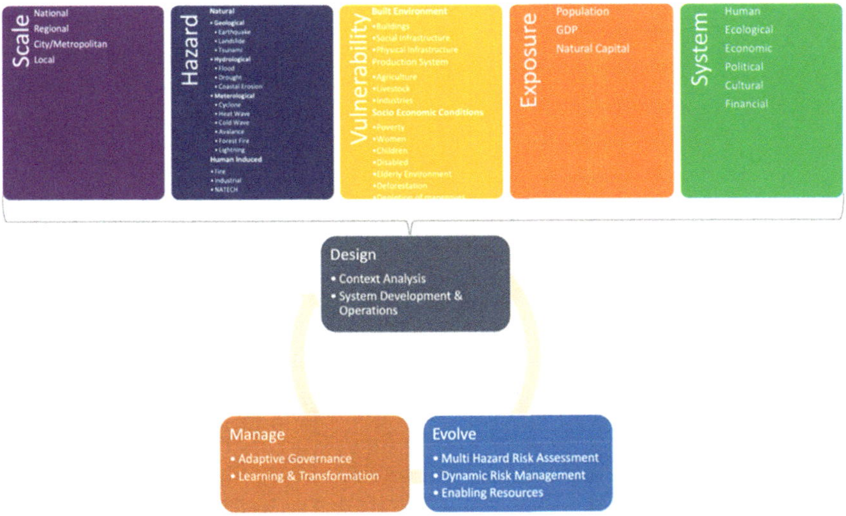

Fig. 1 Schema for systemic risk framework for disasters (SRD) adapted from Mitra and Shaw (2022)

'Manage' dimension highlights the necessity for resources and support to ensure risk readiness and response, ensuring that communities are not only prepared but also equipped to handle systemic threats. Lastly, the 'Evolve' dimension of the SRD advocates for a multi-level governance approach to address disaster and climate risks. This dimension stresses the importance of managing feedback loops effectively, building capacity, and ensuring that communities and systems can adapt and grow in the face of changing risk landscapes.

In conclusion, the Systemic Risk Framework for Disasters (SRD) offers a comprehensive lens through which systemic risks in the disaster risk environment can be viewed and addressed. Addressing these risks necessitates a holistic approach, one that considers not just technical solutions but also political systems, processes, and human values. Effective risk reduction strategies must be all-encompassing, ensuring a comprehensive approach to mitigating systemic risks and building a resilient future.

Subsequently, as Step 2 the Systemic Risk Capability Assessment (SRCA) model is unveiled to localize the SRD framework's application. The SRCA model gauges disaster resilience through two primary facets: Intra System and Inter System resilience, encompassing 4 dimensions and 14 sub dimension in total. The key dimensions include Institutional design, Institutional capacity, System Understanding and Governance. From an Intra Systems perspective there are 5 key systems which include Physical and Infrastructure System, Economic and Finance System, Ecological and Environmental System, Human, Social, and Cultural System and Political and Governance System. Data is gathered via a questionnaire, with respondents across the ecosystem rating and ranking parameters. The model incorporates the Weighted Mean Index (WMI) and Aggregate Weighted Mean Index (AWMI)

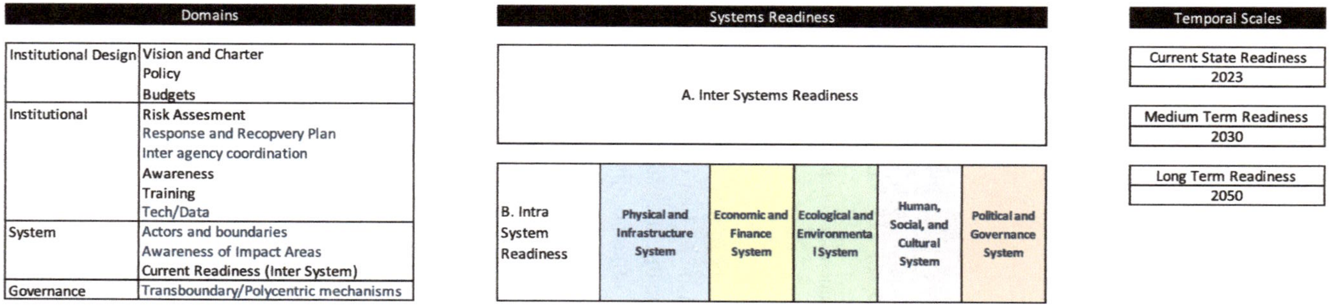

Fig. 2 Schema for SRCA survey domains, systems, and temporal scales

to derive resilience scores. These scores, spanning from 1 to 5, denote readiness for climate-induced disasters across short, medium and long term timeframes. The quality of these outcomes is contingent upon the expertise of the survey participants. Crucially, the SRCA model's results can serve as policy recommendations, aiding in refining disaster management tactics and emphasizing resilience enhancement priorities.

In conclusion, the introduction of the SRD and SRCA marks a significant milestone in our journey towards a safer and more resilient India. By addressing systemic risks holistically, we're not just enhancing our disaster management practices but also taking a step closer to our broader sustainability goals. As we brace for the challenges of the future, it's this comprehensive and integrated approach that will guide our way.

7 Synergizing Resilience for Bilateral Cooperation Between India and Japan in Systemic Risk Governance and Disaster Management

The collaboration between nations, especially in areas of mutual concern like disaster management, can lead to transformative solutions. Given the vulnerabilities both India and Japan face concerning natural disasters, leveraging the SRD and SRCA frameworks bilaterally can be a game-changer. This cooperative initiative, rooted in shared experiences and mutual learning, has the potential to not only address immediate disaster-related challenges but also to shape a resilient future for both countries and set a precedent for global disaster management strategies. Here's how these frameworks can be the cornerstone of a cooperative initiative between India and Japan:

1. **Shared Knowledge and Best Practices**: Japan, with its advanced disaster management strategies, can benefit from the systemic perspective of the SRD and SRCA frameworks. Conversely, India can learn from Japan's technological advancements and infrastructural resilience measures. This mutual exchange of knowledge can lead to the development of more comprehensive and effective disaster management strategies.
2. **Joint Research and Development**: Both nations can invest in joint R&D initiatives, focusing on enhancing the SRD and SRCA frameworks. This can involve developing advanced tools for systemic risk assessment, creating shared databases for systemic disaster events, and devising innovative financial mechanisms tailored for systemic risks.
3. **Capacity Building and Training**: Leveraging the SRCA's capability assessment model, India and Japan can organize joint training programs. These programs can focus on building capacities in both nations, ensuring that officials, responders, and even communities are well-equipped to handle systemic risks.

4. **Technological Collaboration**: Japan's technological prowess can be harnessed to refine the tools and methodologies under the SRD and SRCA frameworks. This can include developing advanced early warning systems, real-time risk assessment tools, and even AI-driven predictive models for systemic risks.
5. **Policy Integration and Harmonization**: Through bilateral cooperation, both nations can work towards integrating their respective disaster management and climate change policies. This harmonized approach can ensure that systemic risks are addressed comprehensively, taking into account the global implications of local actions.
6. **Global Advocacy and Leadership**: Given the global stature of both India and Japan, their bilateral cooperation can set an example for other nations. By showcasing the effectiveness of the SRD and SRCA frameworks, they can advocate for a systemic approach to disaster management at international forums, leading to global policy shifts.
7. **Resilience-Specific Policy and Action Measures**: The collaboration can lead to the development of specific policies that focus on building resilience in critical sectors like infrastructure, agriculture, and healthcare. These policies, backed by actionable measures, can ensure that both nations not only recover from disasters but also thrive in their aftermath.

In essence, the bilateral cooperation between India and Japan, centered around the SRD and SRCA frameworks, can redefine how nations approach disaster management. By focusing on systemic risks and building resilience, this collaboration can not only safeguard the interests of both nations but also provide a roadmap for the world. In an era where global challenges demand global solutions, this India-Japan initiative can be a beacon of hope and leadership.

8 Implications for Global Disaster Risk Management

As the world grapples with an increasing frequency of complex and interconnected disasters, it is imperative to consider the following implications:

1. **Recognition of Systemic Risks**: Foremost among these implications is the acknowledgment of systemic risks as a pivotal challenge within the realm of Disaster Risk Management (DRM). The study posits that contemporary disasters can no longer be perceived as isolated events but rather as outcomes arising from systemic frailties and intricate interdependencies. Grasping this reality constitutes an essential initial stride in addressing the evolving landscape of disaster risks.
2. **Shift Towards Holistic Risk Management**: The conventional DRM approach has traditionally accorded precedence to post-disaster response, relief, and reconstruction at the expense of pre-emptive measures such as prevention and mitigation. Nevertheless, the investigation accentuates the imperative of effecting a

paradigmatic shift toward holistic risk management. This transformative trajectory necessitates the seamless integration of disaster prevention, mitigation, preparedness, response, recovery, and development into a comprehensive and unified strategy.

3. **Incorporating Systemic Risk Assessment**: The research introduces the Systemic Risk Capability Assessment (SRCA) model, a structured framework tailored to assess and enhance disaster resilience at the local level. This model assumes a position of paramount importance in the arsenal of DRM practitioners worldwide, furnishing them with a systematic tool to methodically scrutinize and ameliorate systemic risks within their respective regions.
4. **Global Collaboration and Knowledge Sharing**: Given the inherently transnational character of systemic risks, the imperative of global collaboration looms large. DRM initiatives are compelled to transcend national boundaries and pivot towards a shared comprehension of systemic risks and their far-reaching implications. Consequently, the study enjoins heightened levels of knowledge dissemination, the propagation of best practices, and international cooperation as indispensable facets in efficaciously confronting systemic risks.
5. **Policy and Governance Adaptation**: The dynamic evolution of systemic risks necessitates a continual realignment of policies and governance structures. National and international DRM frameworks must proactively assimilate systemic risk considerations to ensure their currency and effectiveness in addressing nascent challenges.
6. **Research and Capacity Building**: The study underscores the criticality of perpetuated research and capacity building within the ambit of DRM. As systemic risks evolve, the proficiencies and knowledge base of DRM professionals must evolve in tandem. Substantial investment in research endeavors and educational initiatives is pivotal to augmenting the resilience of communities and intricate systems.
7. **Integration of Climate Change Action**: Climate change emerges as a pivotal driver of systemic risks, and its integration into DRM strategies assumes paramount significance. Harmonizing Disaster Risk Reduction (DRR) and Climate Change Adaptation (CCA) endeavours is of paramount importance to address the cumulatively deleterious consequences of climate-induced disasters.
8. **Preparedness for Unpredictability**: Systemic risks are inherently characterized by their intricate nature, imbued with complexity, uncertainty, and ambiguity. DRM strategies must, therefore, accord precedence to preparedness for the mercurial nature of systemic disasters. Early warning systems, adaptive governance frameworks, and versatile response mechanisms represent essential components in this context.

9 Conclusion

In summing up the various insights drawn from the preceding sections, it becomes clear that the battle against systemic risks in disaster management is both complex and critical. The increasing frequency of disasters, fueled by a host of factors including ecological vulnerabilities, climate change, and socioeconomic upheavals, demands an innovative and integrated approach to disaster risk management (DRM). The development of an operational framework towards managing systemic risks emerges as milestone of progress in India's DRM capabilities, offering a structured methodology to understand, evaluate, and manage the layered complexities of systemic risks. This step forward is not just a stride toward bolstering India's resilience but also a step towards achieving broader sustainability objectives.

Furthermore, the symbiotic collaboration between India and Japan through the SRD and SRCA frameworks is set to redefine international disaster management practices. By jointly addressing systemic risks and prioritizing resilience, this partnership could pave the way for a more secure future for both nations and potentially, the world at large. Similarly, The implications for global DRM are profound and manifold. Recognizing systemic risks as a critical concern and shifting towards holistic risk management are fundamental moves towards fortifying global disaster resilience. International collaboration and knowledge sharing, adaptive policy and governance, ongoing research, and capacity building are pivotal to this endeavour.

In essence, the path forward in managing disaster risks lies in a unified, adaptable, and forward-thinking approach. The collaboration opportunities between India and Japan provide a promising template for such initiatives, signalling a shift towards a more resilient and disaster-prepared global community.

References

Berker M (2023) Months of devastation: world battered by rising number of natural catastrophes. https://www.aa.com.tr/en/environment/months-of-devastation-world-battered-by-rising-number-of-natural-catastrophes/2992602

Editor (n.d) Agreements with countries. Retrieved from Disaster Management Division- Ministry of Home Affairs, Government of India: https://ndmindia.mha.gov.in/mou-with-countries

Editor (n.d) Global risks 2033: tomorrow's catastrophes. Retrieved from World Economic Forum: https://www.weforum.org/publications/global-risks-report-2023/in-full/2-global-risks-2033-tomorrow-s-catastrophes/

Forum WE (2020) Half of world's GDP moderately or highly dependent on nature, says new report. Retrieved from https://www.weforum.org/press/2020/01/half-of-world-s-gdp-moderately-or-highly-dependent-on-nature-says-new-report/

Mitra A, Shaw R (2022) Systemic risk from a disaster management perspective: a review of current research. SSRN Electron J

Mitra A, Shaw R (2023) Systemic risk management in India: an analytics perspective. Prog Disaster Sci

Russo A (2020) Half of world's GDP moderately or highly dependent on nature, says new report. Retrieved from World Economic Forum

University (2021) U. N. (n.d.) United Nations University. Retrieved from Interconnected Disaster Risks Report: https://s3.eu-central-1.amazonaws.com/interconnectedrisks/reports/UNU_Interconnected_Disaster_Risks_Report_210908_2022-08-26-221640_jogi.pdf

World Wildlife Forum- Australia (2022) Global wildlife populations fall 69%, Australia records localised extinctions. https://wwf.org.au/news/2022/global-wildlife-populations-fall-69-percent-australia-records-localised-extinctions/

Critical Rare Metal and Collaboration of India and Japan

Neha Mishra and Nabeel Mancheri

Abstract In a zero-carbon scenario, the energy sector's demand for critical minerals might increase by a factor of six by the year 2040. Japan and India have established targets of attaining net zero greenhouse gas emissions by 2050 and 2070, respectively. Nevertheless, their supply chains for critical minerals are currently incapable of fulfilling the ambitious objectives for the clean energy technologies like electric vehicles, solar and wind energy, battery storage, power transmission, and green hydrogen. In response to the supply chain bottlenecks for critical minerals, countries are adopting resilience measures by engaging in cross-country collaborations and global partnerships. India's untapped reserves and rising production potential, in conjunction with Japan's financial, technological, and market capacity, are increasingly serving as examples of the development of such resilient partnerships. Japan and India both face formidable yet complementary challenges: Japan's domestic reserves are virtually nonexistent. India is unable to take advantage of its huge terrestrial rare earth metal reserves due to a lack of funding for research and development and technological innovation. The combined strength of India's reserve and Japan's technology

Ms. Neha Mishra is an Executive Officer for Policy and International Affairs at the Society of Indian Defence Manufacturers (SIDM). Prior to this, she was working as an Associate Fellow at the Centre for Air Power Studies, where she worked on her book project titled "Strategic Use of Rare Earth Elements: Major Powers and Emerging Players". In addition, She is a doctorate candidate at the University of Delhi (2022–2026).

Dr. Nabeel Mancheri is currently serving as the Secretary General at REIA–The Global Rare Earth Industry Association since June 2019. Prior to this, he worked as a Project Manager at Materials Engineering, KU Leuven (2018–2019). He also worked as a Marie Curie Fellow at Leiden University, Institute of Environmental Sciences (2015–2018), and as a Research Fellow at the University of Tokyo (2013–2015). Nabeel also has experience as an Assistant Professor at National Institute of Advanced Studies (2010–2013). He holds a Doctor of Philosophy (Ph.D.) in International Business/Trade/Commerce from Jawaharlal Nehru University (2004–2011).

N. Mishra (✉)
Society of Indian Defence Manufacturers, Confederation of Indian Industry, New Delhi, India
e-mail: mneh3039@gmail.com

N. Mancheri (✉)
The Global Rare Earth Industry Association (REIA), Brussels, Belgium
e-mail: nabeelmnc@gmail.com

R. Shaw and S. R. Choudhury (eds.), *India, Japan and Beyond*,
https://doi.org/10.1007/978-981-97-3282-1_5

has the capacity to significantly enhance human security, environmental sustainability, economic growth, technical innovation, and overall supply chain resilience. Given the above, the purpose of this chapter is to examine the potential of the India-Japan partnership in establishing a sustainable supply chain for critical minerals. The chapter has examined the prospects for India-Japan cooperation in the areas of: (a) diminishing reliance on Chinese imports; (b) fostering self-reliance; (c) attaining energy transition goals; and (d) working towards resilience. The first part of the chapter discusses the role of rare metals in technological innovation, environmental sustainability, and resilience initiatives. The initial section of the chapter examines the significance of rare metals in technological innovations, environmental sustainability, and resilience projects. The next section elucidates India's significant rare earth potential in relation to its reserves, production, and policies. The subsequent part explores Japan's capacity for recycling rare earth elements, despite its lack of reserves. The concluding portion examines the India-Japan partnership in rare metal collaboration, aimed at enhancing resilience and decreasing reliance on China.

Keywords Critical rare metals · Self-reliance · Supply chain resilience · Value chain collaboration · Recycling

1 Introduction

The notion of critical and strategic mineral resources has a long history, dating back over a century to World War I. During this time, the conflicts started with nations facing significant shortages of supplies necessary to support their war efforts. Nevertheless, this idea has been significantly expanded as a result of tremendous advancements in aerospace, metallurgy, and electronics technology. The production of new technologies in the twenty-first century necessitates a substantial supply of raw minerals, leading to a consistent need for mineral resources. It is important to acknowledge that the classification of strategic minerals differs for each country based on factors such as their consumption, production, requirements, availability of alternatives, and technological progress. These classifications are often determined based on two primary variables: vulnerability and criticality.[1]

The critical minerals, whether in their original or refined state, are susceptible to being inaccessible or excessively costly, thereby causing disruptions in their availability. These minerals have a deficiency in substitutability and recycling mechanisms. The global aggregation of extraction and processing activities, the governance structures, and the ecological footprints in resource-rich countries have an adverse effect on the risks associated with availability. Amidst the global supply chain crises in critical minerals, governments are progressively adopting resilience tactics by means of collaboration or partnerships.

[1] Lele and Bhardwaj (2014).

Rare earth elements (REEs) is one such critical metal that plays a pivotal role in various modern technologies, including electronics, renewable energy systems, and defense applications. However, the global supply chain for REEs is vulnerable due to their concentration in a few countries and the complexities of extraction and processing, making their supply chain resilience crucial for maintaining global technological advancement and security. Therefore, focusing on the resilience of REE ecosystems is imperative to mitigate risks associated with supply disruptions and meet the growing demand. This chapter focuses to explore how India's abundant untapped deposits and growing production capacity of REEs, along with Japan's advanced technologies, demonstrate the formation of resilient partnerships that contribute to significant progress in areas such as human security, environmental protection, development, innovation, and resilience. Japan and India have both established targets of attaining carbon neutrality by 2050 and 20,700, respectively. However, their CRM supply chains are presently incapable of achieving ambitious goals for the implementation of solar, wind, battery storage, power transmission, and green hydrogen.[2] The production of rare earths demands a complex and extensive value chain ecosystem, involving multiple processes and plants, to transform reserves into the final product. The activities encompassed in this list are mining, mineral beneficiation, rare earth extraction, separation, oxide refining, metal extraction, alloy manufacture, and obtaining necessary legal approvals.

Presently, mining corporations are actively seeking fresh rare earth element (REE) deposits worldwide that may be commercially developed, while simultaneously looking to recover REEs from waste streams of iron, phosphate, bauxite, uranium, and so on. Many countries are looking at the recycling of rare earth elements (REEs) from pre-consumer scrap, industrial residues, and REE-containing end-of-life products because they lack economically and operationally feasible primary resources within their borders. It is recommended to recycle REE (rare earth elements) because of the issue of maintaining balance. When neodymium is obtained from rare earth element (REE) ores through primary mining, there is an excessive presence of the more prevalent elements, namely lanthanum and cerium. Therefore, recycling neodymium can significantly reduce the total amount of rare earth element ores that need to be harvested.[3]

2 Rare Metals: Innovation, Environment, and Resilience

Each nation's decarbonization objectives necessitate significant quantities of critical raw minerals (CRM), including copper, aluminum, cobalt, lithium, graphite, rare earth elements, and other minerals and metals. In international discourse, CRM is a term used to refer to minerals that are also known as "critical minerals," "critical

---

[2] Isetani et al. (2022).

[3] Binnemans et al. (2013).

strategic minerals," "battery minerals," and "energy transition metals. "In a zero-carbon future, the energy sector's demand for CRM may increase by a factor of six by the year 2040.[4]

Critical minerals refer to mineral resources that are essential inputs in the production process of an economy. The elements germanium, gallium, indium, selenium, tellurium, and other similar elements, known as critical elements, are not usually present in such quantities that would warrant their extraction as primary commodities. Most of these metals are produced as byproducts during the manufacturing and refining processes of primary metals like lead, zinc, copper, aluminum, and tin. An important factor to take into account is the very low proportions at which they are present, even within ore deposits that have the potential to be economically viable. The production process has a significant dependence on technology.[5] Although some minerals are utilised in conventional industries, a considerable portion of them are indispensable for the advanced technical goods required in clean energy, national security, information technology, aviation, and space exploration.

Rare Earth Elements (or REEs), have been a focal point in modern geopolitics due to its numerous strategic applications in both military and civilian domains (see Table 1). Due to their increasing significance in all areas of technological progress, it is unsurprising that countries are striving to safeguard their access to these resources. The global demand for rare earth metals has experienced a significant surge and is projected to reach 315,000 tonnes by 2030. However, this increased demand also brings with it a heightened supply risk. Factors contributing to this risk include China's dominant position in the rare earth supply chain, the global shift towards clean technologies such as electric vehicles, wind power, and solar power, as well as the growing recognition of the economic importance of these metals.[6]

It is worth mentioning that the production of Rare Earth Elements (REEs) was insignificant prior to the 1960s. However, the application of these elements, namely Europium (Eu), was only recognised following the advancement of commercial colour televisions. Since then, rare earth elements (REEs) have been widely utilised in advanced electronics and electronically regulated systems, particularly in the aviation industry and defence technologies. In the 1950s, South Africa, India, and Brazil made first attempts to improve the processing of rare earth ores. However, these efforts did not gain global recognition. The Mountain Pass rare earth deposit, established by Molybdenum Corporation in 1953, was the first western deposit of its kind in the United States. It gained worldwide prominence and surpassed the global production of the rare earth elements (REEs). The Mountain Pass activity significantly boosted the production of Rare-earth Oxides (REO) and successfully kept up with the growing global demand. During the 1980s, the Nuclear Regulatory Commission and the International Atomic Energy Agency implemented a regulation that limited the extraction

[4] Karali and Shah (2022).

[5] Ministry of Mines, "Rare Earths and Energy Critical Elements: A Roadmap and Strategy for India", July 2012, https://mines.gov.in/admin/storage/app/uploads/64a3eae23bee21688464098.pdf. Accessed on 10 September 2023.

[6] Jensen (2022).

Table 1 The seventeen rare earth elements

Rare earth name	Discovery year	Atomic name & number	The chemist who named it	Light/heavy REE	Critical/uncritical	Abundance in the Earth Crust (parts per million)	Usage
Yttrium	1788	Y-39	Sweden-Johan Gadolin	Heavy	Critical		Metal alloys and night vision goggles
Cerium	1803	Ce-58	Europe- John Jacob Berzelius	Light	Excessive	60.0	Automobiles
Lanthanum	1839	La-57	Sweden-Carl Gustaf Mosander	Light	Uncritical	30.0	Optical glasses, night vision goggles
Erbium	1842	Er-68	Sweden-Carl Gustaf Mosander	Heavy	Critical	2.1	Fibre and optic cables machine
Terbium	1843	Tb-65	Sweden-Carl Gustaf Mosander	Heavy	Critical	0.7	Visual display and fuel cells
Ytterbium	1878	Yb-70	Sweden-Jean-Charles Galissard	Heavy	Excessive	2.0	Stainless steels
Holmium	1878	Ho-67	Sweden-Per Teodor Cleve	Heavy	Excessive	0.8	High strength magnets and glass colouring
Scandium	1879	Sc-21	Scandinavia-Lars Fredrik Nilson	Heavy	Critical	16.0	Defence equipment
Samarium	1879	Sm-62	France-Paul-Émile Lecoq de	Light	Uncritical	5.3	Nuclear reactors
Thulium	1879	Tm-69	Sweden-Per Teodor Cleve	Heavy	Excessive	0.3	Lasers, portable, X-ray machines
Praseodymium	1885	Pr-59	Germany-Carl Auer von Welsbach	Light	Uncritical	6.7	Lasers
Neodymium	1885	Nd-60	Germany-Carl Auer von Welsbach	Light	Critical	27.0	Laser ranger finders communication

(continued)

Table 1 (continued)

Rare earth name	Discovery year	Atomic name & number	The chemist who named it	Light/ heavy REE	Critical/ uncritical	Abundance in the Earth Crust (parts per million)	Usage
Dysprosium	1886	Dy-66	France- Paul-Émile Lecoq de	Heavy	Critical	3.8	Permanent magnets
Europium	1886	Eu-63	Europe- Eugène-Anatole Demarçay	Heavy	Critical	1.3	Optical fibres
Gadolinium	1886	Gd-64	Sweden- Johan Gadolin	Heavy	Uncritical	4.0	X-ray and scanning
Lutetium	1907	Lu-71	Austria- Carl Auer von Welsbach and Georges Urbain	Heavy	Excessive	0.4	Petroleum refining
Promethium	1947	Pm-61	America- Jacob A. Marinsky, Lawrence E. Glendenin, and Charles D. Coryell	—	–	10^{-18}	Doesn't exist like earth, but its isotopes are available in the radioactive elements

Source Rose Ragsdale, "Rare earth metals see new medical uses", *Metal Tech News*, June 27, 2020, https://www.metaltechnews.com/story/2020/04/29/tech-metals/rare-earth-metals-see-new-medicaluses/217.html#:~:text=Holmium%20lasers%20emit%20at%202.1,problems%20with%20a%20thulium%20laser. Accessed on 16 November, 2023

of minerals, such as Monazite, which also included thorium and uranium. The levy on these minerals constrained the foundation of the US rare earth market and supply chain. The implementation of this legislation marked the conclusion of the Mountain Pass Era, as it led to a surge in rare earth oxide (REO) production in China, particularly in Baotou, Inner Mongolia, throughout the 1990s.[7]

In 1990s, China effectively capitalised on the diminishing US supply chain in the rare earth market to strengthen its dominance in rare earth elements (REE) and acquire geopolitical power. China emerged as the dominant force in all parts of the rare earth element (REE) supply chains, including production, processing, consumption, and research & development. The Bayon-Obo in Inner Mongolia, China, has emerged as the predominant institution due to its possession of both. Monazite and bastnaesite minerals that are the primary sources of rare earth elements (REEs). In 2010, China, which had a monopoly on the supply chain of rare earth elements (REE), implemented export limitations on crucial REE minerals to Japan. This action was taken in response to the incarceration of a Chinese fishing trawler captain, which had a significant impact on Japan's auto sector. Japan, being the major importer of REEs, was particularly affected by these restrictions.[8] The United States, European Union, and Japan regarded China's resource geostrategy as a significant danger, prompting them to initiate legal action. As a result, the World Trade Organisation ruled that China's embargoes on rare earth element exports were invalid.[9]

The rare earth elements dilemma, which emerged as a consequence of Chinese resource diplomacy during the 2010 event, marked a significant turning point. It compelled the attention of major global powers to safeguard their critical mineral requirements and expand technological demands. The United States, Australia, and Russia initiated the development of their rare earth industries, focusing not only on mining but also on ore processing and REO manufacture. In addition to the major global powers, several other countries with substantial reserves have been actively expanding their production capabilities. These countries include Greenland, Vietnam, Brazil, India, Myanmar, and others. Nevertheless, the incentives and rationale varied among countries in their pursuit of developing their rare earth industry. The primary objective of the major powers was to resist China's resource diplomacy and safeguard their scientific progress, while the emerging nations prioritised their economy and growth. Irrespective of the underlying reason, they have contributed to the establishment of a strategic game using rare earth elements. The significance of Rare Earth Elements (REE) was further amplified by the 2018 trade war between the United States and China, China's subsequent embargo in response,[10] and the recent ban on

[7] Kennedy (2016).

[8] Bradsher (2010).

[9] Ibid.

[10] Ibid.

the exporting of gallium and germanium containing industrial products and materials to the USA.[11]

The importance lies in the impact of these events on the interests of powerful nations, developing economies, and those dealing with the challenge of managing rare earth elements. These elements have been enhancing the actions and efforts of countries heavily reliant on rare earth metals and their collaboration with countries possessing substantial reserves, which has turned the latter into influential players. Countries are participating in resilience initiatives as part of broader global efforts to reduce their reliance on a single country, like China, for rare metals. As an illustration, the Friendshorsing policy of the US government seeks to establish alliances with nations that have similar values, such as the Quadrilateral Security Dialogue (Quad), the Mineral Security Partnership (MSP), and other similar efforts. These objectives entail member countries reducing their dependence on China by establishing their own rare earth mining and processing facilities.[12]

Moreover, the significance of Rare Earth Elements (REE) in the development of clean technology has also increased with escalating environmental concerns and the global commitment to the Paris Agreement's Intended Nationally Determined Contributions (INDCs). These factors are driving the global adoption of clean and developing technologies. Rare earth elements (REEs) have the potential to reduce the cost of infrastructure construction for renewable energy. The seventeen rare earth elements (REEs) possess significant utility potential in improving the infrastructure and conductivity of renewable energy sources for the purpose of power generation. For instance- the Solar and onshore-offshore wind power utilises REEs such as erbium, neodymium, dysprosium, praseodymium, and yttrium for their production.[13]

Nevertheless, the focus of conversations around rare earths internationally revolves around strategies for recycling rare earth elements, methods for minimising or substituting the use of rare earths in products, and the development of alternative products that do not rely on rare earths.

[11] Reuters, "China gallium, germanium export curbs kick in; wait for permits starts", August 1, 2023, https://www.reuters.com/markets/commodities/chinas-controls-take-effect-wait-gallium-germanium-export-permits-begins-2023-08-01/. Accessed on 10 November, 2023.

[12] International Energy Agency, "Critical Minerals Mapping Initiative", October 27, 2022, https://www.iea.org/policies/16093-critical-minerals-mapping-initiative. Accessed on 20 September, 2023.

[13] International Energy Agency, "The Role of Critical World Energy Outlook Special Report Minerals in Clean Energy Transitions", May 2021. https://www.iea.org/reports/the-role-of-critical-minerals-in-clean-energy-transitions.

3 India's Untapped Reserves and Upstream Minining

India's commitment to the energy transition and its objective of attaining net-zero emissions by 2070 have increased the significance of critical minerals. The Indian Parliament has indicated a significant shift in the mining sector and recognised the crucial economic potential of critical minerals with the enactment of the Mines and Minerals (Development & Regulation) Amendment Act, 2023. The goal is to enhance India's involvement in the global commercial sector of critical minerals, hence promoting its clean energy goals. This has occurred concurrently with the advancement of two notable initiatives that seek to attain self-reliance and reduce India's susceptibility in its supply chain. India joined the Mineral Security Partnership (MSP) sponsored by the United States and also officially launched a comprehensive list of 30 critical minerals in 2023.[14]

While India would need to rely on imports for green technology during the next twenty years, it is imperative to optimise the usage of domestic resources to fulfill its long-term needs. India can bolster its readiness by creating the requisite infrastructure for the investigation and exploitation of minerals. India's lack of identified reserves for specific critical minerals necessitates a concentrated effort to build supply networks and acquire international mining assets in order to ensure a reliable and constant supply. Out of these critical minerals, the use of rare earth elements holds great importance in India's goals related to zero-emission automobiles, as well as defence and security technologies. India's principal markets for rare earth elements (REEs) include a wide array of uses encompassing permanent magnets in electricity generators and motors, catalysts for chemical reactions, polishing operations, batteries, electronics, defence technologies, the wind energy industry, aviation, and space exploration. Notable outputs consist of communication and navigation systems, night vision goggles, hybrid and tactical wheeled vehicles, armoured vehicles, laptops, LED monitors/TVs, smartphones, electric drive trains, wind turbines, aircraft components, vehicle components, speakers, steel production, battery anodes, chemical catalysts, glass production, and specialised glass lenses.[15]

The escalating occurrences of global supply chain disruption are hindering India's progress towards achieving an *Atmanirbhar Bharat* and reducing India's ability to compete on a global scale. This demonstrates that the achievement of India's objectives for domestic economic growth and sustainability relies on the investigation and extraction of an extensive range of minerals. India's rare earth production accounts for a mere one percent (2,900 MT) of the global supply, while having the fifth-largest reserves (6.9 million MT) in the world (see Fig. 1). According to a 2022 estimate, a substantial proportion of India's imports, around 14%, consist of critical minerals, including deep-seated minerals like gold, silver, copper, zinc, lead, nickel, cobalt,

[14] Neha Mishra, "India's Potential and Significance in The Critical Minerals Club", News18, https://www.news18.com/opinion/opinion-indias-potential-and-significance-in-the-critical-minerals-club-8268763.html. Accessed on 8 September, 2023.

[15] Ministry of Mines, "Critical Minerals for India", July 24, 2023. https://mines.gov.in/admin/storage/app/uploads/649d4212cceb01688027666.pdf. Accessed on 9 September, 2023.

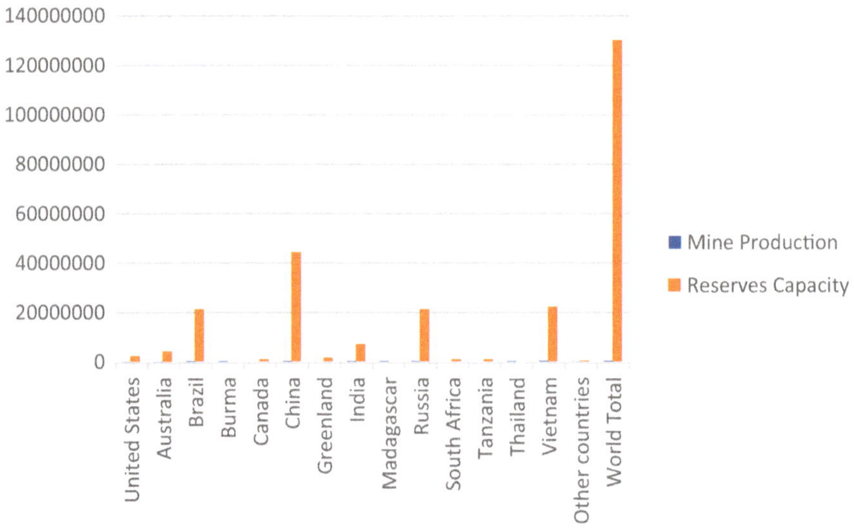

Fig. 1 India's rare earth reserves and production capacity. *Source* United States Geological Survey, "Mineral Commodity Summaries–Rare Earths 2023", http://pubs.usgs.gov/periodicals/mcs2023/mcs2023-rare-earths.pdf. Accessed on 12 September, 2023

platinum group elements (PGEs) and diamonds, worth $8.5 billion and precious metals around $78 billion.[16]

3.1 India's Reserve and Production Capacity

Monazite has traditionally been the main mineral in India that holds the highest viable reserves of light rare earth elements (LREEs). India possesses the largest stocks of monazite, with the most significant concentrations found in the coastal sands of Kerala (see Fig. 2). The estimated amount of monazite, a mineral containing rare earths, discovered in Indian beach sand is 11.93 million metric tonnes, with a rare earth oxide concentration ranging from 55 to 65% (Fig. 3).[17]

The metals and mining sector holds substantial importance in the economic advancement of India, rendering it a crucial constituent of India's growth narrative on both the global and local fronts. It has been asserted that the mining industry can serve as a robust foundation for India's pursuit of achieving its goal of becoming a $5 trillion GDP. In fact, India is a prominent global producer of a diverse range of minerals, encompassing 95 distinct varieties, which may be classified into categories such as minor, atomic, non-metallic, and metallic minerals. Notably, India holds a

[16] Gupta (2022).

[17] Ministry of Mines, "Critical Minerals for India", July 24, 2023. https://mines.gov.in/admin/storage/app/uploads/649d4212cceb01688027666.pdf. Accessed on 9 September, 2023.

Mine	Deposit Kind	State
Chavara Barrier	Beach sand	Kerala
Manavalakurichi beach	Beach sand	Tamil Nadu
Sathankulam	Teri sand	Tamil Nadu
Navaladi-Ovari	Teri sand	Tamil Nadu
Ovari Manapadu	Teri sand	Tamil Nadu
Kuduraimoli	Teri sand	Tamil Nadu
Bhimunipatnam	Beach sand	Andhra Pradesh
Kandivalasa	Beach sand	Andhra Pradesh
Kalingapatanam	Beach sand	Andhra Pradesh
Srikurman	Beach sand	Andhra Pradesh
Bhavanapadu	Beach sand	Andhra Pradesh
Gopalpur	Beach sand	Odisha
Chhatrapur	Beach sand	Odisha
Brahmagiri	Beach sand	Odisha

Fig. 2 Major rare earth deposits in India. *Source* Indian Bureau of Mines, "Indian Minerals Yearbook 2019–Rare Earths", September 2020, https://ibm.gov.in/writereaddata/files/10012020172151RareEarth_2019_AR.pdf. Accessed on 11 September 2023

Fig. 3 Heavy and light rare earth production capacity in India. *Source* Data from Author's Personal Interview of Division Leader, Chavara Mineral Division—Indian Rare Earth Limited (India). October 14, 2023

Element	Light/Heavy	Production Capacity
Cerium	Light	48%
Lanthanum	Light	22%
Samarium	Light	3%
Gadolinium	Heavy	1%
Yttrium	Heavy	0.40%
Dysprosium	Heavy	0.20%

significant position as a leading producer of valuable minerals, including iron ore, chromite, coal, and bauxite.[18] The primary rare-earth public-sector undertakings (PSUs), namely Indian Rare Earth Limited (IREL) and Kerala Rare Earth Limited, have made significant contributions towards augmenting production capacity. It is important to highlight that India is highly dependent on imports, despite their efforts to enhance domestic production capability and the existence of substantial monazite deposits in beach sand. This reliance can be attributed to the specific limitations imposed on mining regulations pertaining to these Public Sector Undertakings

[18] Agarwal (2023).

(PSUs) and the absence of private entities, which has been hampering the India from taking commercial advantage of its mineral strength.[19]

Over the course of recent years, policymakers in India have undertaken many initiatives aimed at reforming the country's mining sector, which could contribute further to yield long-term growth in the rare earth industry. In August 2011, the Ministry of Mines established a Steering Committee to address the current status of Rare Earths and Energy critical Elements in India. The committee, chaired by the Secretary (Mines) and co-chaired by Dr. V. S. Arunachalam, Chairman of the Centre for Study of Science, Technology and Policy (CSTEP) in Bangalore, included members from various organisations such as Department of Science and Technology (DST), Geological Survey of India (GSI), Defence Research & Development Organisation (DRDO), Bhabha Atomic Research Centre (BARC), Institute of Minerals and Materials Technology (IMMT), National Mineral Inventory (NMI), Department of Atomic Energy (DAE), Indian Rare Earth Limited (IREL), and Mr. A. K. Bhandari of Centre for Techno Economic Mineral Policy Options (C-TEMPO) as the Member Secretary. The primary objective of the committee was to prepare a position paper on the aforementioned elements.[20]

In 2015, The Mines and Minerals Development and Regulation (MMDR) Act of 1957 underwent extensive amendments with the aim of implementing various reforms in the mineral sector. These reforms included the introduction of a mandatory auction method for granting mineral concessions, which aimed to enhance transparency in the allocation of mineral resources. Additionally, the amendments established the District Mineral Foundation (DMF) to address the welfare needs of individuals and regions affected by mining activities. Furthermore, the amendments led to the establishment of the National Mineral Exploration Trust (NMET), which aimed to prioritise and promote mineral exploration efforts. Lastly, the amendments also introduced stricter penalties for illegal mining activities.[21]

In 2016, the Atomic Minerals Concession Rules reformed the mining concession standards, stated that the state government could only give mining concessions to private companies if the beach sand they took out had less than 0.75 percent monazite. This is because monazite is a mineral that is important for strategic reasons. Then, in 2017, the Ministry of Mines passed a rule that changed the threshold from 0.75 to 0. This made it illegal for private companies to mine any kind of monazite. The Ministry of Mines sent out another notice in 2018 that banned the private mining and sale of all sand minerals. This was done because all beach sand contains some amount of monazite.[22] During the period from 2017 to 2020, significant attention was directed towards the examination and advancement of Rare Earth Elements (REEs)

---

[19] Akshay (2015).

[20] Ministry of Mines, "Rare Earths and Energy Critical Elements: A Roadmap and Strategy for India", July 2012, https://mines.gov.in/admin/storage/app/uploads/64a3eae23bee21688464098.pdf. Accessed on 10 September 2023.

[21] Ministry of Mines, "Mines and Minerals (Development and Regulation) Act, 1957", March 26, 2015, https://ibm.gov.in/writereaddata/files/07052018100905mmdr%20pdf.pdf. Accessed on 11 September, 2023.

[22] Kukreti (2019).

within the country. The Geological Survey of India (GSI) and Atomic Mineral Division (AMD) collaborated to produce a report titled "Potential Rare Earth Element Deposits of India" in May 2017, which was subsequently submitted to NITI Aayog in New Delhi.[23]

The MMDR Act underwent subsequent amendments in 2016 and 2020 to effectively tackle mining concerns. In November 2020, a strategic plan for augmenting REE exploration in India was jointly presented by GSI and AMD. Additionally, India Exim Bank's working Paper No: 97 in September 2020 emphasised the importance of securing Rare Earth Elements for India.[24] In 2021, another significant round of amendment was introduced to MMDR that aimed to introduce further reforms within the sector, which eliminated the distinction between captive and merchant mines; measures to ensure uninterrupted mining operation through facilitating the transfer of statutory clearance and mineral concessions; and reject the rights of non-auctioned concession holder with unused mining leases. These measures were implemented to ensure that concessions to the private sector are exclusively granted through auction processes.[25]

> Self-reliance is not possible without a strong mining and minerals sector; because minerals and mining are important pillars of our economy.
>
> - Prime Minister Narendra Modi.

As a result of budgetary constraints and a lack of strategic concern over metals and minerals, just 10% of India's geological potential has been explored so far. Authorities' realisation that they needed to update the mining leasing procedure and bring private shareholders back into the mining business was reflected in the 2021 modification to the Mines and Minerals Act (1957).[26] In permitting commercial entities to conduct mineral processing, the MMDR Amendment 2021 offered promise as a precursor to game-changing reforms, with the primary goal of promoting transparency in the auctioning of mines and facilitating increased domestic and foreign investments in the mining industry, thus doubling the current contribution of the mining and mineral sector to GDP by 2025.[27]

Following the suggestions of MMDR 2021, there have been the following initiatives in progress:

(a) In October 2022, Private agencies can also take part in mining sector exploration after obtaining the necessary accreditation from the Quality Council of

---

[23] Ministry of Mines, "Rare Earths and Energy Critical Elements: A Roadmap and Strategy for India", July 2012, https://mines.gov.in/admin/storage/app/uploads/64a3eae23bee21688464098.pdf. Accessed on 10 September 2023.

[24] Ibid.

[25] Ministry of Mines, "Parliament Passes Mines and Minerals (Development & Regulation) Amendment Bill, 2023", August 2, 2023. https://pib.gov.in/PressReleasePage.aspx?PRID=1945102#:~:text=The%20Act%20was%20further%20amended,operations%20even%20with%20change%20of. Accessed on 11 September, 2023.

[26] PIB (2023a).

[27] PIB (2023a).

India-National Accreditation Board for Education and Training (QCI-NABET). Currently, a total of 13 private agencies have obtained accreditation from the central government. There are now 22 government agencies involved in mineral exploration.[28]

(b) In April 2023, CSEP, a not-for-profit policy think tank, released a working paper-49 titled "Assessing the Criticality of Minerals in India" (2023), in which they assessed the importance of 43 non-fuel minerals to the Indian economy and the dangers associated with running out of those minerals, using European Union, albeit with minor modifications. This guided the introduction of the first comprehensive list of 30 critical minerals in July 2023 to ensure that its defence, agriculture, energy, pharmaceutical, and telecommunications sectors have what they need.[29]

The Ministry of Mines has stepped up its efforts to make the discovery of these minerals a national priority, particularly via the employment of the Geological Survey of India (GSI) and other relevant agencies. It has also been mentioned that GSI has conducted over 123 exploration projects in India since 2020. A joint venture company called Khanij Bidesh India Ltd. (KABIL) has developed with investment from three Central Public Sector Enterprises—National Aluminium Company Ltd., Hindustan Copper Ltd., and Mineral Exploration Company Ltd., to identify and acquire critical and strategic mineral assets like lithium, cobalt, and ensure supply side assurance. KABIL has planned to acquire critical minerals from Argentina, Australia, etc.[30] The main issue concerning rare earth elements (REE) relates to the procedure of extracting the individual oxides. The current condition of extraction techniques and facilities for rare earth metals in India has to be consolidated and rectify the existing gaps. India has mining, separation, and refining capabilities for oxide form materials, as well as metal extraction. However, it currently lacks intermediate industrial scale facilities for alloy production, magnet manufacturing, and other similar processes.[31]

4 Japan's Rare Earth Recycling Potential

The availability of critical minerals at competitive costs is crucial for maintaining the global economy and achieving significant progress in high-technology products, sustainable energy technologies, and the commercialization of new inventions. The economic, political, and defence capacities of a nation are primarily influenced by the availability of natural resources. It is crucial to acknowledge that the abundance of natural resources does not necessarily correlate with a nation's degree of development and prosperity. In addition, natural resources are essential not only for promoting the socio-economic advancement of a nation but also for preserving its capacity to

[28] Mishra (2022a, b).

[29] PIB (2023b).

[30] PIBs (2023c).

[31] PIB (2023d).

participate in military conflicts. This is exemplified by Japan, a country that has attained substantial progress and affluence despite its limited natural resources.[32]

Japan's mining resources policy since the 1970s has consisted of the following four main programmes:

- Promoting domestic exploration to ensure the economic viability of domestic mines, which serve as the most reliable source of mining resources.
- Provide assistance for international resource development initiatives and technological collaboration in resource extraction by developing nations, with the aim of ensuring a reliable supply of mining resources from overseas.
- Establishment of a stockpile system for rare metals with the objective of ensuring national economic stability and security.
- Eliminating domestic mine contamination caused by suspended or abandoned mines.[33]

In Japan, rare metals are in dire need for its industrial sectors, particularly utilised in the manufacturing of steel products, aircraft, vehicles, electric and electronic devices, and ships. In July 2009, the Ministry of Economy, Trade and Industry (METI) formulated Japan's first plan to ensure a consistent and reliable supply of rare metals, which highlighted significant obstacles in relation to the acquisition of resources, including scarcity, resource nationalism, and intense rivalry. The technique suggests procuring scarce metals from foreign sources and accumulating them in storage. It proposed the implementation of resource diplomacy tactics in this context. It highlighted the need of establishing strong diplomatic ties with countries that produce valuable resources. The utilisation of Official Development Assistance (ODA) tools is highly advised for the development of infrastructure surrounding mines that includes initiatives such as technological transfers and collaboration on environmental preservation, showcasing Japan's areas of expertise.[34]

After China, the United States, the European Union, India, and Russia, Japan is the sixth greatest GHG emitter worldwide. Japan has a legal obligation to achieve zero net greenhouse gas emissions by the year 2050. To achieve this obligation, Japan has launched several initiatives such as "Green Growth Strategy" the was developed by the Ministry of Economy, Trade, and Industry (METI) in 2020 and the 6th Basic Energy Plan the was authorised by the Japanese Cabinet office on October 22, in 2021, with the goal of increasing the proportion of renewable energy in the country's power mix from its 2019 level of 36–38%.[35]

Diversifying production networks has become a significant approach for manufacturing powerhouses with high raw material consumption, such as Japan. Since the 2010 disruption event for Japan, following the Japanese coast guard's arrest of a Chinese fisherman, Japan has been investing in supply chain resilience. The Chinese informal embargo on rare earth materials in 2010 and export restrictions during the

[32] Randive and Jawadand (2019).

[33] Kikkawa (2013a).

[34] Kikkawa (2013b).

[35] Ibid.

China-Japan Senkaku/Diaoyu Island dispute (2012) had a high negative impact on Japan's automobile industry and its manufacturers, such as Honda, Toyota, Suzuki, and others.[36]

Since then, Japan has been investing in and taking part in a campaign to recycle digital devices like phones and computers to meet their requirements for rare earth materialsand avoid future disruption. In addition, Japan also sought out new partnerships in Mongolia, Myanmar, and other south-central Asian states to get a steady supply of rare earth materials. Japan is also working with Quad partners to strengthen supply chains within the region and commit significant funds to ensure that supply chains continue to diversify through initiatives like developing shared technical standards, 5G diversification and deployment, horizon scanning, and technology supply chains.[37]

4.1 Geopolitics Around REEs: A Wake up Call for Japan

China's 2010 decision to halt the export of REE to Japan prompted extensive research on how China could exploit its market dominance to achieve political objectives. This decision also highlighted China's perception of REE as a valuable resource and the REE sector as a crucial industry, as evidenced by government policies dating back to the 1980s. Over time and through many occurrences, it has been evident that China deliberately employs the use of raw materials. Furthermore, even the assessments of the importance of these minerals are influenced by China's resource policies.[38] Following supply chain disruptions in 2011, Japan has been actively working to ensure the availability of rare earth minerals by initiating mining activities in places outside of China, in order to decrease its dependence on China. In response to the RE crisis, the Japanese government earmarked a sum of US$650 million to procure critical raw materials for the fiscal year 2011.[39]

Furthermore, Japan, along with the United States, European Union, Mexico, India, Brazil, Japan, and Korea, lodged a formal complaint to the World Trade Organisation (WTO) against China in June 2011. The complaint said that China had violated WTO regulations by imposing restrictions on the supply of rare earth raw materials, as well as non-rare earth metals such as tungsten and molybdenum. The countries criticised China's handling of export quotas, permits, minimum pricing, and lack of transparency in publishing measures.[40] Moreover, numerous political representatives shared similar sentiments that Japan must try to invest in supply diversificationwith the aim of obtaining 50% of its RE needs from non-Chinese sources.[41]

---

[36] Bradsher (2010).
[37] Nagy (2022).
[38] Ibid.
[39] Reuters (2011).
[40] Mancheri (2015).
[41] JOGMEC (2012).

Table 2 Japan's REE projects and partnerships

Mine development/ production project	Location country	Partner Countries/ companies	Objective
Mount Weld	Australia	Lynas (Australia) + JOGMEC (Sojitz, Japan)	Financing, offtake agreement, and distribution
Orissa and Manvalakurichi	India	IREL (India) + JOGMEC and Toyota (Japan)	Contract on joint production of rare earths
Dong Pao	Vietnam	Lai Chau-Vimico Rare Earth Joint Stock Company (Vietnam) + Toyota and Tsusho (Japan)	Joint mining and processing project
Brown range	Australia	Sumitomo, Shin-Etsu, Huatai Mining, Jien Mining, Lianyugang Zevu New Materials Sales Co (JFMAG), Sinosteel MECC	MoU for offtake agreement, sales agreement, construction of pilot plant, invesment

Sources Schmid (2019), "Mitigating supply risks through involvement in rare earth projects: Japan's strategies and what the US can learn", Journal of Resource Policy, Volume 63. August 1, 2019, https://doi.org/10.1016/j.resourpol.2019.101457

In 2014, it was demonstrated that RE-based items account for 22% of Japan's overall export value, which is twice as high as the proportion in the U.S. This highlighted the significant importance of RE-based products for the economy of Japan, which was strategically threatened by the supply risks resulting from China's strong market position.[42] After a WTO judgement in August 2014 sided with the United States, the European Union, and Japan, China was ordered to eliminate export taxes and quotas. Since China is still extracting resources for local use, the country cannot claim environmental concerns as a reason for limiting exports.[43] In reaction to a negative verdict, China eliminated the quota system in January 2015 and adopted a stricter licencing system for exporters.[44] In 2016, it was reported that Japan possesses a comprehensive industrial supply chain in the realm of rare earth production and processing, particularly when compared to the United States and the European Unionas a result of the investments Japan made in various rare earth projects in different nations globally during 2013–15 (see Table 2). Significant collaborations involved numerous Japanese trading houses, magnet makers, chemical industries, & technological conglomerates such as Sojitz, Showa Denko, Shin-Etsu, Sumitomo, Toshiba, Daido Steel, and Toyota Tsusho in collaboration with Australia, Canada, India, Jamaica, Kazakhstan, Vietnam, and the United States.[45]

---

[42] Schmid (2019).

[43] WTO (2014).

[44] Mancheri and Marukawa (2016).

[45] Schmid (2019).

The paramount rare earth project for Japan is Lynas' Mount Weld project located in Australia. Sojitz, a prominent Japanese RE importer, has established an offtake arrangement with Lynas to ensure a steady supply from Lynas' Mount Weld mining operation. Sojitz, in collaboration with the state agency JOGMEC (Japan Oil, Gas, and Metals National Corporation), granted a total of US$250 million in the form of loans and equity.[46] In 2016, JOGMEC and Sojitz reached an agreement to restructure the debt of Lynas, aiming to prevent its collapse. This was achieved by deferring fixed repayments and reducing the interest rate.[47] Japan's political backdrop is critical for understanding its goals of direct access to rare-earth production and a diverse supply structure, as the Japanese government uses its specialised agency JOGMEC to ensure that it has access to the raw materials it need. JOGMEC is funded by a wide variety of initiatives, involving total expenditures from both politicians and corporations on a massive scale.[48]

In addition to these initiatives in the wake of the RE crisis of 2010–2011, a considerable investment has been made in rare earth projects. However, the overall effect on the global RE market is small as China still dominates in RE mining and processing. Thus, Japan's attention has also been directed towards the manufacturing of secondary raw materials. The recycling of rare earth elements (REE) at the end-of-life (EoL) phase, such as from batteries, magnets, and lamps, along with the replacement of primary raw materials with secondary raw materials, which would reduce overall reliance on external supply sources. Nevertheless, the issue of price competitiveness and the quality of secondary raw materials continues to pose a difficulty. Additional tactics such as substituting and stockpiling RE have also demonstrated their effectiveness.[49]

The Japanese government has also been moving forward with support for countries that prioritise domestic manufacturing, diversify overseas production away from China towards ASEAN and other countries, and utilise digital technologies to strengthen the global supply chains of Japanese companies. Japan's Ministry of Economy, Trade, and Industry (METI) is also promoting subsidies for domestic investment promotion projects for supply chain measures and subsidies for projects supporting the diversification of overseas supply chains.[50] Moreover, Japan has been offering subsidies to domestic companies with manufacturing operations in China,

---

[46] Sojitz, "Agreement to Form Strategic Alliance with Lynas, an Australian Rare Earths Company, Concerning Rare Earths Supply for the Japanese Market and Supporting the expansion of Lynas Rare Earths Project", November 24, 2010. https://www.sojitz.com/jp/news/docs/101124_e.pdf. Accessed on 10 October, 2023.

[47] Reuters, "Lenders rescue rare earths producer Lynas from collapse", October 26, 2016, https://www.reuters.com/article/lynas-corp-debtrenegotiation-idUSL4N1CV674.

[48] Lifton (2016).

[49] Smith and Eggert (2016).

[50] Centre for Strategic & International Studies. "Building Resilient Global Supply Chains: The Geopolitics of the Indo-Pacific Region". February 19, 2021. https://www.csis.org/analysis/building-resilient-global-supply-chains-geopolitics-indo-pacific-region. Accessed on 25 August 2022.

either to repatriate their supply chains or relocate them by setting up bases in Southeast Asian countries, such as Vietnam, Indonesia, and Laos, and now increasingly shifting towards South Asian countries like India and Bangladesh.[51]

5 India-Japan Rare Metal Collaboration Towards Resilience

The rare metal collaboration between India and Japan dates back to early 2011, with the Indo-Japan Joint Statement signed between the Department of Atomic Energy of India (DAE) and the Ministry of Economy, Trade, and Industry of Japan (METI). As part of the agreement, the DAE designated Indian Rare Earths Limited (referred to as IREL) and METI designated the Toyota Tsusho Corporation (referred to as TTC), as the partner companies to carry out the separation and refining of rare earth. In order to further enhance their long-standing cooperative relationship, both parties decided to expedite the establishment of a joint venture firm between them.[52] Following the commitment, India and Japan established a wholly owned subsidiary in India called Toyotsu Rare Earths India Pvt Ltd. to process and purify rare earth oxides.

It is worth noting that the majority of India's REE production comes in the form of chlorides and oxides, which have a relatively low market value. India faces obstacles due to limited funds, technology, infrastructure, and policies to produce rare earths in metal and alloy form for the manufacturing of high-value products. As a result, India's dependence on imports of rare earth elements (REE) persists. To advance the domestic mining capacity of India, the Toyota Tsusho Corporation has achieved significant advancements by engaging in the purification and subsequent exportation of rare earth chemicals as part of Japan's efforts to decrease its reliance on China, with an objective to reach a total of 4,000 metric tonnes annually.[53]

With the aim of enhancing Rare Earths refining and downstream activities in India, the IREL (India) Limited extended its agreement with M/s. Toyotsu Rare Earths (India) Pvt. Ltd., and M/s. Toyota Tsusho Corporation, Japan recently.[54]

In addition, the Indian Government has launched a 'Deep Ocean Mission' to investigate mineral resources in the Indian Ocean, in which a potential collaboration with Japan, known for their proficiency in deep-sea mining, could prove advantageous. Similarly, the Japan Oil, Gas and Metals National Corporation (JOGMEC) devised

---

[51] S-RM. "Asia Pacific: The Future Of "Made in Asia" Supply Chains". September 17, 2022. https://insights.s-rminform.com/supply-chains-asia-pacific. Accessed on 30 November 2022.

[52] Ministry of External Affairs, "MoU on the Cooperation in the Rare Earths Industry in India", November 16, 2012. https://www.mea.gov.in/Portal/LegalTreatiesDoc/JP12B0296.pdf. Accessed on 20 November, 2023.

[53] Bhandari and Faridi (2022).

[54] India Frontline, "IREL India commitment to boost the rare earth sector renewed", February 10, 2022. https://indiafrontline.com/irel-india-commitment-earth-sector/. Accessed on 20 November, 2023.

a method for extracting nickel and cobalt from the ocean floor through a process of crust excavation.

In April 2022, the Japanese government chose eight projects for the "Programme for the Supply Chain Resilience in the Indo-Pacific Region" using around $8.6 million USD from the third supplementary budget for fiscal year 2021. Indian initiatives account for six out of the total. Initially, these initiatives will implement a pilot plan to support Japanese businesses in enhancing their supply chain transparency, efficiency in logistics, effectiveness in trade procedures, and then will focus on the diversification of their product portfolio. These projects have the potential to significantly strengthen India's supply chain resilience and industrial competitiveness by bridging the gap between Japan and India.[55]

In addition to the bilateral collaborations, India and Japan have been engaging in significant trilateral and multilateral initiatives such as:

a-The Supply Chain Resilience Initiative (SCRI) was officially launched by India, Japan, and Australia with the aim of developing robust supply chains in the Indo-Pacific region. The initiative is driven by the desire to decrease reliance on China and has the potential to be expanded through mutual agreement. The objective is to enhance and expand the current bilateral frameworks such as the ASEAN-Japan Economic Resilience Action Plan and India-Japan Industrial Competitiveness Partnership, with the aim of attracting foreign direct investment in the region.[56] The primary objective of the SCRI is to ensure that global supply networks remain unaffected by pandemics and other unforeseen catastrophic events, commonly referred to as "black swan" events. Given its ability to enhance communication among governments sharing common concerns regarding economic security in the Indo-Pacific region, the SCRI is gaining prominence as a significant subject of discussion among several states. The SCRI serves as a significant balancing force to China's increasing influence, particularly due to the Belt and Road Initiative.[57]

b- Quadrilateral Security Dialogue (Quad) group, whose objective is to reduce the region's reliance on China by establishing a free and open Indo-Pacific and developing strong supply networks. The Quad members demonstrated efficient communication and coordination during the outbreak, which was evident in their diplomatic endeavours to create and execute a vaccine. To advance their climate-focused partnership, they have chosen to strengthen the resilience of clean energy supply chains, particularly in areas such as clean hydrogen. Additionally, they plan to establish a Quad green shipping network. The Quad members have reached a consensus to collaborate on crucial technologies including biotechnology, genome sequencing,

---

[55] Mancheri et al. (2019).

[56] Suneja (2021).

[57] Dr. Jagannath Panda, "PacNet #31 – The Structural Limits of the Supply Chain Resilience Initiative", Pacific Forum, https://pacforum.org/publication/pacnet-31-the-structural-limits-of-the-supply-chain-resilience-initiative. Accessed on 25 March, 2023.

and genome manufacturing. Additionally, they have decided to establish a semiconductor supply chain initiative and establish an Open RAN policy alliance to facilitate the widespread and diverse implementation of 5G.[58]

6 Concluding Remarks

India's unexploited reserves, along with Japan's technological prowess and recycling capabilities, suggest a promising and extensive path towards achieving supply chain resilience. There are numerous prospects for partnership between India and Japan in the fields of semiconductors, robust supply chains, rare-earths extraction, space technology, artificial intelligence, and quantum computing, as recently highlighted by RBI Governor Shaktikanta Das during his speech at the Tokyo Chamber of Commerce and Industry. He also said that, "Our collaboration has the potential to be further enhanced in the field of human resource", which reflects the confidence in the future holding boundless opportunities to enhance India-Japan interactions for the betterment of citizens and the global community.[59]

References

Akshay KVL (2015) India not realising potential of rare earth industry. The Economic Times. https://economictimes.indiatimes.com/industry/indl-goods/svs/metals-mining/india-not-realising-potential-of-rare-earth-industry/articleshow/54940321.cms?from=mdr. Accessed 10 Sept 2023

Agarwal BB (2023) Government ministering the growth of the metal and mining sector. The Times of India. https://timesofindia.indiatimes.com/blogs/voices/government-ministering-the-growth-of-the-metal-and-mining-sector/. Accessed 10 Sept 2023

Bhandari A, Faridi S (2022) Partnering with Japan for critical minerals supply. Gateway House. https://www.gatewayhouse.in/partnering-with-japan-for-critical-minerals-supply/. Accessed Nov 2023

Binnemans K et al (2013) Recycling of rare earths: a critical review. J Clean Produc. https://doi.org/10.1016/j.jclepro.2012.12.037. Accessed 10 Oct 2023

Bradsher K (2010) Amid tension, china blocks vital exports to Japan. The New York Times. https://www.nytimes.com/2010/09/23/business/global/23rare.html#:~:text=HONG%20KONG%20%E2%80%94%20Sharply%20raising%20the,wind%20turbines%20and%20guided%20missiles. Accessed 20 Feb 2023

Bradsher K (2010) Amid tension, china blocks vital exports to Japan. The New York Times

Gupta DD (2022) Mineral exploration: new India & green dreams. Retrieved from The Times of India: https://timesofindia.indiatimes.com/blogs/voices/mineral-exploration-new-india-green-dreams/

---

[58] Mishra (2022a).

[59] Free Press Journal, "India, Japan Can Collaborate In Chips, Rare-Earth & AI, Says RBI Chief", November 9, 2023. https://www.freepressjournal.in/business/india-japan-can-collaborate-in-chips-rare-earth-ai-says-rbi-chief. Accessed on 20 November, 2023.

Isetani A, Shimizu A, DeWit A, Shaw R (2022) Indo-Japanese collaboration on energy security and critical raw materials (CRM), vol 20, no 18, 5th edn

Jensen PD (2022) Collective action to save our one and only Earth. Grundfos. https://www.grundfos.com/about-us/news-and-media/news/collective-action-to-save-our-one-and-only-earth. Accessed 20 Nov 2023

JOGMEC (2012) Annual Report. Japan Oil, Gas and Metals National Corporation. https://www.jogmec.go.jp/content/300075892.pdf

Karali N, Shah N (2022) Bolstering supplies of critical raw materials for low-carbon technologies through circular economy strategies. Energy Res Soc Sci 88:102534

Kennedy JC (2016) Rare earth production, regulatory USA/international constraints and Chinese dominance: the economic viability is bounded by geochemistry and value chain integration. In: Rare earth industry. Edinburgh Elsevier Inc, pp 37–55

Kikkawa T (2013a) Why has Japan revised the mining act? -historical and contemporary background. Hitotsubashi J Commer Manag 47:33–42

Kikkawa T (2013b) Why has Japan revised the mining act?-historical and contemporary background. Hitotsubashi J Commer Manag 47:33–42

Kukreti I (2019) Environment ministry stops clearance to mining of beach sand minerals. Retrieved from Down-to-Earth: https://www.downtoearth.org.in/news/mining/environment-ministry-stops-clearance-to-mining-of-beach-sand-minerals-63999

Lele A, Bhardwaj P (2014) Strategic materials: a resource challenge for India. Institute for Defence Studies & Analyses (IDSA), Pentagon Press, New Delhi, p 221

Lifton J (2016) Lifton challenges the WSJ editorial on Honda no longer using heavy rare earths. InvestorIntel 20.07.2016. https://investorintel.com/sectors/technologymetals/technology-metals-intel/whoever-covers-rare-earths-wall-street-journalcannot-see-forest-trees/

Mancheri N, Marukawa (2016) Rare earth elements: China and Japan in industry. Trade and Value Chain ISS CCRS No.17. http://web.iss.u-tokyo.ac.jp/kyoten/research/issccs/no17.html

Mancheri N (2015) World trade in rare earths, Chinese export restrictions, and implications. Resour Policy 46(2):262–271. https://doi.org/10.1016/j.resourpol.2015.10.009

Mancheri N et al (2019) Effect of Chinese policies on rare earth supply chain resilience. Resour Conserv Recycl 142:101–112. Accessed 20 Aug 2023

Mishra A (2022) Centre allows 13 private agencies to explore minerals, Shall boost mineral production. Swarjya. https://swarajyamag.com/news-headlines/centre-allows-13-private-agencies-to-explore-minerals-shall-boost-mineral-production

Mishra N (2022) QUAD's quest to overcome dependence on China in critical and emerging technologies. Bharat Shakti. https://bharatshakti.in/quads-quest-to-overcome-dependence-on-china-in-critical-and-emerging-technologies/. Accessed 28 March 2022

Nagy S (2022) Supply chain resilience in the Indo-Pacific: building mutual beneficial dependency. Canadian Global Affairs Institute. https://www.cgai.ca/supply_chain_resilience_in_the_indo_pacific_building_mutual_beneficial_dependency. Accessed 25 Aug 2022

PIB (2023a) The mines and minerals (development and regulation) amendment act, 2021. Retrieved from Ministry of Law and Justice: https://pib.gov.in/PressReleasePage.aspx?PRID=1945102#:~:text=The%20Act%20was%20further%20amended,operations%20even%20with%20change%20of

PIB (2023b) Union Minister Pralhad Joshi unveils list of "Critical Minerals for India". Retrieved from Ministry of Mines: https://pib.gov.in/PressReleaseIframePage.aspx?PRID=1936083

PIB (2023c) Thirty critical minerals list released. Ministry of Mines. https://pib.gov.in/PressReleasePage.aspx?PRID=1942027#:~:text=These%20minerals%20are%20Antimony%2C%20Beryllium,%2CZirconium%2C%20Selenium%20and%20Cadmium. Accessed 5 Sept 2023

PIB (2023d) Mining of rare earth elements. Department of Atomic Energy. https://pib.gov.in/PressReleasePage.aspx?PRID=1914305#:~:text=As%20such%20in%20a%20finished,are%20non%2Dexistent. Accessed 10 Sept 2023

Randive K, Jawadand S (2019) Strategic minerals in India: present status and future challenges. J Miner Econ. https://doi.org/10.1007/s13563-019-00189-0

Reuters (2011) Japan budgets $650 mln for rare earths, resources. 05.01.2011. http://www.reuters.com/article/japan-china-rareearth-idUSTOE70404220110105

Schmid M (2019) Mitigating supply risks through involvement in rare earth projects: Japan's strategies and what the US can learn. J Resour Policy 63. https://doi.org/10.1016/j.resourpol.2019.101457.

Smith BJ, Eggert RG (2016) Multifaceted material substitution: the case of NdFeB, 2010–2015. JOM 68(7):1964–1971

Suneja K (2021) India, Japan, Australia launch supply chain resilience initiative, may be expanded based on consensus. The Economic Times. https://economictimes.indiatimes.com/news/economy/foreign-trade/india-japan-australia-launch-supply-chain-resilience-initiative-may-be-expanded-based-on-consensus/articleshow/82276307.cms?utm_source=contentofinterest&utm_medium=text&utm_campaign=cppst. Accessed 27 Aug 2022

WTO (2014) China-measures related to the exportation of rare earths, tungsten, and molybdenum. Reports of the Panel, WT/DS431/R, WT/DS432/R, WT/DS433/R, 26, March, Geneva. https://www.wto.org/english/tratop_e/dispu_e/431_432_433r_e.pdf

Industry and Innovation

Consideration of Risks and Resilience Leading to New Policy Framework for Micro, Small, and Medium Enterprises (MSMEs) in India

Tomo Kawane and Rajib Shaw

Abstract Micro, small and medium enterprises (MSMEs) significantly contribute to Indian economy, and they provide livelihoods of the largest numbers of people. Climate change, disasters and other socio-economic and well as geo-political risks pose significant threats to the MSMEs, especially to those enterprises in the informal sector. In this study, vulnerabilities of MSMEs are assessed through qualitative meta-analysis to synthesize qualitative findings. The study highlights that gender disparity among the enterprises, regional disparities in terms of their growth as well as formal registration, and their adaptation to digitization, urbanization, disasters and climate change impacts are to be considered in the framework of risks, resilience, and recovery strategies of MSMEs. A policy framework is proposed to enhance entrepreneurial ecosystem in India through: (1) creating self-assessment avenues for MSMEs, (2) imparting educational opportunities for MSMEs against economic distress due to secondary perils, (3) nurturing collective disaster-preparedness for the MSME sector, and (4) exploring new forms of assistance for the MSME sector. Implications for further research are also drawn to support entrepreneurial resilience.

Keywords MSMEs · India · Entrepreneurship · Climate change · Resilience

1 Introduction

Since long, in India, the presence of Micro, Small, and Medium Enterprises (MSMEs) in the Indian economy has been acknowledged as the key determinant to sustain the economic growth of the country. Common characteristics of the dominant presence of micro enterprises and their precariousness in terms of sustainable business practices and stable financial base and support of the overall MSMEs have been pointed out in preceding studies on the MSME sector in India. Micro-managing characteristics with

T. Kawane (✉) · R. Shaw
Graduate School of Media and Governance, Keio University, Fujisawa, Kanagawa 252-0882, Japan
e-mail: kawane@sfc.keio.ac.jp

R. Shaw and S. R. Choudhury (eds.), *India, Japan and Beyond*,
https://doi.org/10.1007/978-981-97-3282-1_6

insufficient financial support because of the MSMEs being small in their business scale have been acknowledged in various countries throughout Asia and the Pacific.

For a country like India, an emerging economy that holds the highest number of population in the world, the most crucial challenge is to create entities that offer employment opportunities, contributing to the growth of its economy. The value and worth of the MSMEs have been largely discussed from the viewpoint of socio-economic as well as political priorities of India as a nation, initially by promoting small-scale industries. The contribution of the MSMEs to the Indian economy amounts to 30% of India's GDP (2019–20). The MSMEs have contributed to 49.5% of India's export (2020–21). The MSMEs have a share of 24% in the service sector, and 36.9% in the manufacturing sector in terms of the GDP ratio of the MSME Sector (2019–20). Micro enterprises in India are the most dominant enterprises among different types of enterprises. Micro enterprises make up 99.4% of all the MSMEs, 0.52% as small enterprises and just 0.01% belong to medium enterprises. Regarding employment opportunities, 97% of the employment in MSMEs comes from the micro sector (Ministry of Micro, Small and Medium Enterprises, Government of India 2022).

Social enterprises with frugal innovation ready to take risks are appreciated and promoted. Yet, this is merely one aspect of the MSME sector in the Indian economy where the skewed development and regional baggage of environmental issues and various impacts of climate change have been experienced more frequently in recent times. On the other hand, social groups characterized by caste, class, gender, and in some cases religion, would give cross-cutting cultural implications on the growth and performance of the MSME sector in India.

What has been missing is giving due consideration to entrepreneurial resilience. To understand the factors that could establish entrepreneurial resilience, the impact of the entrepreneurial mindset of entrepreneurs could be reviewed in terms of the ecosystem of the MSMEs. The ecosystem of MSMEs has been posing continuous challenges, not always in the form of primary perils but as secondary perils caused by climate change. Environment-driven economic disruptions are not always fully insured. Awareness on the part of both policymakers and the MSME sector needs to be better informed for formulating practical steps for disaster preparedness. This aspect is beyond the past policy interventions to develop its MSME clusters based on its traditional local-grown industries, especially home-based cottage industries, as an effort to ameliorate social disparities and economic integration of weaker sections of society.

In this study, the authors aim to offer a discussion point to review the framework of studying the MSME sector in India in the context of climate change, entrepreneurial resilience, risk management as well as disaster preparedness.

2 Methodology

The study is a systemic review of the literature on MSMEs in India to assess their vulnerabilities. It has used qualitative meta-analysis to synthesize qualitative findings from distinctive features of case studies anticipating to obtain an understanding of contextual factors or shedding light on reasons for certain phenomena that usually cannot be investigated through quantitative studies. The insights gained through qualitative meta-analysis will be useful in identifying influential variables or patterns as well as deriving a meta-causal network. Contextualized case examples can be studied and aggregated to obtain a larger picture that can be generalized, significantly contributing "for academics in terms of theory development and for practitioners in terms of evidence-based management or entrepreneurship" (Hansen et al. 2022: 7). An overview of MSMEs in India is carried out with the use of secondary data from annual reports of the Ministry of Micro, Small and Medium Enterprises, Government of India, as well as other relevant data from government offices. The available published government data on MSMEs is further compared with the ongoing registration of MSMEs under the Government of India digital UDYAM platform by referring to the official websites.

The Scopus search of the keywords, "MSME" and "India" together generated 217 document results, all of which have been published in English. Out of those, 16 documents deal with "Entrepreneurship" (14 documents) and "Entrepreneur" (2 documents). There are 18 documents regarding sustainability such as "Sustainability" with 8 documents, "Sustainable Development" with 6 documents, "Sustainable Goals" with 2 documents, and "Sustainable" with 2 documents. However, the most crucial term, "Risk" was only found in "Risk Management" (2 documents) and "Credit Risk" (2 documents). Other notable terms were "Innovation (10 documents)", "Manufacturing (13 documents)}", "Manufacture (10 documents)", and "Lean Manufacturing (8 documents)". Those 16 documents dealing with the subject of entrepreneurship or entrepreneur do not talk about risk or risk management at all. Further, it was found that those dealing with the theme of sustainability of the MSMEs rather focused on the issues related to operational managerial challenges or financial constraints, being silent on environmental risks.

The authors, therefore, embarked on a further search of relevant references through ResearchGate, Mendeley, Google Scholar, and international publications by the United Nations, the United Nations Industrial Development Organization (UNIDO), the International Labor Organization (ILO), the World Bank, the World Economic Forum, and so on. To understand previous studies on various disaster impacts on the MSMEs, case examples from other countries that highlight the vulnerability of MSMEs were also taken into consideration in the process. For the meta-analysis, the inclusion of all available studies (including grey literature) is intended, since highly cited articles tend to be found easier and faster than less cited articles: here lies the point of publication bias that tends to give more preference to statistically driven research publications (Hansen et al. 2022: 11).

Table 1 Classification of the MSMEs in India

Classification of enterprises	Micro	Small	Medium
Manufacturing	Investment < Rs. 10 million	Investment < Rs. 100 million	Investment < Rs. 500 million
Services	Turnover < Rs. 50 million	Turnover < Rs. 500 million	Turnover < Rs. 3.5 billion

Source Ministry of Micro, Small, and Medium Enterprises, Government of India (2022)

3 Overview of MSMEs in India

3.1 The Definition of MSMEs

The Government of India announced a new definition of MSMEs to expand the scope of supporting MSMEs. The definition of MSME got stipulated in the Small and Medium Enterprise Development Act of 2006, with separate definitions for micro, small, and medium enterprises. The classification of the MSMEs in India was revised in July 2020 and based on investments and turnover of the enterprise (Reserve Bank of India 2020). Ministry of Micro, Small and Medium Enterprises (2022) classified micro enterprises as enterprises with investment less than Rs. 10 million which is around 0.12 million US Dollars, since 1 US Dollar is equivalent to 82.46 Indian Rupees as per the average RBI reference rate for the month of December 2022 (Reserve Bank of India 2022). and turnover of less than Rs. 50 million (0.6 million US Dollars). Small enterprises are classified as enterprises with investments less than Rs. 100 million (1.2 million US Dollars) and turnover of less than Rs. 500 million (6 million US Dollars). And Medium enterprises are classified as enterprises with investments less than Rs. 500 million (6 million US Dollars) and turnover of less than Rs. 3.5 billion (42 million US Dollars). The classification is summarized below (Table 1).

3.2 Registered MSMEs in India

In the financial year 2021–22, the Ministry of Micro, Small, and Medium Enterprises, Government of India, reported that 63.4 million enterprises exist in the MSME sector in India, employing 110 million people and contributing to offer 46% of the total employment in India (Ministry of Micro, Small, and Medium Enterprises, Government of India 2022). In India, the MSME sector has the second largest share of employment following the agriculture sector. As on 30 December 2022, the number of registered MSMEs in India was 13,024,581. This showed that the number of registered MSMEs is around 20% of the total MSMEs. The rest of them have not yet registered themselves or could be in the process of registration.

Regarding the registered MSMEs, Table 2 shows their composition according to the classification of enterprises.

The major issue of the MSMEs in India has been the huge informal sector. The government policy for the MSME sector has encouraged MSMEs to get themselves registered under UDYAN of the Ministry of Micro, Small and Medium Enterprises, Government of India, publishing its guidebook in 23 official languages of India which is available from their official website of the Ministry of MSMEs of Government of India (Office of the Development Commissioner, Ministry of Micro, Small and Medium Enterprises, Government of India 2021). The data after the introduction of UDYAN MSME registration shows the skewed development in terms of the number of registered MSMEs in India as shown in Table 3.

Prepared by Authors

Source MSME Dashboard, Micro, Small, and Medium Enterprises, Government of India https://dashboard.msme.gov.in/Udyam_Statewise.aspx, accessed on 30 December 2022

Out of 37 States and Union Territories as mentioned above in Table 3, the top 12 states in terms of the registered number of MSMEs have 81.4% of all the MSMEs in India, including so-called BIMARU states (Bihar, Madhya Pradesh, Rajasthan, and Uttar Pradesh) that are considered to be recording inferior growth regarding industrialized development. Those top 12 states are Maharashtra, Tamil Nadu, Uttar Pradesh, Gujarat, Rajasthan, Karnataka, Madhya Pradesh, Bihar, Punjab, Haryana, Telangana, and West Bengal.

It is notable here that the two Indian states, namely, Maharashtra and Gujarat, have one-fourth of the registered MSMEs in India. Further, when we look at each classification of MSMEs for the two states, it is notable to mark that 30% of the country's registered medium enterprises are located here: (1) Micro enterprises: Maharashtra (2,404,892 enterprises) and Gujarat (954,420 enterprises) together have 3,359,312 enterprises (26.77% of the total number of registered micro enterprises); (2) Small enterprises: Maharashtra (63,130 enterprises) and Gujarat (48,643 enterprises) together have 111,773 small enterprises in total (25.67% of the total number of registered medium enterprises); and (3) Medium enterprises: Maharashtra (7,090 enterprises) and Gujarat (4,872 enterprises) together have 11,962 medium enterprises in total (30% of the total number of registered medium enterprises in India).

Table 2 Composition of the registered MSMEs in India

Classification of enterprises	Number of registered MSMEs	Percentage of registered MSMEs in total MSMEs
Micro enterprises	12,549,488	96.35
Small enterprises	435,261	3.34
Medium enterprises	39,832	0.31
Total	**13,024,581**	**100.00**

Source MSME Dashboard, Micro, Small and Medium Enterprises, Government of India https://dashboard.msme.gov.in/Udyam_Statewise.aspx, accessed on 30 December 2022

Table 3 State-wise registered MSMEs in India

Sl. No	State/UT name	Registered MSMEs	Percentage to all India registered MSMEs	Micro	Small	Medium
1	Maharashtra	2,475,112	19.00	2,404,892	63,130	7,090
2	Tamil Nadu	1,372,870	10.54	1,330,613	39,010	3,247
3	Uttar Pradesh	1,095,098	8.41	1,058,720	33,734	2,644
4	Gujarat	1,007,935	7.74	954,420	48,643	4,872
5	Rajasthan	990,521	7.61	962,810	25,810	1,901
6	Karnataka	768,587	5.90	738,356	27,792	2,439
7	Madhya Pradesh	607,845	4.67	588,433	18,130	1,282
8	Bihar	517,719	3.97	506,842	10,314	563
9	Punjab	475,544	3.65	457,209	16,830	1,505
10	Haryana	447,499	3.44	424,111	21,477	1,911
11	Telangana	419,137	3.22	400,076	17,146	1,915
12	West Bengal	418,165	3.21	395,875	20,417	1,873
13	Andhra Pradesh	378,150	2.90	361,197	15,673	1,280
14	Delhi	353,137	2.71	325,933	24,324	2,880
15	Kerala	317,109	2.43	303,826	12,328	955
16	Odisha	271,564	2.09	262,734	8,249	581
17	Jharkhand	219,630	1.69	214,107	5,176	347
18	Chhattisgarh	175,102	1.34	167,233	7,096	773
19	Assam	171,689	1.32	165,903	5,385	401
20	Jammu and Kashmir	160,718	1.23	157,740	2,791	187
21	Uttarakhand	122,741	0.94	118,664	3,752	325
22	Himachal Pradesh	73,786	0.57	71,070	2,440	276
23	Manipur	40,104	0.31	39,681	407	16
24	Goa	25,345	0.19	24,301	936	108
25	Chandigarh	22,563	0.17	21,183	1,237	143
26	Puducherry	20,440	0.16	19,721	664	55
27	Tripura	16,990	0.13	16,483	476	31
28	Mizoram	9,487	0.07	9,348	132	7
29	Nagaland	9,272	0.07	9,146	117	9
30	Andaman and Nicobar Islands	7,784	0.06	7,581	199	4
31	Dadar and Nagar Haveli	7,436	0.06	6,846	503	87
32	Meghalaya	6,989	0.05	6,708	252	29

(continued)

Table 3 (continued)

Sl. No	State/UT name	Registered MSMEs	Percentage to all India registered MSMEs	Micro	Small	Medium
33	Arunachal Pradesh	5,680	0.04	5,488	173	19
34	Ladakh	4,732	0.04	4,663	67	2
35	Sikkim	4,005	0.03	3,904	91	10
36	Daman and Diu	3,604	0.03	3,179	360	65
37	Lakshadweep	492	0.004	492	0	0
Total		**13,024,581**		**12,549,488**	**435,261**	**39,832**

According to the National Sample Survey (NSS) 73rd round conducted under the Ministry of Statistics & Program Implementation in 2015–2016, it was recorded that the distribution of non-agriculture MSMEs in India showed that 36% of the MSMEs engaged with trading, while 33% of the MSMEs with services, and 31% with manufacturing (MSME Annual Report 2022:25). Regarding the number of employees, the MSME workers are hired most in services (35%), followed by trade (33%), and manufacturing (32%) (MSME Annual Report 2022: 28). The Top 5 Industrial sectors of registrations are—Food Products, Textile, Apparel, Fabricated Metal Products and Machinery and equipment (Press Trust of India 2022).

3.3 Ownership of the MSMEs: Disparities of Vulnerable Social Groups

The ownership of MSMEs is male-dominated. Nearly 80% (79.63%) of the MSMEs are owned by men, which is nearly four times of the remaining 20.44% of MSMEs owned by women. As for regional distribution, 51% of MSMEs are in rural areas, while 49% of them are in urban areas. In both rural and urban areas, the dominance of male ownership is prominent; it is reported as 77.76% of MSMEs in rural areas and 81.58% of MSMEs in urban areas are owned by men. That leaves only 22.24% of MSMEs in rural areas and 18.42% of MSMEs in urban areas in the hands of women (Ministry of Micro, Small and Medium Enterprises, Government of India 2022). The MSME sector in India has been characterized by a skewed gender ratio in terms of the ownership of the three different scales of enterprises, namely, micro, small, and medium enterprises. Majority of micro-enterprises have been owned by women (around 80%) and this is the sector that has significantly contributed to creating employment opportunities for women in India. Thus, the scale of the industry has been considerably influenced by gender-factor and this part requires urgent attention.

Securing finance to sustain an enterprise has been often posed as the most acute problem in micro-sectors owned by women.

The financial constraint of women-owned 'very small enterprises' (WVSE) [emphasis ours] segment in India with less than 20 employees and turnover range remains between Rs.1,000,000 and 5,000,000. They have not been able to be addressed by banks and microfinance institutions require the topmost attention. Over 70 percent of WVSEs are engaged in seven activities, namely, textiles manufacturing and wearing apparel, handicrafts manufacturing, food, and grocery retail, jewelry and accessories retail, beauty salons and personal services, and eateries (Soni 2021). In the MSME sector in India, 72% of the women belong to the micro sector compared to 29% for male owners: while male ownership has been almost equally divided among the sizes of enterprises, marking 35% of ownership in medium enterprises, 35% in small enterprises, and the lowest 29% in micro enterprises, only 9% of the women are employed in the medium sector. It is also noted that 66.27% of all the MSMEs are owned by socially backward communities, while only 24.97% belong to medium enterprises (Ministry of Micro, Small and Medium Enterprises, Government of India 2022:27). The compiled data of the Government of India suggests social disparities and unequal development according to gender and socially backward communities, leading to the skewed regional distribution of the MSMEs. The micro sector in India has marked its vulnerability in the composition of the dominance of socially disadvantaged classes.

4 Literature Review

4.1 Entrepreneurial Mindset: Emphasis on Self-Employment

Golla (2018) represented idealized features of the MSME sector in India as a catalyst for creating a better ecosystem that enhances rural development through poverty reduction while achieving economic growth and development through generating employment opportunities. The study concluded that institutional framework and policy specifications under Micro, Small, and Medium Enterprises Development (MSMED) Act of 2006, played a vital role to help MSMEs grow. Over those 8 years from the year 2006–7 to 2014–15, the total number of units showed a compound annual growth rate (CAGR) of 19.42%, while employment showed a growth rate of 18.82%, and market value of fixed assets recorded growth rate of 30%.

Javalgi and Todd (2011), based their study on the 150 small and medium enterprises in India: they emphasize that it is entrepreneurial orientation,commitment to internationalization, and the ability to leverage human capital influence that matters for the internationalization of their business, focusing on the factors that influence the success of Indian SMEs to expand internationally. Gurtoo (2009) argues that risk-taking as well as the learning ability of entrepreneurs could be derived through interactions with the market, enabling them to navigate their behavior and strategy: entrepreneurs seek benefits and find opportunities while playing with the demand and supply of the market. It is adequate training, confidence, and risk-taking ability of the entrepreneurs that would enable them to attain sustainable economic growth. In such processes, entrepreneurs strive for social amalgamation to achieve larger representation while monitoring to defer unjust exploitation.

Thus, entrepreneurs are entrusted to function as a combined social force, as an enabling factor for enhancing the socio-economic development of India to ameliorate poverty by creating employment opportunities. Moreover, an entrepreneurial mindset is acknowledged as a driving force for the MSME sector to materialize internationalization.

4.2 Risk Management and MSMEs

Panigrahi (2012) studies the risk management practices of small and medium enterprises in India and acknowledges the highlight of risk management as the survival of a business entity that greatly owes to "its capabilities to anticipate and prepare for the change (Panigrahi 2012: 59)" rather than merely reacting to such change. It is necessary that "the risks are consciously taken with complete knowledge and clear understanding so that it can be measured to help in mitigation (Panigrahi 2012: 59)." Risk taking is "failure prone", so risk management is to reduce the risks to a reasonable and manageable level on an on-going basis. While acknowledging that risk-taking is "failure prone", risk management is regarded to reduce the risks to a reasonable and manageable level, on an ongoing basis. Risks faced by MSMEs that are discussed by Panigrahi can be classified as follows in Table 4.

In this context, it is clear that risk management for the MSME sector in India is perceived as the subject of the internal operational management of the enterprises. Risk factors affecting the operation, the market, and the credit scenes of the MSME sector in India still need to be discussed further from the viewpoint of environmental factors.

4.3 Climate Change and MSMEs

Kapitsa (2020) researched climate change and its impact on MSMEs and acknowledged that it resulted in enormous economic and human costs, hitting the poorest and most vulnerable worst. Kapitsa maintains that the problems of environmental degradation, disaster risks, and socio-economic development need to be simultaneously addressed while governments and civil society groups are undertaking policies and measures to reduce mass poverty and social inequality. Further, with regard to surviving disasters and transforming adaptation measures, Kapitsa recommended disaster-based insurance, though the difficulty of buying commercial insurance schemes for households and businesses in low-income countries without support

Table 4 Risks faced by Micro, Small, and Medium Enterprises (MSMEs) according to the type, relevant factors, inherent risks and impacts/effects

Type of risk	Factors	Inherent risks	Impacts/effects
Operational risk	Constitution of business entity	Lack of professionalism Over-dependence on one or two key personnel	Limiting the funds mobilization efforts and leveraging capacity
	Most of the MSMEs are small players	Tough competition	Facing tough competition from the bigger players
	Limited growth prospects	Qualified personnel gain experience and change employment Unskilled employees	High employee turnover Wastage of manpower The additional cost in training and knowledge updating Affecting continuity Lowering productivity
Market risk	Inadequate margin	High input cost	Not able to raise their price
	The scale of business is much small	Not able to dictate terms to their customers	Low collection in account receivables
Credit risk	Little financial resources	Poor in leveraging the financial structure	Not able to go for technological advancement
	Collateral security	Family home as security against the amount borrowed	Catalyzing to avoid default
	Microfinance through a repayment incentive structure	Adopting profit-centered approach	SMEs will pay fair risk-adjusted cost of capital
	Bank lending with the knowledge of prospects of venture	Risks facing their business	

Source Prepared by Authors (in reference to Panigrahi (2012))

from extended family or the government was pointed out at the same time. In continuation to this discussion, Kapitsa points out that MSMEs in developing countries are regarded from the perspective of productivity and competitiveness, where leading issues would be considered as transformation, employment, and social protection that the MSMES would bring to society.

Swiss Re Institute (2019) observed that the issue of under-insurance has already become "a theme across the advanced and emerging economies", urging insurers to improve their risk modeling capabilities for "Secondary Perils" with an enhanced product range for insurance. It also indicates the potential of private–public partnership in building "socio-economic resilience through their investment activities (Swiss Re Institute 2019:1)". According to their overview of the state of industry insured against disasters, they emphasize that the significant impact of "Secondary

Perils" need to be taken into consideration in terms of insurance coverage. Quoting the data of the year 2018, insured losses from catastrophes in 2018 amounted to 85 billion US Dollars, much beyond the average of the previous 10 years of 71 billion US Dollars. (Swiss Re Institute 2019: 1) Combined with the natural catastrophe loss of 2017, it amounted to 219 billion US Dollars, which was recorded as the highest insured loss till then over a two-year period. The peak loss of the global catastrophe protection gap of 199 billion US Dollars in 2017 came down to 80 million US Dollars in 2018, recording less than half of the 2017 gap.

Considering their available data before the COVID-19 pandemic, it is noteworthy to mark the observation of the Swiss Re Institute (2019) that the accumulated losses from the secondary peril disaster events made the year the fourth costliest year for the industry as recorded in 2019. The combined insured losses resulting from natural catastrophe events in 2017 and 2018 amounted to 219 billion US Dollars, the highest figure of the losses ever for a consecutive two-year period. The previous two-year high in 2018 prices was 2011–2012 at 207 billion US Dollars. How to make the MSME sector sustainable against disaster risks by enhancing disaster preparedness is crucial in consideration of the increasing impacts of secondary perils in regional contexts.

4.4 Digitalization Through Information and Communication Technology (ICT) Introduction

Another point of concern regarding the MSME sector is the degree of Information and Communication Technology (ICT) introduction in their day-to-day business operations. The ICT introduction to the MSME sector needs to consider how to accept and adapt the hurdles such as literacy level and educational qualifications of the enterprise owners. There should be further consideration for digitalization with various means of adaptation as a key of surviving strategies of the MSMEs.

Enhancing employment opportunities has been marked as one of the most important contributions of the MSMEs. It has also been acknowledged that it is the entrepreneurial mindset of the founders, owners, and/or managers that have been deciding the direction of the business of the enterprises. Of late some studies have acknowledged the importance of ICT introduction to operational management for the MSMEs to make a significant transformation regarding its expansion of the market, preparedness for disruptions due to disasters and climate change, or enhanced market reach. It is notable to mention here, that such initiatives of technological enhancements greatly depend on the leadership role of the entrepreneurial mindset of the founders, owners, and key professionals of concerned enterprises that have a final say about the adaptation and introduction of digitalization processes.

Das et al. (2020) acknowledge that technology adaptation makes SMEs competitive and enables their sustainable growth in their paper titled "Technological Adaptation and Survival of SMEs: A Longitudinal Study of Developing Countries" on the impact of technological environments upon the sustainability of the

SMEs in developing nations, focusing on the period between 2012–16, carried out through secondary resources from the Global Competitive Index Report (GCIR) and the Global Entrepreneurship Monitor (GEM). They claim that the main focus of preceding research on the technological adaptation of the SMEs has been on specific technology-related policies of the concerned country of the research target to examine the effectiveness or impact of policy regarding one particular dimension of the research. The market competitiveness of the SMEs requires awareness of market change as well as keeping pace with technological advances on their own so that they can be at par with global standards, but "the sustainable SMEs in developing countries are strongly dependent on technological environments that are resilient and adaptive to the high level of technological volatility at present [emphasis ours] (Das et al. 2020: 69–70)." In their study, Institutional Capabilities (Factor 1) were measured in terms of 5 sub-parameters, namely, comparative variables: 1)Availability of latest technologies (Factor loading 0.91), 2) Firm-level technology absorption (Factor loading 0.89), 3) FDI and technology transfer (Factor loading 0.75), 4) Individuals using internet (Factor loading 0.72), 5) Fixed broadband Internet subscription (Factor loading 0.71), have been found to have significantly impacted on the sustainability of the SMEs, while External Capabilities (Factor 2) measured in terms of a comparative variable, International Internet bandwidth (Factor loading 0.36) was not found significant. They point out there is a research gap to assess the overall technological-environmental factors of entrepreneurship sustainability of SMEs in developing countries. Further, they caution that the impact on the sustainability of business in developing nations needs to be studied in consideration of (1) the lack of structurally defined technological environments and (2) some other dimensions of the environment. It is in this context that cross-cultural factors upon the influence of the technological environment also need to be reviewed in further study on the MSME sector in India to seek "a more entrepreneurial technological environment for business (Das et al. 2020: 71)."

4.5 Key Findings from Literature Review

The entrepreneurial mindset has been valued and encouraged as a driving force for contributing to the employment generation of the country, promoting self-employment to ameliorate poverty by creating employment opportunities as well as exploring international business opportunities. MSME policies in India have centered around fostering opportunities for entrepreneurs to start their businesses while existing entrepreneurs have been encouraged to receive skill training to be able to maximize their business potential. The risk management for the MSME sector in India tends to be interpreted as internal operational management of the enterprises concerned with those risk factors affecting the operation, the market, and the credit scenes of the MSME sector. The threats to the entrepreneurial ecosystem because of environmental factors and sustainability have been hardly acknowledged in the study of the MSME sector in India. Disaster preparedness of the MSMEs has been considered in terms of credit risks or insurance. No mention of the impacts of secondary perils in regional contexts is the focus of the research on Indian MSMEs. Further, urban development with socio-economic impact as well as political implications and environmental risks with considerable economic distress due to secondary perils of late needs to be studied with the degree of involvement in digitalization of the enterprises on the part of the entrepreneurs in the MSME sector in India.

5 Discussion

The focus of risk management for MSMEs centered around operational managerial means, counting on the initiatives, changes, as well as modifications on the part of the stakeholders: MSMEs, financial institutions, as well as government agencies. What is missing is to conceptualize risks beyond the stakeholders. Risk factors for the MSME sector in India today go beyond the subject matter of internal operational managerial issues, demanding a comprehensive review to reflect upon the increasing threats in its entrepreneurial ecosystem.

Disasters and environmental hazards compel us to opt for a broader framework where practical challenges linger in the aftermath of disasters and natural calamities. The changing global scenario regarding the role of India has been also taken into consideration regarding disaster risk reduction in the context of India's G20 presidency (Shaw and Kishore 2023). The relationship between the MSMEs in India and the environment has been considered from the framework of environmental compliance, pollution control as well as energy efficiency as the key elements of "zero effect" for the industry (Ministry of Micro, Small and Medium Enterprises, Government of India 2012: 8). Other environmental factors such as climate change and disasters need to be taken into consideration.

The study of the MSMEs needs to overcome the dichotomy of rural–urban classifications prominent in the MSME policy approach in India. Such dichotomy may lead to a misinterpretation of rural locations with urban characteristics if (1) incorrect standards of applied services and unmet requirements of local public goods are used; (2) funds are allocated for such developments and programs designed for rural areas; (3) failure on the part of politicians failing to recognize the aspiration of the populace no longer remain rural and agrarian, having same aspirations as found in urban areas. The linear development and transformation from rural to urban may not explain fully how the growth of the MSMEs will be anticipated in coming years in consideration of the urbanization trend.

The study of the MSME sector in India starts with facing the reality of its huge informal sector. On average, 20% of MSMEs are formally registered, while unregistered MSMEs exist four times more than those registered. But this ratio varies greatly among the states within India as well as among the regions within a state. Therefore, the entrepreneurial ecosystem needs to be studied in consideration of a region-specific approach for data-based planning for effective policies.

While the existing MSME policy framework needs to be reviewed in consideration of external factors that pose risks to the enterprises due to climate change, disasters, as well as urbanization, the digitalization factor needs to be reviewed on how it has been impacting the MSMEs. Social distancing and prolonged closure of offices and schools accelerated digital usage in society. Yet there is "a phenomenon that has received scant attention in the peer-reviewed disasters literature to date", where consumers preferred to opt for online shopping, creating opportunities for entrepreneurship with a surge of new small businesses (Chang et al. 2022). Infectious disease, as one of the highest impact risks in 2021, as well as the digital divide have marked the overall risk landscape, widening the gap between north and south as well as urban and rural, in going through a digital transformation accelerated by COVID-19, causing severe disparity due to age and gender factors (Shaw and Kishore 2023:1).

What is important is to understand how entrepreneurs face risks and be resilient enough to find and adapt recovery strategies in the MSME sector in India. Figure 1 depicts that an entrepreneurial mindset co-exists with risks and opportunities. Institutional support implies government agencies and financial institutions support entrepreneurs who are prone to face risks such as climate change and disasters, yet can balance opportunities and risks. This figure indicates that an entrepreneurial mindset could exist at various moments, with different degrees of opportunities and risks. The support of financial institutions could also come from domestic financial players to achieve the financial inclusion of MSMEs by offering simplified access to credit for entrepreneurs of small businesses (Press Trust of India 2022).

Figure 2 shows the key factors which affect MSMEs in India: gender disparity, regional disparities, digitalization, urbanization, and principal and secondary perils. Addressing these factors will help us to oversee some possible recovery routes as well as the means of risk mitigation.

In the policy framework for the MSMEs, more attention needs to be drawn toward the striking absence of consideration over environmental factors due to climate change. Therefore, the following suggestions are proposed:

1. **Creating Self-Assessment Avenues for MSMEs**: Climate change has induced the MSMEs prone to be affected by unexpected weather conditions and susceptible to silent localized disasters. Vulnerability factors and the inability to

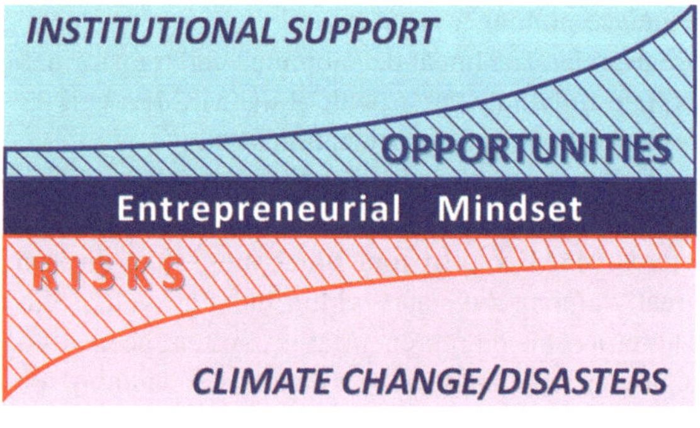

Fig. 1 An ecosystem for entrepreneurial mindset with risks and opportunities. *Source* Authors

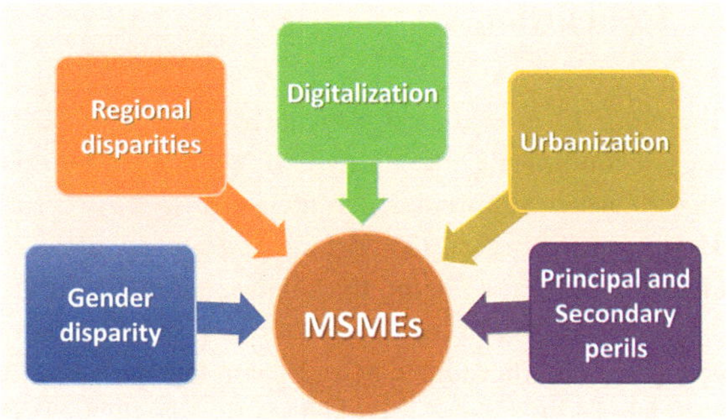

Fig. 2 Conceptual diagram of the study. *Source* Authors

ensure disaster preparedness require the MSMEs to have their means of evaluating resilience. Measure for self-assessment of risks would help them become resilient MSMEs, planning ahead of disasters.

2. **Imparting Educational Opportunities for MSMEs Against Economic Distress Due to Secondary Perils**: Secondary perils of disasters have compelled many MSMEs to face constant threats of vulnerabilities beyond their control due to climate change. Restructuring insurance applicable to probable economic distress due to secondary perils needs to be structured. Such initiatives with the support of local, regional, and national government agencies can create climate data collectives by engaging communities and educational institutions as co-learners and co-creators. Creating hubs of citizen groups who learn about climate change and its impact upon their lives locally will create the pool of "improved risk data at the individual level" that will ensure "expanded modeling capabilities" where "greater emphasis will be placed on current as well as future risks with improved educational impact on personal risk, leading to further expanded coverage of the more insured population (Pande 2022)."
3. **Nurturing Collective Disaster-preparedness for the MSME Sector**: Trade-based collective disaster-preparedness drills for MSMEs can be explained in practical vernacular forms. Assessing existing threats and anticipating future risks will help MSMEs make their managerial planning as well as plan for incentives for the MSMEs to enhance their preparedness against natural and human calamities.
4. **Exploring New Forms of Assistance for the MSME Sector**: Sector-specific support in consideration of the characteristics of the entrepreneurs with regional implications is required. It will generate effective and workable stakeholder-specific recovery measures to ameliorate vulnerabilities of the MSMEs owned by particular communities, considering gender factors (Sugaraj and Salve 2014) and taking into account of region-specific impacts being manifested due to climate change.

6 Conclusion

The MSME sector has significantly contributed to enhancing opportunities for the socio-economic development of the country in various parts of the world. The MSME sector has been appreciated as the major force in creating employment opportunities. This point is especially important for a country like India where the informal economy takes the burden of being the major source of employment. Climate change-bound disasters are environmental threats. Rapid urbanization has been impacting MSMEs, too. However, these have neither been fully documented, nor reflected in the MSME policy in India. Health-related hazards, like the COVID-19 pandemic, have impacted the world so extensively, though variedly, that they have rendered many governments to consider ameliorative measures to save the MSMEs in their countries. For India, too, it is inevitable to reflect upon existing conceptual frameworks and parameters for measuring the growth and development of MSMEs. Though types of disasters and their impacts would vary according to regions and countries, there has been a common thread to understand the underlying theme: to understand the inherent vulnerability factors of the MSME sector. However, in the context of the increasing impact of climate change, it is paramount to research yet another dimension of the vulnerability of the MSMEs in India, which necessitates further research.

7 Declaration of Competing Interest

The authors declare that they have no known competing financial interests or personal relationships that could have appeared to influence the work reported in this paper.

Funding The authors acknowledge the funding from India Japan Laboratory of Keio University as well as JSPS Kaken C Grant-in-Aid (Number: 23K11610).

References

Chang SE, Brown C, Handmer J, et al. (2022) Business recovery from disasters: Lessons from natural hazards and the COVID-19 pandemic. Intern J Disast Risk Red 80:1–13. https://doi.org/10.1016/j.ijdrr.2022.103191

Das S, Kundu A, Bhattacharya A (2020) Technology adaptation and survival of SMEs: a longitudinal study of developing countries. Technol Innov Manag Rev 10(6):64–72. https://timreview.ca/sites/default/files/article_PDF/TIMReview_2020_June%20-%206.pdf

Golla SK (2018) Growth and performance of micro small and medium enterprises in India after MSMED Act 2006. Intern J Res Manag Econ Comm 08(1):52–59. https://www.researchgate.net/publication/322654613_Growth_and_Performance_of_Micro_Small_and_Medium_Enterprises_in_India_after_MSMED_Act_2006

Gurtoo A (2009) Policy support for informal sector entrepreneurship. J Devel Entrep 14(2):181–194. https://doi.org/10.1142/S1084946709001223

Hansen C, Steinmetz H, Block J (2022) How to conduct a meta-analysis in eight steps: a practical guide. Manag Rev Quart 72:1–19. https://doi.org/10.1007/s11301-021-00247-4

Javalgi RG, Todd PR (2011) Entrepreneurial orientation, management commitment, and human capital: the internationalization of SMEs in India. J Bus Res 64(9):1004–1010. https://doi.org/10.1016/j.jbusres.2010.11.024. https://www.sciencedirect.com/science/article/pii/S0148296310002535

Kapitsa LM (2020) Climate change and micro, small and medium enterprises. MGIMO Rev Intern Relat 13(4):216–231. https://doi.org/10.24833/2071-8160-2020-4-73-216-231

Ministry of Micro, Small and Medium Enterprises, Government of India (2012) MSME Policy Launch of Draft Consultation Paper. http://www.dcmsme.gov.in/MSME_Policy_Consultation.pdf#:~:text=Any%20MSME%20Policy%20for%20India%20should%20take%20into,Ministry%20of%20Corporate%20Affairs%20and%20Ministry%20of%20Finance

Ministry of Micro, Small and Medium Enterprises, Government of India (2022) Annual Report 2021–22. https://msme.gov.in/sites/default/files/MSMEENGLISHANNUALREPORT2021-22.pdf

Ministry of Micro, Small and Medium Enterprises, Government of India (2022) MSME DASHBOARD. https://dashboard.msme.gov.in/Udyam_Statewise.aspx (Accessed on 30 December 2022)

Office of the Development Commissioner, Ministry of Micro, Small and Medium Enterprises (2021) Benefits of Taking Udyam Registration. https://udyamregistration.gov.in/docs/Benefits_of_UR.pdf

Office of the Development Commissioner, Ministry of Micro, Small & Medium Enterprises (2019) Up-scaled Entrepreneurship and Skill Development Programme: a sub-component under "Development of MSMEs" vertical of Office of DC (MSME) schemes (As amended on 20.11.2019). https://msme.gov.in/sites/default/files/Up-Scaled-ESDP-Guidelines.pdf

Pande M (2022) Bizarre or the new normal? The challenge of understanding secondary perils. SwissRe. https://www.swissre.com/risk-knowledge/risk-perspectives-blog/challenge-of-understanding-secondary-perils.html

Panigrahi AK (2012) Risk Management in Micro, Small and Medium Enterprises (MSMEs) in India: a critical appraisal. Asia Pac. J Marketing Manag Rev 1(4):59–72. https://www.researchgate.net/profile/Ashok-Panigrahi/publication/323394128_Risk_Management_in_Micro_Small_and_Medium_Enterprises_MSMEs_in_India_A_Critical_Appraisal/links/5a958764aca27214056934c4/Risk-Management-in-Micro-Small-and-Medium-Enterprises-MSMEs-in-India-A-Critical-Appraisal.pdf

Press Trust of India (2022, December 15) Kinara Capital Adds 15 New Branches; Reaches 125 Branches Across 100 plus Cities in India to Serve MSME Sector. Press Trust of India. https://www.ptinews.com/pti/kinara-capital-adds-15-new-branches-reaches-125-branches-across-100-plus-cities-in-india-to-serve-msme-sector/58984.html (accessed on 02 January 2023)

Reserve Bank of India (2020) Notification vide no. RBI/2020–2021/26 FIDD.MSME & NFS.BC.No.4/06.02.31/2020–21. https://msme.gov.in/sites/default/files/RBI-NOTIFICATION.pdf

Reserve Bank of India (2022) Reference rate archive. https://rbi.org.in/scripts/ReferenceRateArchive.aspx

Shaw R, Kishore K (2023) Disaster risk reduction and G20: a major step forward. Prog Disast Sci 17:100274, ISSN 2590–0617. https://doi.org/10.1016/j.pdisas.2023.100274

Soni S (2021) Women's entrepreneurship day: why only 20% of India's over 6 crore MSMEs are led by women? https://www.financialexpress.com/industry/sme/msme-eodb-womens-entrepreneurship-day-why-only-20-of-indias-over-6-crore-msmes-are-led-by-women/2372108/

Sugaraj JM, Salve PS (2014) Constraints/challenges faced in a journey of women entrepreneurship in the development in western Maharashtra. IOSR J Econ Fin 3(2). IOSR Journals:53–60. https://doi.org/10.9790/5933-03215360

Swiss Re Institute (2019) Natural catastrophes and man-made disasters in 2018: "secondary" perils on the frontline. *Sigma,* No 2. https://www.swissre.com/dam/jcr:c37eb0e4-c0b9-4a9f-9954-3d0bb4339bfd/sigma2_2019_en.pdf

United Nations Department of Economic and Social Affairs (2019) World Urbanization Prospects: the 2018 Revision (ST/ESA/SER.A/420). New York: United Nations Department of Economic and Social Affairs/The Population Division. https://population.un.org/wup/Publications/Files/WUP2018-Report.pdf

United Nations Industrial Development Organization (2020) Guidance for Micro, Small, and Medium Enterprises Business Recovery in the Wake of the COVID-19 Pandemic. Geneva: UNIDO. https://www.unido.org/sites/default/files/files/2020-05/Guide%20for%20MSMEs%20in%20wake%20of%20COVID-19_EN%20FINAL%202_0.pdf

Van Wely M (2014) The good, the bad and the ugly: meta-analyses. Human Reproduct 29(8):1622–1626. https://doi.org/10.1093/humrep/deu127, https://academic.oup.com/humrep/article/29/8/1622/2914052

Renewable and New Energy: Potential Collaboration Between Japan and India

Yoshiro Kaku and Yuta Hikichi

Abstract Prime Minister Modi's declaration in 2021 that India will achieve carbon neutrality by 2070 has transformed the energy and environmental landscape in India and is attracting attention from around the world. In light of this, this chapter analyzes the background of such a declaration and trends surrounding energy and environment in India, including renewable energy, biomass, thermal power, and clean mobility, as well as potential collaborations between Japan and India in these fields.

Keywords Climate change · Carbon neutrality · Renewable energy · Thermal power · Biofuel · Clean mobility · India-Japan relationship · Sustainability

1 Introduction

In 2023 India just became the most populated country in the world. Also, the success of lunar landing and exploration project "Chandrayaan 3" in August and the G20 Summit in September 2023 are fresh in our memories. On the other hand, India is currently the world's third largest energy consumer after China and the United States, and further growth in energy consumption is estimated as its population increases (IEA 2023a, b, c). Therefore, in the fields of energy and climate change, as seen in the stance of International Energy Agency (IEA), India has been receiving increasing attention (IEA 2021a, b).

In 2021, India's words in relation to energy and climate action came into the spotlight. Most notably, at COP26 in Glasgow, UK, Prime Minister Modi stunned the world by announcing "Panchamirt—Five Commitments," including that "India will achieve carbon neutrality (CN) by 2070." (Press Information Bureau n.d.)

Y. Kaku (✉)
NEDO, New Delhi, India
e-mail: kaku-yoshiro@nedo.in

Y. Hikichi
Embassy of Japan, New Delhi, India

R. Shaw and S. R. Choudhury (eds.), *India, Japan and Beyond*,
https://doi.org/10.1007/978-981-97-3282-1_7

Here, questions arise as to why India made such a declaration. Admittedly, India's intention to tackle climate change, such as reducing greenhouse gas (GHG) emissions, cannot be denied at all. However, India's traditional stance on climate change is that of developing countries. Just as India has insisted on "common but differentiated responsibilities" in the international arena, India's stance that "developed countries are primarily responsible for climate change" remains unchanged. (PTI 2021).

In fact, it should be noted that India included an absolute amount of CO2 emission target in the Five Commitments announced at COP26, by saying it will reduce total planned carbon emissions by 1 billion tonnes by 2030 (Press Information Bureau n.d.). But, this absolute amount target is not included in India's revised NDC submitted in 2022 (Press Information Bureau n.d.). Rather than such an absolute amount target, the inclusion of an emission intensity target in this NDC, as described below, suggests that India is pursuing climate change countermeasures in a manner consistent with its desire for continuing economic growth (Press Information Bureau n.d.).

In this chapter, we will analyze the background of why India has declared it will achieve CN by 2070, by grasping the current circumstances surrounding energy and environment in India, and highlight India's potential for clean energy in each of the following sectors:

Clean thermal power
Renewable energy
Biomass
and Clean mobility

Also, we will discuss the current status and future trends of cooperation between Japan and India in each of these sectors. While India is also focusing on hydrogen and ammonia, these are analyzed in a separate chapter, and will not be discussed in this chapter.

2 Background of India's CN Declaration

Here, we will consider the reasons behind India's declaration of CN (factors other than climate change countermeasures).

First, Govt. of India has an intention to make the most of domestic energy and to reduce imports of resources and fuels to the maximum. India depends on foreign countries for around 75% of its crude oil and around 50% of its natural gas. India places importance on reducing such dependence, from the viewpoint of energy security and preventing the outflow of national wealth (IEA 2021a, b). In fact, on Independence Day in August 2021, prior to COP26, Prime Minister Modi declared that India will achieve energy independence by 2047, and mentioned that India would work on biomass, electrification of mobility, renewable energy, and green hydrogen (Press Information Bureau n.d.).

In addition, in terms of domestic energy, it is important to mention coal-fired power. Coal is an important domestic energy source because India is around 70%

self-sufficient in coal (IEA 2021a, b). Moreover, since many workers are engaged in coal mines in the country, the use of coal energy is a sensitive issue with regard to securing employment. Actually, in India's National Electricity Plan announced in May 2023, coal-fired power generation plans to increase by approximately 50 GW in absolute terms over the next 10 years (while its overall share will be smaller) (Central Electricity Authority—Ministry of Power of India 2023).

According to Mr. Alok Sharma, President of COP26, which was held in Glasgow, UK, and where Prime Minister Modi declared CN to be achieved by 2070, the Govt, of India objected to the words "phase out of coal-fired power" in the draft joint declaration, resulting in the words "phase down of coal-fired power" (Harvey and Mason 2021). This is another important glimpse into India's basic stance to energy policies.

Secondly, it may be aimed at encouraging investment and technology from overseas. India generally intends to solve social problems with new technological innovations, and regarding technologies that India does not possess now, it has a policy of promoting domestic production by attracting technology from overseas (the Production-Linked Incentive (PLI) scheme is designed to promote domestic production). Indeed, in PM Modi's Five Commitments speech, he said, "India expects developed countries to quickly provide $1 trillion in climate finance" (Press Information Bureau n.d.). The $1 trillion in climate finance, which developed countries promised to provide to developing countries at COP21, has been repeatedly claimed by developing countries that their commitments have not been met. Because of these circumstances, India is expected to attract foreign investment and technology transfer in the clean energy sector by declaring carbon neutrality.

Third, India emphasizes measures to deal with air pollution, health hazards, etc. India has one of the highest levels of air pollution in the world, especially in the northern part of the country, particulary in fall (IQAir 2023). One of the causes of air pollution is the burning of rice straw and other agricultural waste in the surrounding areas, and so it is expected that biomass resources such as rice straw will be used as energy. Also, Govt, of India intends to promote the intruduction of clean mobility to replace diesel and fuel-inefficient vehicles (IQAir n.d.). Moreover, in rural areas of India, where firewood is being used for cooking traditionally, health hazards to women and children who are engaged in cooking have also been reported. In regards to this issue, the Govt. of India is making efforts to promote clean cooking through the PMUY scheme for supplying LPG and other products to rural households, which is highly appreciated by the IEA (IEA 2023a, b, c; Press of information Bureau n.d.).

Against these backgrounds, India is promoting the introduction of clean energy to achieve carbon neutrality. In light of this, the India's revised NDC includes the targets of "Reduce the carbon intensity (GHG/GDP) by 45% from 2005 levels by 2030" and "50% non-fossil energy by 2030," as well as the "LiFE—Lifestyle For Environment," a lifestyle change movement promoted by Prime Minister Modi to combat climate change (Press of information Bureau n.d.). In addition, the National Electricity Plan mentions that the share of renewable energy such as solar power will increase rapidly in the future (Press of Information Bureau n.d.). (Graph 1) Thus, India is strongly promoting initiatives related to clean energy.

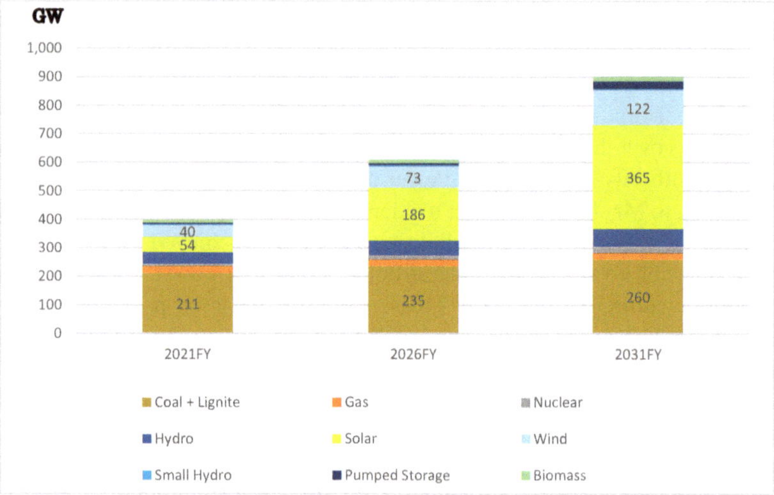

Graph 1 Projected installed capacity. *Source* "National Electricity Plan" (Central Electricity Authority 2023)

3 Existing Frameworks and Policies for Promoting Collaboration Between Japan and India in Energy Field

As this chapter discusses Japan-India collaborations in the energy and environment sector, we would like to mention some general frameworks and policies for promoting such collaborations.

Since Govt. of India presently intends to keep utilizing coal-fired power, it supports the concept of "realistic energy transition" and "various pathways to achieve CN," advocated by Japan in the "Asia Energy Transition Initiative (AETI)" and other initiatives (Ministry of Economy, Trade and Industry 2021). Since India is calling for such an energy transition that takes into account the reality of Asia, India is an extremely important partner for Japan in terms of spreading this concept around the world.

In addition, Japan-India Energy Dialogue has been held 10 times, and concrete discussions have been held on Japan-India cooperation (Ministry of External Affairs of India 2022). Moreover, the Japan-India Clean Energy Partnership (CEP) was announced at the Japan-India Summit Meeting in March 2022 and served as a basic framework for further Japan-India cooperation in various energy fields (Ministry of Economy, Trade and Industry 2022). Based on this, it is expected that Japanese companies with carbon–neutral technologies will make various efforts in India's growing energy-related market.

Next, we would like to introduce some programs to support the activities of Japanese companies in the energy field. First, NEDO (New Energy and Industrial Technology Development Organization, under Ministry of Economy, Trade and Industry, Govt. of Japan) has a program to give funding to projects to demonstrate Japanese companies' technologies overseas (NEDO n.d.). To date, NEDO has conducted more than 10 demonstration projects in India, mainly in the sectors of energy conservation and renewable energy (NEDO n.d.).

The Joint Crediting Mechanism (JCM) is also one of such programs. Negotiations have been underway between Govt. of Japan and India to conclude JCM (Ministry of Environment of Japan and Ministry of Environment, Forest and Climate Change of India 2023). If JCM is concluded, it will be possible for Japanese companies to utilize support schemes by NEDO or Ministry of Environment of Japan, and to allocate a portion of the credits generated in India to Japan. Japanese companies are strongly requesting the introduction of JCM (Press Information Bureau n.d.).

The following sections will perform analysis for each individual field's current circumstance and Japan-India collaborations.

4 Clean Thermal Power

As noted above, India's prioritization of energy security is clear. Thermal power sector has many domestic stakeholders.

As well as India is the world's fifth largest producer of coal, it is the world's second largest coal consumer after China. India's consumption of coal is accounting for 14% of its global consumption, and coal power generation accounts for 74% of India's domestic coal consumption. As IEA projects, India will continue to be a major player in boosting coal demand globally with growing by 3.5% per year until 2026 (it is projected that coal power generation will grow by 9% only in 2023), while China's coal consumption begins to trend downward (IEA 2023a, b, c).

According to India's Ministry of Power (MoP), as of May 2023, fossil fuel-derived thermal power generation accounted for 56.8% of installed capacity in India (coal: 50.7%, gas: 6%, and diesel: 0.1%). Govt. of India has set a goal is to reduce the share of coal-derived power generation capacity to about 30% of this total by 2030 (Press Information Bureau n.d.).

On the other hand, India's demand growth has led to an increase in generation capacity over the past decade until 2023, and the country has stated that it will add 23 GW of new coal power plants by 2030. Thus, the absolute amount of coal power generation is expected to continue to increase even after 2030, and perhaps as long as it remains a developing country until 2047, as India has stated. Specifically, as MoP notes, as well as coal power generation will not be decommissioned until 2030, it will continue to be used after 2030 while facilities will be refurbished, modernized, extended in service life, and operated in two shifts mode to promote grid integration of solar and wind energy (IEA 2023a, b, c; Press Information Bureau n.d.; *Dispatch of Coal to thermal power and coal consumption/off-take during last ten years* n.d.; Press Information Bureau n.d.; Central Electricity Authority 2023).

India considers before 2030 as the short-term phase, 2030–2047 as the medium-term phase, and 2047–2070 as the long-term phase when discussing decarbonization.

It can be inferred that when the expected operation period of coal power plants is estimated at 40 years, and when they are installed by 2030, it can be consistent with India's goal of achieving CN by 2070. If PM Modi's party wins the general election in 2024, it would be possible to suppress discussion on the abolition of coal power plants and whether or not the 2030 target would be achieved until 2029, when PM Modi's third term would be completed (Press Information Bureau n.d.).

With regard to natural gas, the Govt. of India has advocated a gas-based economy, proposing to increase the share of gas in the country's energy mix from the current approximately 6 to 15% by 2030. On the other hand, against the backdrop of soaring gas prices, the current utilization rate of gas-fired power generation is low at around 12%, and many of gas-fired power plants have become stranded assets. However, the role of gas as a bridge fuel for future decarbonization is not small. As will be discussed later, there is room for the introduction of gas-fired power plants to play a role as stabilizing the output for renewable energy and as in-house power generation for fertilizer plants, etc. (Press Information Breau n.d.).

India, as a developing country, will work to reduce GHG emissions per its GDP by the massive introduction of renewable energy until 2047. However, as mentioned above, we assume that India could gradually be put in a tougher position internationally with regard to its stance on GHG reductions, in view of the fact that India, overtaking China, will become the country with the largest emissions and still need to build more coal-fired power plants.

Therefore, as Govt. of India takes efforts to meet electricity demand as well as reduce its GHG emissions, while adding and replacing coal-fired power generation, it is considered a major issue which should be addressed immediately to increase the efficiency of thermal power generation, in addition to the introduction of renewable energy (Press Information Bureau n.d.).

With these backgrounds in mind, we will review India's efforts toward cleaner thermal power generation, including higher efficiency thermal power generation, by introducing some collaborations with Japanese companies, in the following paragraphs.

First, we would like to mention ultra-supercritical (USC) technology. One of the Japan-related companies that have already introduced this technology in India is Toshiba J&W Power Systems. Toshiba J&W Power Systems has a plant for manufacturing turbines in Chennai and has received orders for USC turbines from UP State Power Generation Corporation. Also, L&T-MHI Power Boilers Pvt. Ltd. and L&T-MHI Power Turbine Generators Pvt. Ltd., in which Mitsubishi Heavy Industries (MHI) and Mitsubishi Electric Corporation are participating, have completed the construction of NTPC's USC power plant in Madhya Pradesh (LARSEN&TOUBRO n.d.; *Toshiba Wins Rs. 3436 crore EPC Contract for Harduaganj Ultra-supercritical Thermal Power Project from Uttar Pradesh Rajya Vidyut Utpadan Nigam Ltd (UPRVUNL) in Lucknow, India.* n.d.).

As well as the USC technology, the combined cycle gas turbine technology has been put into practical use. This technology has a significant contribution to CO2 reduction compared to USC technology, which has a limited reduction effect, as combined cycle gas turbine's CO2 emissions per unit of electricity generated are approximately half those of traditional coal power generation. However, gas power generation is not so prioritized in India, because India relies on imports of gas for power generation, and cheaper domestic gas is preferentially supplied for fertilizer and mobility sectors.

In addition to power plants, it is important to improve the energy efficiency in steel mills, cement plants and fertilizer plants as well. For example, fertilizer plants generate power by using the gas generated in the plant. Japan has achieved gas turbines with very high power generation efficiency for gas power generation, and this is a possible area where Japan can contribute (Press Information Bureau n.d.).

Another new technology is hydrogen co-firing and ammonia co-firing, which are in the demonstration stage. These technologies can be introduced simply by modifying existing facilities. For example, ammonia co-firing is a process in which air is mixed with coal fines, which are then mixed with ammonia for combustion. IHI Corporation, together with Kowa Company Ltd. and Adani Power, has started a feasibility study on ammonia co-firing in coal boilers as part of a NEDO demonstration project, with the aim of achieving 20% co-firing. In addition, IHI and GE have signed an MoU for conducting joint research and demonstration on ammonia gas turbines. As for hydrogen co-firing, MHI is demonstrating technology for 30% mixed combustion in gas turbines, with the aim of achieving 100% in the future (*Development of Co-Firing technology of pulverized coal and ammonia for suppressing NOX generation* n.d.; *Adani Power To Co-Fire Green Ammonia At Its Mundra Plant For A Sustainable Future* n.d.; *GE and IHI Sign Memorandum of Understanding to Develop Gas Turbines that Can Operate on 100% Ammonia* n.d.).

In addition, the Govt. of India is keen to utilize Carbon Capture, Utilization and Storage (CCUS) to capture CO_2 emitted from industrial activities, and utilize it to produce valuable products or store it underground. NITI Aayog published a research report in November 2022 titled "Carbon Capture, Utilization and Storage—Policy Framework and its Deployment Mechanism in India." According to this report, deep saline aquifers distributed in almost all regions and basaltic rock formations unevenly distributed in the western part of the country are considered promising. However, there are many challenges due to the lack of geological data and the long distance from the CO_2 source. One of the CCUS-related technologies, methanation is a promising option, where CO_2 and hydrogen are converted to methane by using a catalyst. Hitachi Zosen and IHI possess technology for methanation (Press information Bureau n.d.; NITI Aayog 2022; Tuli, Hemant and Deepak 2023; *Methanation Equipment | Hitachi Zosen Corporation* n.d.; *IHI Launches Methanation System to Produce Fuel from Carbon Dioxide and Hydrogen* 2022).

Moreover, Govt. of India has a strong interest in waste-to-energy technology, which will contribute not only to increasing electricity generation capacity, but also to reducing waste volume. It is critical because there is a growing shortage of landfill sites for waste in India, particularly in large cities. Hitachi Zosen has already received orders for a total of three waste-to-energy plants in Madhya Pradesh and Maharashtra (*Received third order for the EfW plant in India | Hitachi Zosen Corporation* 2021).

PSUs such as NTPC and Coal India Ltd (CIL), as mentioned above, are under pressure to introduce renewable energy in order to improve their own in-house energy mix, which will be discussed below in the section on renewable energy.

Finally, many of coal power plants have not installed Flue Gas Desulfurization (FGD) equipment to reduce SOx emissions. Although the MoP has set a timeframe for the installation of FGD in 600 plants close to cities, only 22 plants had been installed as of January 2023. Since the related guidelines were drawn up in 2015, postponement of the installation has been repeatedly suggested by MoP and Central Electricity Agency (CEA), due to high project costs. This is said to be partly because the recent geopolitical situation is affecting the import of certain components (Ministry of Power of India 2022a, b; Press Information Bureau n.d.).

JBIC and Japanese private banks have signed a loan agreement with NTPC for the introduction of FGDs (Behl 2022).

5 Renewable Energy

India aims to increase its total installed capacity from non-fossil based energy sources to at least 50% by 2030 (*500GW nonfossil fuel target | Government of India | Ministry of Power* n.d.; Press Information Bureau n.d.). It has decided to publicly offer 50 GW of renewable energy projects annually from FY2023 to FY2027, and apparently its intention is to keep electricity prices low by increasing market transparency and ensuring a stable supply of demand through policy measures. Such an increase in renewable energy is to reduce the dependence on energy imports and increase energy self-sufficient rate (Press Information Bureau n.d.).

As for demand generation, since 2018, the Renewable purchase obligation (RPO) has been imposed on the operators with transmission and distribution licenses. RPO targets for 2030 are set by Ministry of New and Renewable Energy (MNRE) in the categories of wind, hydro, and other sectors, respectively. Similarly, there is an obligation to secure storage for wind and solar power (Ministry of Power 2018; Ministry of Power 2022).

In addition, coal power plants on which RPO is imposed on, except on-site power plants, has become obligated to generate a certain percentage of their electricity from renewable energy sources. Coal power plants commencing operations after 2023 are obligated to generate at least 40% of their thermal generation capacity from renewable energy sources (Press Information Bureau n.d.; Ministry of Power of India 2023).

Purchase demand for renewable energy is being maintained through the series of public procurement programs such as RPO, rooftop solar program, and PM KUSUM. Similarly, in order to stably increase renewable energy supply, efforts are being made to lower the barriers to entry for power generators, such as the renewable energy bank system and solar parks, as described below.

The Electricity (Promoting Renewable Energy Through Green Energy Open Access) Rules 2022, established in June 2022, allows renewable energy producing companies to deposit renewable energy for a specific period of time. At the same time, consumers will proceed with the procurement of green electricity as stipulated in the Green Energy Open Access Rules. In particular, the conditions for suppliers continue to be eased, for example, by lowering of the minimum transaction amount from 1 MW to 10 kW (*Notification of "the Electricity (Promoting Renewable Energy Through Green Energy Open Access) Rules 2022."* 2023; *ANNEXURE Electricity (promoting renewable energy through Green Energy Open Access) Rules 2021* n.d.).

However, DISCOMs, which adjust the supply and demand of electricity, are obligated to purchase renewable energy while suppressing thermal power generation to reduce oversupply during the monsoon season when demand for cooling and agriculture is low but wind power generation is high. Also, they compensate for wind power fluctuations during peak demand. Since such a structure makes DISCOMs' cost burden heavier, some states are leaving this open access scheme (Ayallore 2022).

Similarly, in the context of reducing the mismatch between supply and demand, in addition to the above, storage technologies such as batteries and hydrogen ammonia are essential. From such perspective, the Confederation of Indian Industry (CII), with the support of the Ministry of Power (MoP), has issued a report that considers hydrogen production, storage, and solar modules as a critical part of India's supply chain (Press Information Bureau n.d.; Confederation of Indian Industry & EY n.d.).

The following sections provides an overview of three categories of renewable energy: solar, wind, and hydropower. As of July 2023, the installed capacity in India is 70 GW of solar, 44 GW of wind, 5 GW of small hydro, 47 GW of large hydro, 10.2 GW of biomass cogeneration, and 1 GW of waste-to-energy (*Renewable energy in India—Indian Power Industry investment* n.d.).

5.1 Solar Power Generation

In general, the northwestern part of India is said to be suitable for solar power generation due to the high intensity of sunlight. So the purchase price of solar power has been successfully reduced, making it a competitive energy source. ReNew Power and Mitsui Corporation have signed India's first long-term round-the-clock power purchase agreement (*Mitsui to invest in a 1,300 MW renewable energy project in India—MITSUI & CO., LTD.* n.d.).

As a mechanism to stabilize supply, the Govt. of India has established the Solar Energy Corporation of India Ltd (SECI) and other entities to provide public procurement for solar power, offering procurement through bidding and ensuring stable royalty payments to operators. In addition, in order to encourage the entry of power generators, MNRE asks states to develop solar parks and obtain legal clearance for land, roads, transmission systems, and other infrastructure (*Solar Energy Corporation of India Limited (SECI), a Government of India enterprise, under Ministry of New and Renewable Energy* n.d.).

In addition, because solar power generation fluctuates across the day, it cannot be operated round the clock without storage. Therefore, battery storage and pumped hydro are needed in order that solar power plants directly connect to the grid and provide a stable supply. Moreover, in some cases, renewable energy power producers need to develop substations as well to connect to the grid, which is said to be a challenge because it requires a large investment.

Other government measures to increase renewable power supply include promoting of roof-top solar, which installs solar power on the rooftops of homes and public facilities. Also, PM KUSUM promotes the supply of solar-derived electricity to irrigation pumps used to secure water for agricultural use. PM KUSUM, in particular, not only contributes to increasing the ratio of renewable energy, but also benefits rural communities, which account for the majority of the nation's population (Press Information Bureau n.d.).

5.2 Wind Power Generation:

According to the International Renewable Energy Agency (IRENA), India ranks fourth in the world in terms of wind power generation capacity and sixth in terms of its electricity generation. On the other hand, more than 90% of its wind power generation potential is below 260 W/m^2, the lowest category. So, India has only a few suitable sites for wind power, except for offshore Tamil Nadu and other areas (*Country rankings* n.d.; International Renewable Energy Agency n.d.).

Many of the wind power plants in India were installed before the year 2000, and subMW class wind power plants have been installed in suitable locations for wind power generation. The "National repowering & life extension policy for wind power projects" was announced in December 2023, which aims to increase power generation capacity by replacing some of the turbines and other components of these inefficient wind power plants. The wind power capacity for the whole country is 25.4 GW, and by state: 7.4 GW in Tamil Nadu, 4.7 GW in Gujarat, 3.4 GW in Maharashtra, and 3.0 GW in Karnataka (Ministry of New and Renewable Energy of India 2023).

MNRE is the lead ministry for promoting offshore wind power generation. It will bid for projects worth 4 GW per year in Tamil Nadu and Gujarat for three years starting in FY2022, on the condition that the government will provide power transmission from the EEZ sea area where the power is generated and the offshore pooling substation. Thereafter, project bids equivalent to 5 GW per year will be conducted until 2030. A separate study estimates the offshore wind potential to be 70 GW for the two states (Press Information Bureau n.d.; *Evaluation of Wind Energy in India* n.d.).

The feed-in tariffs, as in the case of other energies, had set a certain price for the sale of electricity until 2017, but the bidding price has been even lower since the competitive bidding process began. On the other hand, Non-performing assets amounted to Rs. 600 crores as of 2021, and Rs. 14,247 crores as of 2022, partly because they are not paid by DISCOMs.

5.3 Hydroelectric Power Generation

Hydropower can be divided into three categories: large-scale hydropower, pumped storage, and small-scale hydropower using the flow of rivers.

Regarding a typical large hydropower, we should note financial issues surrounding it. The debt/equity ratio for its project is 70:30, but many projects require the use of short-term debt that cannot cover the project period. Therefore, 10 out of 16 projects have been suspended for financing reasons, as of 2019 (PRS Legislative Research n.d.).

As for pumped storage, while it is estimated that India has a potential of 96 GW as of 2021, only 8 locations nationwide are currently operational with a total of 3.3 GW. However, it is expected to serve as storage to mitigate output fluctuations of solar and wind power. For example, Greenko and ReNew Power have each submitted bids for BOO (Build-Own-Operate) pumped storage projects based on the peak generation unit price issued by SECI (Nova 2023; Saurenergy 2020).

Japan has a track record in pumped storage power generation through ODA. For example, Japan has provided loans of approximately 18 billion yen since 2006, to install 900 MW of pumped storage power generation and related transmission facilities in the Purulia Pumped Storage Project in West Bengal. Japanese companies such as Toshiba and J-Power are involved in this project (*Hydropower in India—Event*—IEA n.d.).

6 Biomass

India is a big agricultural country with high biomass energy potential. According to IEA, bioenergy could provide 130 million tonnes of crude oil equivalent by 2040 in India, or about 15% of its total energy available at that time, if India pursued various policies (IEA 2021a, b).

According to PwC, India has the following potential for bioenergy production (Our take—PwC 2023).

Parameter	Quantity
Power generation potential	Annual power generation potential of 208 billion units (BU) from 28 GW
Additional bagasse-based cogeneration potential	Annual power generation potential of 65 BU from 14 GW
Bioethanol production potential from agri-waste	From sugarcane (1 G): • Sugar/sugar syrup—9,523.52 KLPD • B molasses—24,843.98 KLPD • C molasses—13,309.27 KLPD From rice (2G)—897.25 KLPD From maize (2G)—395.67 KLPD

(continued)

(continued)

Parameter	Quantity
Compressed biogas production potential	From cattle dung—38,981.25 TPD From municipal solid waste—4,853.98 TPD From paddy straw—16,377.03 TPD

Since India attaches great importance to domestic resources, the utilization of its biomass energy is crucial from the viewpoint of energy security.

The use of biomass will also contribute to mitigate air pollution, as pointed out as above. The burning of rice straw and other agricultural waste after harvest is said as one of the causes of heavy air pollution in fall in northern India, including Delhi. The reduction of air pollution is expected, if agricultural residues can be used as energy without being burned.

In addition, the benefits to farmers are important factors when discussing biomass energy. If previously discarded or incinerated agricultural residues are purchased by energy suppliers as their resources, additional income will be generated for farmers. Farmers, who make up a large portion of India's population (more than 50%), have a strong social influence, and this is also an important point in view of biomass energy (Press Information Bureau n.d.).

Due to such a big potential of biomass, Govt. of India has been implementing policies to promote biomass energy, and, in response, efforts by state and private companies have also been promoted. First, the Ministry of Petroleum and Natural Gas (MoPNG) formulated the "National Biofuels Strategy 2018," which promotes domestic production of biofuels and the blending of gasoline with ethanol, and revised the strategy in 2022 (Ministry of Petroleum and Natural Gas of India 2018). In November 2022, the Ministry of New and Renewable Energy (MNRE) formulated the "National Bioenergy Programme," which has an incentive program totaling Rs 8.58 billion to support the production of biogas from waste and pellet and briquettes for power generation (Press Information Bureau n.d.).

In addition, in an important recent development, Prime Minister Modi announced the establishment of the Global Biofuels Alliance (GBA) at a side event of the G20 Summit (chaired by India) in September 2023 (Press Information Bureau n.d.). Already, 19 countries, including the United States and Brazil, and 12 international organizations have agreed to join the alliance. While GBA's specific activities are currently unclear, it is said to include promotion of joint research and development through pilot projects, knowledge sharing, technical support, development of standards and regulations, identification of key issues, formation of markets for trade activation, formation of capacity building and awareness, and mapping and database of information. Minister Puri, the Minister for Petroleum and Natural Gas, who is in charge of the GBA, mentioned that GBA would contribute to the development of international standards in the field of sustainable aviation fuel (SAF) in order to make India a major production and export hub for the sector.

With the overall trends described above in mind, we would like to review and analyze the current situation in three main areas of biomass energy, by introducing Japan and India collaborations.

6.1 Bioethanol, Etc.

Bioethanol has been described as one of the most successful examples in the field of biomass energy in India recently.

In National Biofuels Strategy 2018, India set a target of "20% ethanol in gasoline, and 5% biodiesel in diesel by 2030." (Ministry of Petroleum and Natural Gas of India 2018) In December 2021, the Govt. of India lowered the GST for ethanol used in gasoline blends from 18 to 5% (Press information Bureau n.d.). With these policies, the 10% ethanol blend (E10) was achieved in June 2022, 5 months earlier than the target at that time (Press Information Bureau n.d.). In response to it, in May 2022, the Govt. of India revised its National Biofuels Strategy and decided to bring forward the 20% ethanol blend (E20) target to by FY 2025 from by FY 2030 (Press Information Bureau n.d.).

Here, we would like to mention the supply and demand situation of bioethanol. According to NITI Aayog, C molasses have been mainly used for raw materials until now, because it is cheaper than other raw materials. Recently, however, the diversification of raw materials including B molasses has progressed.

In addition, NITI Aayog estimates that the demand for ethanol in the future will be 10.2 billion liters in FY 2025 (the year aiming to achieve E20), while the supply of ethanol from molasses will be 7.6 billion liters and that from grains will be 7.4 billion liters. Therefore, ethanol is projected to meet the demand in that year (NITI Aayog and MoPNG 2021).

On the other hand, it is necessary to increase the ethanol supply capacity in order to promote ethanol mixing further than E20. Therefore, advanced bioethanol such as 2nd generation (2G) ethanol is important. As 2G ethanol is still a new field, efforts to demonstrate it are progressing with government support. In this context, we would like to introduce the 2G ethanol plant by IOCL (Indian Oil Company ltd.) in Panipat, Haryana, which started commercial operation in August 2022 after demonstration (Press information Bureau n.d.). It is a state-of-the-art plant called "One of a kind and the first in Asia" that can produce about 3,000 L of ethanol per year from 200,000 tons of rice straw per year. The technology behind the plant is provided by Indian company Praj Industries (Praj Industries 2022a, b). Founded in 1985, Praj has developed a variety of technologies related to biomass, including biofuels and bioplastics, as well as an R&D center that is considered the most advanced in Asia. The technology related to the production of 2G ethanol is a cellulosic ethanol production technology called "enfinity™". This technology can process various types of raw materials including rice straw, wheat straw, bagasse, corn residue, and soft wood to produce multiple biofuels such as bioethanol, bioCNG, and biochemicals (Praj Industries 2022a, b).

Efforts are also underway for automobile manufacturers to promote bioethanol blending as a vehicle fuel. Minister Gadkari, Minister for Roads, Transport and Highway, has strongly encouraged automobile manufacturers to produce flex-fuel engines, and Toyota Kirloskar, in particular, has shown efforts to respond to this. First, in October 2022, the company announced the Corolla Altis Hybrid, India's first ethanol-compatible flex-fuel hybrid vehicle ("Launch of Toyota's First of Its Kind Pilot Project on Flexi-Fuel Strong Hybrid Electric Vehicles (FFV-SHEV) in India," n.d.). Next, in August 2023, Toyota Kirloskar introduced the world's first electric flex-fuel vehicle in the Innova model (*"Toyota Kirloskar Motor Unveils Prototype of the World's First BS 6 (Stage II) Electrified Flex Fuel Vehicle, Contributing Towards Realising National Goals,"* n.d.). This car can use up to 100% ethanol as an alternative fuel to gasoline, and is equipped with both a flex-fuel engine and an electric powertrain that can use ethanol and improve fuel efficiency.

While it is desirable for India to promote flex-fuel engine vehicles, one major issue is the high tax rate for flex-fuel engine vehicles. In August 2023, the Indian Sugar Mills Association (ISMA), the sugar industry association, requested that the GST for flex-fuel vehicles be reduced to 5%, the same as that for EVs (Business Standard 2023).

6.2 Biogas

The Govt. of India is also focusing on biogas. As a policy to promote it, MoPNG launched the SATAT (Sustainable Alternative Towards Affordable Transportation) program in 2018. Under the program, off-taker oil companies offer EOIs (Expressions of Interest) to biogas suppliers, with the goal of installing 5,000 compressed biogas plants in India and producing 15MMT of compressed bio gas (CBG) per year by 2025. Under this scheme, 9,019 MT of CBG was produced until August 2022 (Our take—PwC 2023).

MNRE also supports biogas production plants using waste materials, including agricultural residues, through the National Bioenergy Programme (Press Information Bureau n.d.). Under this scheme, incentives of Rs 4 per 4,800 kg/day of production of CBG from new plants and Rs 3 per 4,800 kg/day of production of CBG from existing plants are provided (Press Information Bureau n.d.).

Maruti Suzuki India Ltd. is one of the leading companies to promote the utilization of biogas. Maruti Suzuki has focused on compressed natural gas (CNG) vehicles, and is now working to mix biogas with them. Biogas is produced by artificially generating raw gas from collected cow dung to refine automotive fuel (which is CN fuel). As a result, this will contribute to the reduction of atmospheric methane emissions and contribute to the reduction of GHG. In addition, the residue after the generation of biogas can be used as organic fertilizer, with the aim to contribute to the government's policy of promoting organic fertilizer. In August 2022, Maruti Suzuki signed a memorandum of understanding (MOU) with the National Dairy Development Board (NDDB), an Indian government agency, for the implementation of a biogas demonstration project (Suzuki starts biogas demonstration project in India 2022; *NDDB, Suzuki R & D Center inks MoU to provide clean energy solutions and organic fertilisers | nddb.coop n.d.*). In September 2023, Suzuki Motor's subsidiary, NDDB and Banas Dairy signed a contract worth Rs 2.3 billion to establish 4 biogas production plants in Gujarat using cow dung and other raw materials. Each facility plans to produce 1.2 ton/day of biogas, which can provide fuel for 680 vehicles/day (Suzuki concludes a Three-Party agreement for the biogas demonstration project in India 2023; *NDDB, Suzuki & Banas Dairy to setup first four dung based CBG plants in Gujarat's Banaskantha district | nddb.coop n.d.*).

In October 2023, Air Water Inc. launched a basic study on biogas in India utilizing the NEDO program (Inspection Project aimed at Building a Bioenergy Supply Chain in India Selected as a NEDO Project~Aiming to contribute to a decarbonized society through the manufacturing of cow manure-derived energy~ 2023). This project aims to demonstrate a local supply chain model that uses raw material such as cattle dung and agricultural waste, which are unused resources in India, to produce biomethane by using the company's methane fermentation technology and use it as regional CN energy. This technology is called semi-dry methane fermentation system technology, which is compact and adaptable to a variety of raw materials. Through this research, the company aims to advance to a demonstration project in India after a couple of years.

6.3 Biomass Co-Firing

The Govt. of India has promoted biomass co-firing. First, the National Electricity Plan 2018 set a target of 10 GW of installed biomass power capacity by 2022. To achieve this goal, co-firing using bagasse from sugar mills has contributed greatly. In the National Electricity Plan revised in 2023, a target of 15.5 GW of biomass power generation capacity for 2031 is mentioned (Press Information Bureau n.d.).

In addition, the Ministry of Power (MoP) issued a notification to make it mandatory to co-fire 5% of biomass in thermal power plants, which was planned to be enforced from October 2022 (Ministry of Power of India 2020). However, MoP decided to postpone the enforcement date, to make it mandatory from April 2024, and to raise the mandatory level to 7% from April 2025. Until May 2023, 1,64,976 MT of biomass was co-burned at 47 coal-fired power plants (Press Information Bureasu n.d.).

Here, we would like to introduce the efforts of an Indian company, Punjab Renewable Energy Systems Pvt. Ltd (PRESPL). The company handles agricultural residue supply chain management in India (*About PRESPL n.d.*). In August 2021, Mitsui Corporation invested about 400 million yen in this company (*Mitsui invests in PRESPL; India's Leading Biomass Supply-Chain Management Company—MITSUI & CO., LTD n.d.*). Taking advantage of its strong network with farmers, PRESPL has developed a wide range of businesses, including the collection and supply of agricultural residues such as rice straw, and the production of fuel including briquets and pellets for thermal power co-firing using agricultural residues as raw material.

6.4 Conclusion of Biomass

Against the backdrop of India's enormous biomass potential as described above, Govt. of India and state and private companies are focusing on biomass energy. With regard to biomass energy, it has been pointed out that agricultural residues and cattle dung, which can be used as raw materials, are widely dispersed, and that there are seasonal fluctuations in the supply of these materials. It is important to efficiently establish such raw material supply chains and to process and concentrate them for use as energy. There is also a lot of space for Japanese companies with its technologies to contribute in this field, so further efforts are expected to progress.

7 Clean Mobility

This section provides an overview of electric/hybrid mobility, driving automation, smart logistics, and drones.

7.1 Electric Mobility/Hybrid Mobility

7.1.1 EV Policy

India's goal is to have 80% of its two-wheeled and three-wheeled vehicles and 30% of its four-wheeled vehicles be EVs by 2030.

The Faster Adoption and Manufacturing of Electric (and Hybrid) Vehicles in India (FAME) was launched in 2015 to encourage early adoption of electric vehicles (EVs). In addition, Phase-II of FAME Scheme (FAME II) commenced from 1st April 2019. Other manufacturing and purchasing promotion measures have strongly encouraged EVs, such as incentives under the PLI scheme for auto parts and Advanced Chemistry Cell (ACC) batteries, and a reduction in GST from 12 to 5%. As for charging stations, 4,873 charging stations have been installed in the country under the FAME scheme (PRS legislative Research 2023; Press Information Bureau n.d.).

Notably, the Govt, of India supports the concept of interchangeable batteries and battery swapping, and the Ministry of Heavy Industries has discussed battery standardization for the same. Battery swapping is to reduce the initial installation cost of EVs by using them as a shared asset.

Honda Motor Company Ltd., for example, has completed a demonstration project in Karnataka and has started manufacturing and selling interchangeable batteries for EV rickshaws. The batteries can be used not only as a power source for normal driving, but also as an emergency evacuation power source and storage for off-grid renewable energy generation (*Honda Begins Sales of "Honda Power Pack Exchanger e:" Battery Swapping Station—Gachaco, a battery sharing business venture, begins operation of the first unit of Honda Power Pack Exchanger e: in Japan*—2022; Auto E 2022).

7.1.2 EV Player

The majority of EVs sold in India are two-wheelers. The top four companies located in Tamil Nadu account for 70% of EV two-wheelers production in India, with Ola Motor Electrics, the top company, accounting for 40% of its production.

Honda plans to unveil two models of EVs in FY2024: a model for medium-range vehicles and a two-wheeled model with a swapping battery, with the goal of producing 1 million EVs by 2030.

Maruti Suzuki is developing a four-wheeled EV and plans to begin production of an SUV-type vehicle in the fall of 2024, with an expected price of over 4 million yen, which is believed to be aimed at higher-income class segment. Maruti Suzuki is aiming to convert 15% of its production in India to EVs by 2030, and intends to use India as a production base to sell its EVs to Japan and Europe.

7.1.3 Hybrid Vehicles

Compared to EVs, hybrid vehicles in India has less preferential treatment.

Toyota Kirloskar Motor is currently seeking tax breaks for hybrids and flex-hybrids from the Govt. of India. The current tax rate for hybrid vehicles is 43%, compared to 48% for regular gasoline vehicles. From a consumer's perspective, it is a small discount relative to the environmental benefit (Livemint 2023).

7.1.4 Fuel Cell Vehicles

As India is promoting hydrogen production under the "National Green Hydrogen Mission," the use of hydrogen in India is expected to continue to be an important topic in the future. Toyota Kirloskar Motor has modularized its fuel cells and supplied two units to Ashok Leyland to manufacture commercial fuel vehicles in India, and is currently working on prototyping and feasibility studies (Ministry of New and Renewable Energy of India n.d; *Toyota Kirloskar Motor Supplies Hydrogen Fuel Cell Module for a Proto Examination and Feasibility Study purposes, Contributing to Nation's Energy Self-reliance and Carbon Neutral Goals* 2023).

7.2 Operation Optimization and Automation, Smart Logistics

To reduce traffic congestion and GHG emissions from transportation sector, optimization of vehicle operation, as well as overall logistics, is an effective approach.

In India, where total transportation demand continues to increase, it is necessary not only to introduce EVs, but also to introduce operational optimization initiatives (Igliński and Babiak 2017).

In October 2023, Panasonic started a demonstration project in Delhi in cooperation with ETO Motors to promote the use of e-rickshaws as a means of last-mile transportation between Delhi Metro stations and destinations. For this project, Panasonic introduces a system to optimize the operation of EV operators and drivers (*Demonstration Project to realize Efficient Operation Support by IT System for Last Mile Transportation launched in India | press release | NEDO* 2023).

The Unified Logistics Interface Platform (ULIP) is a logistics platform led by National Industrial Corridor Development Corporation Ltd (NICDC) under the PM Gati Shakti scheme. Under this scheme, ULIP will serve as a platform for more than 30 logistics systems overseen by seven ministries and agencies. The integration of the systems is expected to ensure seamless operations (Press Information Bureau n.d.).

7.3 Drone

The policy implications of drones in India are mainly in the area of security, especially for border security. In addition to that, however, the industry promotion aspect is also becoming more and more important. In India, the Drone Act of 2021 has been established, with new rules regarding licensing requirements and flight areas. These rules include drone cabs, which weigh 500 kg or less (Press Information Bureau n.d.; *Digital Sky* n.d.).

In urban areas of India, traffic congestion is a social problem, and there are options for using drones for mobility, including emergency transportation. For example, IndiGo, one of the leading Indian airline companies, has concluded an agreement with Archer Aviation and is considering the introduction of eVTOL in India (Sharma 2023).

SkyDrive, a Japanese company, plans to establish a subsidiary with Suzuki Motors for the purpose of manufacturing "flying cars". The company plans to begin manufacturing eVTOLs on Suzuki's factory premises in the first half of 2024. SkyDrive is considering introducing "flying cars" in the Indian market. Elsewhere, SkyDrive has already signed a memorandum of understanding (MOU) for cooperation with IIT Hyderabad's Institute of Automated Driving Technology, with the common goal of creating a market for logistics drones (Autocar India New Desk 2023).

8 Conclusion

India is becoming increasingly important in the energy and environment area because of its continued economic growth. In addition, as India supports the concept about the realistic energy transitions in Asia and various pathways to achieve carbon neutrality, advocated by the Govt. of Japan, it is necessary for Japan and India to further collaborate in spreading this concept to the world. On the other hand, India is a country rich in renewable energy resources and has made remarkable progress in solar, wind, biomass, clean mobility, and other related sectors. While many Japanese companies have long developed advanced technologies in the energy and environmental fields, it is expected that more business and projects including demonstration and joint research will come between Japan and India.

References

500GW nonfossil fuel target I Government of India I Ministry of Power (n.d.). https://powermin.gov.in/en/content/500gw-nonfossil-fuel-target

ANNEXURE Electricity (promoting renewable energy through Green Energy Open Access) Rules, 2021 (n.d.). Retrieved January 4, 2024, from https://static.pib.gov.in/WriteReadData/specificdocs/documents/2021/aug/doc2021816111.pdf

About PRESPL (n.d.). https://prespl.com/about-us/about-prespl

Adani Power To Co-Fire Green Ammonia At Its Mundra Plant For A Sustainable Future I Media-Releases I Newsroom I Home (n.d.). Adani Power. https://www.adanipower.com/newsroom/media-releases/Adani-Power-to-co-fire-Green-Ammonia-at-its--Mundra-Plant-for-a-sustainable-future

Auto E (2022). Honda Power Pack begins battery swapping at HPCL petrol stations in Bengaluru. ETAuto.com. https://auto.economictimes.indiatimes.com

Autocar India News Desk (2023) Suzuki, SkyDrive sign deal to develop flying cars. Autocar India. https://www.autocarindia.com

Ayallore C (2022) A case for strengthening the framework on banking of renewable energy—environmentality. https://environmentality.cprindia.org/blog/a-case-for-strengthening-the-framework-on-banking-of-renewable-energy

Behl M (2022). Power plants may delay installing FGDs for few years more. The Times of India. https://timesofindia.indiatimes.com

Business Standard (2023) Sugar body ISMA seeks 5% GST on flex-fuel vehicles in line with EVs. https://www.business-standard.com/economy/news/sugar-body-isma-seeks-5-gst-on-flex-fuel-vehicles-in-line-with-evs-123082800712_1.html

Central Electricity Authority, Ministry of Power of India (2023) National Electricity Plan (volume I Generation). https://cea.nic.in/wp-content/uploads/notification/2023/06/NEP_2022_32_FINAL_GAZETTE_English.pdf

Central Electricity Authority, Ministry of Power of India (2023) FLEXIBILISATION OF COAL FIRED POWER PLANT. https://cea.nic.in/wp-content/uploads/tprm/2023/03/Report_21022023.pdf

Confederation of Indian Industry and EY (n.d.). Global champions for advancing renewable energy innovation and manufacturing. 4th International Conference & Exhibition on 'AatmaNirbhar Bharat for Clean Energy Transition'. https://static.pib.gov.in/WriteReadData/specificdocs/documents/2023/sep/doc2023915253001.pdf

Country rankings (n.d.). https://www.irena.org/Data/View-data-by-topic/Capacity-and-Generation/Country-Rankings

Demonstration Project to realize Efficient Operation Support by IT System for Last Mile Transportation launched in India | press release | NEDO (2023). https://www.nedo.go.jp/english/news/AA5en_100453.html

Development of Co-Firing technology of pulverized coal and ammonia for suppressing NOX generation | Technical Information | Technology | IHI Corporation (n.d.). IHI Corporation. https://www.ihi.co.jp/en/technology/techinfo/contents_no/1199400_13586.html

Digital Sky (n.d.). https://static.pib.gov.in/WriteReadData/specificdocs/documents/2022/jan/doc202212810701.pdf

Diespatch of Coal to thermal power and coal consumption/off-take during last ten years (n.d.). Retrieved January 4, 2024, from https://coal.gov.in/sites/default/files/2021-01/Generation-of-Thermal-Power-from-Raw-Coal-and-Coal-Consumption-during-last-ten-years.pdf

Evaluation of Wind Energy in India (n.d.). https://prsindia.org/policy/report-summaries/evaluation-of-wind-energy-in-india#:~:text=The%20commercially%20exploitable%20potential%20of,of%20the%20commercially%20exploitable%20potential

GE and IHI Sign Memorandum of Understanding to Develop Gas Turbines that Can Operate on 100% Ammonia | 2022FY | News Articles | IHI Corporation (2023). IHI Corporation. https://www.ihi.co.jp/en/all_news/2022/resources_energy_environment/1198148_3488.html

Harvey F, Mason R (2021) Alok Sharma 'deeply frustrated' by India and China over coal. The Guardian. https://www.theguardian.com/environment/2021/nov/14/alok-sharma-deeply-frustrated-by-india-and-china-over-coal

Honda Begins Sales of "Honda Power Pack Exchanger e:" Battery Swapping Station—Gachaco, a battery sharing business venture, begins operation of the first unit of Honda Power Pack Exchanger e: in Japan – (2022). https://global.honda/en/newsroom/news/2022/p221025eng.html

Hydropower in India—Event—IEA (n.d.). https://www.iea.org/events/hydropower-in-india

IEA, World Energy Outlook 2021 (2021). https://www.iea.org/reports/world-energy-outlook-2021

IEA (2021). World Energy Outlook 2021. https://iea.blob.core.windows.net/assets/1de6d91e-e23f-4e02-b1fb-51fdd6283b22/India_Energy_Outlook_2021.pdf

IEA (2023). A Vision for Clean Cooking Access for All. https://iea.blob.core.windows.net/assets/75f59c60-c383-48ea-a3be-943a964232a0/AVisionforCleanCookingAccessforAll.pdf

IEA (2023). World Energy Outlook 2023. https://www.iea.org/reports/world-energy-outlook-2023

IEA (2021). India and the IEA enter new phase of closer collaboration with signing of Strategic Partnership Framework—News—IEA. https://www.iea.org/news/india-and-the-iea-enter-new-phase-of-closer-collaboration-with-signing-of-strategic-partnership-framework https://www.iea.org/news/india-and-the-iea-enter-new-phase-of-closer-collaboration-with-signing-of-strategic-partnership-framework

IEA (2023). Coal 2023—Analysis and forecast to 2026. https://iea.blob.core.windows.net/assets/a72a7ffa-c5f2-4ed8-a2bf-eb035931d95c/Coal_2023.pdf

IEA (n.d.). Coal Information : Overview—Production. https://www.iea.org/reports/coal-information-overview/production https://www.iea.org/reports/coal-information-overview/production

IHI Launches Methanation System to Produce Fuel from Carbon Dioxide and Hydrogen | 2022FY | News Articles | IHI Corporation (2022). https://www.ihi.co.jp/en/all_news/2022/resources_energy_environment/1198178_3488.html

IQAir (2023). World Air Quality Report—Region & City PM2.5 Ranking. https://www.greenpeace.org/static/planet4-india-stateless/2023/03/2fe33d7a-2022-world-air-quality-report.pdf

Igliński H, Babiak M (2017) Analysis of the potential of autonomous vehicles in reducing the emissions of greenhouse gases in road transport. Procedia Engineering 192:353–358. https://doi.org/10.1016/j.proeng.2017.06.061

Inspection Project aimed at Building a Bioenergy Supply Chain in India Selected as a NEDO Project~Aiming to contribute to a decarbonized society through the manufacturing of cow

manure-derived energy~ | News Releases | What's New | AIR WATER INC (2023). https://www.awi.co.jp/en/news/release/release-4297897468737482 5048.html
International Renewable Energy Agency (n.d.). ENERGY PROFILE INDIA. https://www.irena.org/-/media/Files/IRENA/Agency/Statistics/Statistical_Profiles/Asia/India_Asia_RE_SP.pdf
LARSEN&TOUBRO (n.d.). Annual report 2020–2021—POWER BUSINESS. https://investors.larsentoubro.com/pdf/AR-2021-22/Power%20Business%20-%20MDA%20LNT%20AR20-21%20.pdf
Launch of Toyota's first of its kind pilot project on Flexi-Fuel Strong Hybrid Electric Vehicles (FFV-SHEV) in India (2022). https://www.toyotabharat.com/news/2022/launch-of-toyotas-first-of-its-kind-pilot-project-on-flexi-fuel-strong-hybrid-electric-vehicles-ffv-shev-in-india.html
Livemint (2023). Tax cut: Toyota lobbies government reduction in hybrid-car taxes to 37% | Mint. Mint. https://www.livemint.com
Methanation Equipment | Hitachi Zosen Corporation (n.d.). https://www.hitachizosen.co.jp/english/business/field/electrolytic-hydrogen/methanation.html
Ministry of Economy, Trade and Industry of Japan (2021). Minister Kajiyama announed the Asia Energy Transition Initiative (AETI). https://www.meti.go.jp/english/press/2021/0528_002.html
Ministry of Economy, Trade and Industry of Japan (2021) Announcement of the India-Japan Clean Energy Partnership (CEP). https://www.enecho.meti.go.jp/en/category/others/cep/20220319.html
Ministry of Environment, Forest and Climate Change of India (2023). Aide Memoire on the Joint Crediting Mechanism between the Ministry of the Environment of Japan and the Ministry of Environment, Forest and Climate Change of the Republic India. https://www.env.go.jp/content/000121764.pdf
Ministry of External Affairs of India (2022). Announcement of the India-Japan Clean Energy Partnership(CEP). https://www.mea.gov.in/Portal/ForeignRelation/Japan_-_Bilateral_Brief_MEA_Website_Oct_2023.pdf
Ministry of New and Renewable Energy of India (n.d.). National Green Hydrogen Mission. https://mnre.gov.in/national-green-hydrogen-mission/#:~:text=The%20National%20Green%20Hydrogen%20Mission,efficiently%20transport%20and%20distribute%20hydrogen
Ministry of New and Renewable Energy (2023). National Repowering & Life Extension Policy for Wind Power Projects Reg. https://cdnbbsr.s3waas.gov.in/s3716e1b8c6cd17b771da77391355749f3/uploads/2023/12/20231213187429 6229.pdf
Ministry of Power of India (2018). Long term growth trajectory of Renewable Purchase obligations (RPOs) for Solar and Non-solar for a period of three years i.e. 2019–20 to 2021–2022—regarding. https://powermin.gov.in/sites/default/files/webform/notices/RPO_trajectory_2019-22_Order_dated_14_June_2018.pdf#:~:text=In%20continuation%20to%20the%20earlier%20Order%20dated%2022nd,excluding%20consumption%20met%20from%20hydro%20sources%20of%20power.
Ministry of Power of India (2020). Revised Policy for Biomass Utilization for Power Generation through Co-firing in Coal based Power Plants. https://powermin.gov.in/sites/default/files/Revised_Biomass_Policy_dtd_08102021.pdf
Ministry of Power of India (2022a). Annual Report2022–2023. https://powermin.gov.in/sites/default/files/uploads/MOP_Annual_Report_Eng_2022-23.pdf
Ministry of Power of India (2022b). Renewable Purchase Obligation (RPO) and Energy Storage Obligation Trajectory till 2029–30—regarding. https://powermin.gov.in/sites/default/files/Renewable_Purchase_Obligation_and_Energy_Storage_Obligation_Trajectory_till_2029_30.pdf
Ministry of Power of India (2023). Draft Notification on Renewable Generation Obligation—Regarding. https://powermin.gov.in/sites/default/files/webform/notices/Draft_Notification_on_Renewable_Generation_Obligation_0.pdf
Mitsui invests in PRESPL; India's Leading Biomass Supply-Chain Management Company—MITSUI & CO., LTD (n.d.). https://www.mitsui.com/jp/en/topics/2021/1241716_12171.html
Mitsui to invest in a 1,300MW renewable energy project in India—MITSUI & CO., LTD (n.d.). https://www.mitsui.com/jp/en/release/2022/1243209_13406.html

NDDB, Suzuki & Banas Dairy to setup first four dung based CBG plants in Gujarat's Banaskantha district | nddb.coop (n.d.). https://www.nddb.coop/node/2415
NDDB, Suzuki R & D Center inks MoU to provide clean energy solutions and organic fertilisers | nddb.coop (n.d.). https://www.nddb.coop/node/2350
NEDO (n.d.). NEDO GLOBAL ACTIVITIES. Retrieved January 4, from https://www.nedonewdelhi.in/globalactivities.aspx
NITI Aayog & Ministry of Petroleum and Natural Gas of India (2021). Roadmap for Ethanol Blending in India 2020–2025, https://www.niti.gov.in/sites/default/files/2021-06/EthanolBlendingInIndia_compressed.pdf
NITI Aayog (2022). Carbon Capture Utilization and Strage (CCUS)—Policy Framework and its Deployment Mechanism in India. https://www.niti.gov.in/sites/default/files/2022-11/CCUS-Report.pdf
Ministry of Power of India (2023) Notification of "the Electricity (Promoting Renewable Energy Through Green Energy Open Access) Rules, 2022." (2023, June 6). https://www.acma.in/uploads/ciculer-attachement/Electricity%20(Promoting%20Renewable%20Energy%20through%20Green%20Energy%20Open%20Access)%20Rules,2022.pdf
Nova (2023) Hydropower in India: An update. https://www.orfonline.org/expert-speak/hydropower-in-india-an-update
Our take—PwC (2023). https://www.pwc.in/assets/pdfs/fuelling-indias-future-with-bioenergy.pdf
PRS Legislative Research (2023). Standing Committee Report Summary -Promotion of Electric Vehicles in the Country. https://prsindia.org/files/policy/policy_committee_reports/SCR_Summary-EV.pdf
PRS Legislative Research (n.d.).Hydro Power. https://prsindia.org/policy/report-summaries/hydro-power
PTI (2021). India being lectured, pressured on environmental issues by developed countries: PM. Deccan Herald. https://www.deccanherald.com/india/india-being-lectured-pressured-on-environmental-issues-by-developed-countries-pm-1054963.html
Praj Industries (2022). Enfinity Technology—Praj Industries. https://www.praj.net/businesslines/2g-ethanol/enfinity-technology/
Praj Industries (2022). IOCL's 2G Ethanol Bio-Refinery based on Praj's technology inaugurated by Prime Minister Modi, https://www.praj.net/wp-content/uploads/2022/08/20220816-Press-Release-IOCLs-2G-Ethanol-Bio-Refinery-based-on-Prajs-technology-inaugurated-by-Prime-Minister-Modi.pdf
Press Information Bureau (2022). January 4, 2024 https://static.pib.gov.in/WriteReadData/specificdocs/documents/2022/jan/doc202212810701.pdf
Press Information Bureau (n.d.). Retrieved January 4, 2024, from https://pib.gov.in/PressReleasePage.aspx?PRID=1768712
Press Information Bureau (n.d.). Retrieved January 4, 2024, from https://pib.gov.in/PressReleaseIframePage.aspx?PRID=1847812
Press Information Bureau (n.d.). Retrieved January 4, 2024, from https://pib.gov.in/PressReleasePage.aspx?PRID=1746070
Press Information Bureau (n.d.). Retrieved January 4, 2024, from https://pib.gov.in/FeaturesDeatils.aspx?NoteId=151483&ModuleId%20=%202
Press Information Bureau (n.d.). Retrieved January 4, 2024, from https://pib.gov.in/PressReleaseIframePage.aspx?PRID=1928750
Press Information Bureau (n.d.). Retrieved January 4, 2024, from https://pib.gov.in/PressReleasePage.aspx?PRID=1883921
Press Information Bureau (n.d.). Retrieved January 4, 2024, from ttps://pib.gov.in/Pressreleaseshare.aspx?PRID=1808263
Press Information Bureau (n.d.). Retrieved January 4, 2024, from https://pib.gov.in/PressReleasePage.aspx?PRID=1943391
Press Information Bureau (n.d.). Retrieved January 4, 2024, from https://pib.gov.in/PressReleasePage.aspx?PRID=1947384

Press Information Bureau (n.d.). Retrieved January 4, 2024, from https://pib.gov.in/PressReleaseIframePage.aspx?PRID=1944386

Press Information Bureau (n.d.). Retrieved January 4, 2024, from https://pib.gov.in/PressReleasePage.aspx?PRID=1967167#:~:text=India's%20current%20share%20of%20Gas,creating%20a%20Gas%20based%20economy

Press Information Bureau (n.d.). Retrieved January 4, 2024, from https://pib.gov.in/PressReleaseIframePage.aspx?PRID=1911550

Press Information Bureau (n.d.). Retrieved January 4, 2024, from https://pib.gov.in/PressReleaseIframePage.aspx?PRID=1975033

Press Information Bureau (n.d.). Retrieved January 4, 2024, from https://pib.gov.in/newsite/erelcontent.aspx?relid=186572

Press Information Bureau (n.d.). Retrieved January 4, 2024, from https://pib.gov.in/PressReleaseIframePage.aspx?PRID=1943363

Press Information Bureau (n.d.). Retrieved January 4, 2024, from https://pib.gov.in/PressReleasePage.aspx?PRID=1668137

Press Information Bureau (n.d.). Retrieved January 4, 2024, from https://pib.gov.in/PressReleasePage.aspx?PRID=1879865

Press Information Bureau (n.d.). Retrieved January 4, 2024, from https://pib.gov.in/FeaturesDeatils.aspx?NoteId=151141&ModuleId%20=%202

Press Information Bureau (n.d.). Retrieved January 4, 2024, from https://pib.gov.in/PressReleaseIframePage.aspx?PRID=1900216

Press Information Bureau (n.d.). Retrieved January 4, 2024, from https://pib.gov.in/PressReleaseIframePage.aspx?PRID=1913789

Press Information Bureau (n.d.). Retrieved January 4, 2024, from https://pib.gov.in/PressReleasePage.aspx?PRID=1947375#:~:text=The%20resolution%2C%20inter-alia%2C%20mandates%20that%20any%20generating%20company,and%20supply%20renewable%20energy%20equivalent%20to%20such%20capacity

Press Information Bureau (n.d.). Retrieved January 4, 2024, from https://pib.gov.in/PressReleseDetailm.aspx?PRID=1909269

Press Information Bureau (n.d.). Retrieved January 4, 2024, from https://pib.gov.in/Pressreleaseshare.aspx?PRID=1739835

Press Information Bureau (n.d.). Retrieved January 4, 2024, from https://pib.gov.in/PressReleseDetail.aspx?PRID=1832708

Press Information Bureau (n.d.). Retrieved January 4, 2024, from https://pib.gov.in/PressReleasePage.aspx?PRID=1874209

Press Information Bureau (n.d.). Retrieved January 4, 2024, from https://pib.gov.in/PressReleaseIframePage.aspx?PRID=1947505

Press Information Bureau (n.d.). Retrieved January 4, 2024, from https://pib.gov.in/PressReleaseIframePage.aspx?PRID=1942506

Press Information Bureau (n.d.). Retrieved January 4, 2024, from https://pib.gov.in/PressReleasePage.aspx?PRID=1864095

Received third order for the EfW plant in India | Hitachi Zosen Corporation (2021) https://www.hitachizosen.co.jp/english/newsroom/news/release/2021/20210921_001418.html

Renewable energy in India—Indian Power Industry investment (n.d.). https://www.investindia.gov.in/sector/renewable-energy

Saurenergy (2020). Greenko, ReNew Power win SECI's 1.2 GW Hybrid Plus Storage Tender. https://www.saurenergy.com/solar-energy-news/greenko-renew-power-win-secis-1-2-gw-hybrid-plus-storage-tender

Sharma S (2023) IndiGo parent InterGlobe & Boeing-backed Archer Aviation plan to launch electric air taxi in India in 2026. The Indian Express. https://indianexpress.com

Solar Energy Corporation of India Limited(SECI), a Government of India enterprise, under Ministry of New and Renewable Energy (n.d.). https://www.seci.co.in/view/publish/tender?tender=all

Suzuki concludes a Three-Party agreement for the biogas demonstration project in India (2023). https://www.globalsuzuki.com/globalnews/2023/0906.html

Suzuki starts biogas demonstration project in India (2022). https://www.globalsuzuki.com/globalnews/2022/0831.html

Toshiba Wins Rs. 3436 crore EPC Contract for Harduaganj Ultra-supercritical Thermal Power Project from Uttar Pradesh Rajya Vidyut Utpadan Nigam Ltd (UPRVUNL) in Lucknow, India (n.d.). https://toshiba-india.com/pr-Toshiba-Wins-Rs.3436-crore-EPC-Contract-for-Harduaganj-Ultra-supercritical-Thermal-Power-Project-from-Uttar-Pradesh-Rajya-Vidyut-Utpadan-Nigam-Ltd-(UPRVUNL)-in-Lucknow-India.aspx

Toyota Kirloskar Motor Supplies Hydrogen Fuel Cell Module for a Proto Examination and Feasibility Study purposes, Contributing to Nation's Energy Self-reliance and Carbon Neutral Goals (2023). https://www.toyotabharat.com/news/2023/tkm-supplies-hydrogen-fuel-cell-module-for-a-proto-examination-and-feasibility.html

Toyota Kirloskar Motor Unveils Prototype of the World's First BS 6 (Stage II) Electrified Flex Fuel Vehicle, contributing towards realising national goals (2023, August 29). https://www.toyotabharat.com/news/2023/tkm-unveils-prototype-of-the-worlds-first-bs-6-stage-ii-electrified-flex-fuel-vehicle-contributing-towards-realising-national-goals.html

Tuli B, Hemant M, Deepak Y (2023) Assessing India's CO_2 storage potential: a critical analysis of what lies beyond the theoretical potential. CEEW. https://www.ceew.in/publications/how-can-india-scale-carbon-capture-storage-projects-co2-storage-potential-sequestration

Expectations and Challenges for Jointresearch and Development on Hydrogen Energy Between India and Japan

Takuma Otaki and Rajib Shaw

Abstract India and Japan have incentive to deepen their cooperation in the hydrogen field because Indian government is trying to actively in implementing hydrogen in society, Japan has already developed technologies in significant areas for India, India has a high R&D (research and development) capability and has a very structured roadmap for R&D in the hydrogen area. To deepen cooperation between India and Japan in the hydrogen area, the first short-term policy measures needed are a special IP (intellectual property) protection environment in the hydrogen area and the early establishment of a bilateral credit scheme. Once these two measures are realized, it is expected that many R&D projects and actual projects will be structured between the two countries and that many connections will be established between the parties involved.

Keywords India and Japan · Hydrogen energy · Research and development · Intellectual property · Joint crediting mechaanism

1 Background

Hydrogen technologies are attracting whole world as a new technology which have potential for reducing greenhouse gases gas (GHG) emissions. Hydrogen enables load shifting of fluctuating renewable energy sources. I improves the flexibility of the entire energy system and applies low carbon energy to areas where it has been difficult to utilize low carbon energy sources (Otto et al. 2022; Schrotenboer 2022). India and Japan are no exception and both nations are actively working toward the social implementation of hydrogen technology. Japan, in particular, was the first country in the world which announced a national Hydrogen strategy, and as an early investor, has long been developing technologies in the hydrogen field (METI 2017). Indian

T. Otaki (✉) · R. Shaw
Graduate School of Media and Governance, Keio University, Fujisawa, Japan
e-mail: takuma1004@gmail.com

R. Shaw and S. R. Choudhury (eds.), *India, Japan and Beyond*,
https://doi.org/10.1007/978-981-97-3282-1_8

government issued National Green Hydrogen in January 2023, and active discussions on the use of Hydrogen in society are proactively discussed (MNRE 2023). This chapter will discuss areas of high potential for deepening cooperation between India and Japan in the hydrogen area, specific ways to cooperate, and challenges to deepening cooperation.

2 Strong Hydrogen Utilization Potential in India

The potential for the use of hydrogen technology is particularly high in India. With the high demand for energy in all areas and the high potential for renewable energy, there is an inherent need for hydrogen to enable a shift in energy loads. India is becoming the world's third largest country in terms of energy consumption, about Fossil fuels cover more than 70% of Primary Energy demand as of 2019, and the share of fossil fuels is projected to continue to exceed 70% in 2040. Particularly According to the IEA report, fossil fuel consumption is expected to increase as India's economy grows, particularly with respect to oil, which is projected to be imported twice as much in 2040 as it was consumed in 2018 (IEA 2021). However, externally, India has been clear about its policy to reduce GHGs as much as possible, in fact Indian government declared to achieve the target of reducing their GDP emissions intensity of GDP by 45% by 2030, compared to the 2005 level (MoEFCC 2023). Currently, the most emphasized concrete measure to achieve this goal is the introduction of renewable energy. India's Ministry of Renewable Energy has stated that it will expand its renewable energy capacity to 450 GW by 2030 (MNRE 2021). In July 2023, the amount of renewable energy capacity has been reached to 176.49 GW (Wind power: 43.7 GW, Solar Power: 70.1 GW, Biomass/Co-generation: 10.2 GW, Small Hydro Power: 4.94 GW, Waste To Energy: 0.55 GW, Large Hydro: 46.85 GW) (Invest India 2023). India is oriented toward reducing GHGs for society as a whole by rapidly increasing renewable energy, which is unstable in output, and the amount of renewable energy installed is increasing sufficiently, although not necessarily progressing as targeted. As shown above, the potential for utilizing hydrogen is very high in India from the perspective of responding to the ever-increasing energy needs and the rapidly increasing and unstable output of renewable energy.

In addition, there is a high demand for industrial applications of hydrogen in India. Harnessing Green Hydrogen, published by NITI Aayog, a government research institute in India, lists (a) Ammonia, (b) Steel, (c) Heavy Duty Vehicles (HDVs), and (d) Power Sector as areas with high potential for utilizing hydrogen from renewable energy sources produced in India (NITI Aayog and RMI 2022).

(a) Ammonia

Ammonia is an important raw material for fertilizers, and India is the world's third largest fertilizer importer. The amount of subsidies related to fertilizer imports in FY2022 has reached USD 20billion in India. (Volza 2022; MoCF 2022). Therefore, the replacement to the domestic production of ammonia has been serious matter for the Indian government from perspectives of food security and trade balance.

(b) Steel

The use of hydrogen in the steel sector is also identified in the National Green Hydrogen Mission as an effective means of reducing the use of fossil fuels. And, Ministry of Steel was assigned to the role for driving adoption of Hydrogen in the steel sector, identifying and facilitating pilot projects for use of Hydrogen in steel production. It is desired to gradually replace the hydrogen already used in the steelmaking process with hydrogen derived from renewable energy sources, and to realize steel production from renewable energy sources at a time when economic rationality can be achieved through technological progress (MNRE 2023).

(c) HDVs

Indian players are also very active in the HDVs space: in September 2023, Indian Oil Corporation (IOC) ran India's first hydrogen bus in Faridabad, near Delhi. The Minister for Petroleum and Natural Gas, Hardeep Singh Puri, also conveyed his hope Hydrogen could be a transition fuel from fossil fuels (The Economic Times 2023). And there is already examples of Japanese corporation cooperating with Indian company' initiatives in this area already in existence. In January 2023, Toyota Kirloskar Motors announced that they supply their fuel cell modules to Ashok Leyland, a major Indian commercial vehicle manufacturer (The Economic Times 2022).

(d) Power

In India, above 70% of electricity (kWh basis) in FY2020 was supplied from coal power generation sources (IEA 2022). And generation capacity of coal power is projected to increase till 2040 in India, so it is crucially important for Indian society to install technologies which mitigates GHG emissions from coal power plant. Ammonia co-firing technology will be an important technology for India, which needs the continued use of coal power plants to ensure energy security and to cover its increasing energy demand due to economic growth. Fortunately, cooperation between India and Japan has already started in this area. Adani and IHI Corporation have begun demonstration testing of an ammonia and coal co-firing power plant (IHI 2022). A very realistic energy transition would be achieved if hydrogen produced from renewable energy sources in India could be converted to ammonia, stored, and burned within India to reduce the use of coal.

3 Hydrogen-Related Technologies Developed by Japan, an Early Investor

Japan formulated its Sixth Basic Energy Plan, which calls for that Hydrogen and Ammonia to increase to about 1% of the power supply mix by FY2030 (METI 2021). Additionally Japanese government has revised the Basic Hydrogen Strategy as a guideline to show strong initiatives government and the private sector should share a common vision and be strongly committed to the early realization of a hydrogen society for achieving carbon neutrality by 2050. Japanese government reaffirmed that the goal was set to introduce up to 3 million tons of hydrogen per year by 2030 and approximately 20 million tons per year by 2050. A goal of 12 million tons of hydrogen per year to be introduced by 2040 was also set. The meeting also decided to introduce12 million tons per year for the introduction of Hydrogen by 2040. In particular, to encourage Japanese companies to take the risk of being first movers, an incentive program that offers a subsidy to bridge the price difference between the foreign price of hydrogen and ammonia and the parity price of the current fuel is needed to establish a hydrogen value chain with foreign countries (METI 2023). Additionally, in August 2023, the Japanese government unveiled its "Green Transformation" plan, pledging to issue 20 trillion JPY in green bonds in order to fund a practical energy transition. The policy further declares that the funding required to construct a supply chain for ammonia and hydrogen will come from the Japanese government's FY2024 budget (CAS 2023).

Japan's R&D and demonstration projects in areas of interest to India are also progressing very well. Japan has long been conducting advanced research in India's focus areas of (a) Ammonia, (b) Steel, (c) HDVs, and (d) Power Sector, all of which have great potential to be emphasized by India and Japan (Otaki and Shaw 2023).

(a) Ammonia

The most significant ammonia-related R&D effort in Japan has been the JERA-led demonstration project for the development of ammonia co-firing technology and the composition of the supply chain. Chiyoda Corporation, JERA, and Tokyo Electric Power Company are creating new ammonia production catalysts and performing technology demonstrations relevant to the establishment of a fuel ammonia supply chain as part of their efforts to establish a reliable ammonia supply chain (JERA 2022). In addition, the development of ammonia-fueled vessels has been actively pursued in the past in order to transport ammonia internationally with fewer GHG emissions. The development of ammonia-fueled vessels, where the goal was to have them in commercial service as soon as possible by 2028, was pushed forward to achieve higher goals, with additional R&D budget allocated to promote the project so that it could be put into practical use sooner. In July 2023, the public solicitation for additional R&D implementers was launched for (1) the development of technology to remove nitrous oxide (N_2O) from the exhaust gas of ammonia-fueled engines, which was a particular challenge, and (2) technology development related to safety measures when ammonia fuel is supplied to ships (MLIT 2023).

(b) Steel

Japan aims to reduce CO_2 emissions in the steelmaking process by up to 50% or more by applying hydrogen reduction technology to the current blast furnace (blast furnace hydrogen reduction technology) and establishing technology to directly reduce low-grade iron ore with hydrogen, with the aim of achieving carbon neutrality in the steelmaking process from 2021. In September 2023, the world's first 22% reduction in CO2 emissions was accomplished in a test using a tiny test blast furnace thanks to the invention of low-carbon technology that uses external hydrogen and CO2 contained in blast furnace exhaust gas. The conceptual design of the test furnace has been completed, and basic process studies and experiments are being conducted to develop elemental technologies (NEDO 2023a).

(c) HDVs

In Japan, there is funding available for the development of HDVs in the domains of commercial vehicles, ships, and airplanes, in that order. There are 123 fuel cell buses in operation as of October 2022, and talks regarding fuel cell-powered commercial vehicles are already progressing in the commercial stage (METI 2022). Ship demonstration tests are currently in progress. Ammonia-fueled ships and hydrogen-fueled centrifuges are both being researched in Japan. The development of not only ammonia vessels but also hydrogen fueled vessels running on liquefied hydrogen is progressing as planned, and basic designs for the development of multi-cylinder engines, marine hydrogen fuel tanks, and hydrogen fuel supply systems are in progress (NEDO 2023b). Development is also underway for aircraft. R&D for the development of a hydrogen aircraft is being conducted with a focus on core technology development and weight reduction of key structural components. As of October 2023, development is progressing as planned, including improved burner design, prototype production of hydrogen pumps, valves, and heat exchangers, weight reduction design of components for development of liquefied hydrogen fuel storage tanks, heat insulation performance, and vibration resistance characteristics evaluation (NEDO 2023c).

(d) Power Sector

In the power sector, efforts are underway to implement hydrogen derived from renewable energy in society. Green chemical plants and large-scale alkaline water electrolyzers are being shown by Asahi Kasei Corporation and JGC Holdings Corporation. Energy demand conversion and utilization technology has been created by Yamanashi Prefectural Bureau of Public Enterprises, Tokyo Electric Power Holdings, Toray Industries, Hitachi Zosen Corporation, Siemens Energy Corporation, and Miura Industry Co. utilizing a PEM-type water electrolyzer (Otaki and Shaw 2023). In addition to this, Toyota Motor Corporation (TMC) has announced the development of a water electrolysis system, which is expected to increasingly accelerate the discussion of hydrogen production to energy use. TMC announced that it has developed a new water electrolyzer that produces hydrogen by electrolyzing water, appropriating the fuel cell technology used in Toyota's MIRAI fuel cell vehicle, and start operation in March 2023 at Denso Corporation's Fukushima plant as a place to implement the technology to promote its widespread use in the future (TMC 2023).

4 High Potential and Challenges for Technological Cooperation Between India and Japan

Japan is making steady progress in R&D in areas of particular focus in India, and there is great potential for Japan to contribute to the diffusion and development of efficient hydrogen technology in India by deepening technical cooperation between India and Japan.

Concern over intellectual property protection is one of the psychological obstacles preventing Japanese businesses from investing in India. 45.9% of small and medium-sized businesses in a 2013 study of 113 Japanese businesses operating in India stated they were having some issues. The results were broken down as follows: extremely serious (5.4%), very problematic (8.1%), problematic (10.8%), and slightly problematic (21.6%) (Sato 2017). Additionally, it takes an average of 6.7 years to secure a patent in India, making it a very challenging procedure for one of the few ways to safeguard corporate secrets (JETRO 2022). Though India and Japan have previously established the Patent Prosecution Highway (PPH), a framework for the rapid evaluation by the patent authorities of both nations of the procedures for obtaining patents in areas agreed upon by both nations, it does not encompass hydrogen-related technologies (MoC 2019). If intellectual property protection is properly established in India and this fact is communicated and understood not only in Japan but also in the rest of the world, it is expected that India's R&D potential will be fully utilized, and hydrogen-related technologies will be created.

5 R&D Cooperation Between India and Japan Beginning

As mentioned in Sect. 4, while many Japanese companies are concerned about intellectual property, some of the earliest cases of joint research with India have begun to be identified. In December 2023, Mitsui Mining & Smelting (MMS) signed a memorandum of understanding for joint research with the Indian Institute of Technology (IIT) Delhi and opened an R&D center in the Research and Innovation Park on the IIT Delhi campus. The MOU announced that Japanese researchers and IIT Delhi researchers will work together to develop electrode materials for use in water electrolyzers. At the MOU signing ceremony, MMS introduced expensive precious metals in the electrolysis cells incorporated in the electrolysis stack make the whole system very expensive. And MMS explained that joint research with IIT Delhi will help reduce the price of water electrolysis equipment, which is currently a hurdle to introduce. MMS expressed the desire to contribute to the acceleration of social implementation of Hydrogen-related equipment by lowering the price of electrolysis equipment, which is currently a hurdle for its introduction. MMS cited the following four factors in selecting India as a research base: (1) high competitiveness in renewable energy, (2) potential for low-carbon hydrogen production, (3) strong policy incentives for hydrogen production, and (4) India's globally competitive research track record. The research center is expected not only to conduct research on new materials, but also to perform a research function on the implementation of research results in Indian society, he said. Professor Rangan Banerjee of IIT Delhi addressed the signing ceremony, expressing his gratitude for the opportunity to conduct joint research with Mitsui Kinzoku Mining, and his hope to conduct research that will lead not only the energy issues between Japan and India, but also the global discussion on hydrogen (JETRO 2023). This is a very ideal project, and one that is expected to increase in the future, as India and Japan develop hydrogen technology not only for social implementation in India and Japan, but also to contribute to the world.

6 India's Growing R&D Capacity

India's R&D capacity is very high. India ranks third globally in terms of the number of PhDs given in science and engineering (almost 25,000), behind the United States and China. The number of publications in SCI journals has increased, and the country is expected to rank third globally in 2022 (MST 2022). In addition, the Ministry of New and Renewable Energy announced its "R&D ROADMAP FOR GREEN HYDROGEN ECOSYSTEM IN INDIA" in October 2023, and more and more research and development in the hydrogen field is expected (MNRE 2023b). In this roadmap, hydrogen-related technologies belonging to Hydrogen Production, Hydrogen Storage, Hydrogen Transport, and End-use Application are classified into short-term (0 to 5 years), medium-term (0 to 8 years), and long-term (0 to 15 years) development periods according to the maturity level of each technology, and the R&D plans are systematically organized and published. The roadmap also includes a broad range of technologies that are currently being developed around the world. In particular, it compares the status of development in India with that in the rest of the world for technologies that are considered necessary for development in India. For example, the development of materials necessary to reduce the production cost of MMS's water electrolyzer, as described in Sect. 5, is listed as a top priority among Hydrogen Production's development items. Thus, in India, government agencies, private companies, universities, and other research institutions have cooperated to visualize their focus areas in a single roadmap. Based on this roadmap, Japanese companies and research institutions with hydrogen-related technologies have a concrete image of joint research with India, and by introducing equipment developed by Japanese companies to Indian universities and research institutions, where there are many extremely talented researchers, and by conducting joint development, it is expected that highly competitive hydrogen-related technologies will be developed in the world. To this end, it is believed that providing reassurance to

Japanese stakes that their intellectual property will be properly protected in India will accelerate the research and development of new technologies in India and strongly contribute to deepening cooperation between the two countries in the hydrogen field.

7 Carbon Credit Scheme to Facilitate Implementation of Joint Projects

The early establishment of the Joint Crediting Mechanism (JCM) between India and Japan is also expected to contribute strongly to the promotion of research and development and diffusion of hydrogen technology. The main purpose of the JCM is to reduce and absorb GHG emissions through the diffusion of excellent decarbonization technologies, products, services, systems, and infrastructure to partner countries and the implementation of measures in those countries, and to quantitatively evaluate the results of these measures. Japan has signed JCM agreements with 28 countries by December 2023, and has provided more than 200 public supports to promote JCM projects by Japanese companies, including allocation of subsidies for equipment installation, feasibility studies (FS), and funding for demonstration projects by METI and other related organizations. The JCM project is currently underway (METI 2023b). A JCM scheme has not yet been established between India and Japan, and discussions between the two governments are ongoing. As of right now, talks are moving forward to sign a memorandum of cooperation on the establishment of JCM. An aid memoire was signed in March 2023 in New Delhi, India, by the Ministry of Environment of Japan and the Ministry of Environment, Forests, and Climate Change of India, confirming the intention to establish JCM The signing of the Memorandum of Understanding is scheduled to take place in New Delhi, India (MoE 2023). At the 15th Japan-India Foreign Ministers' Strategic Dialogue held in July 2023, the Japanese side requested cooperation for the early conclusion of JCM, but as of December 2023, JCM between the two countries has not been concluded (MoFA 2023). According to a report compiled by Keidanren, Japan's leading business organization, Japanese companies have e xpressed a need to form and implement a wide range of JCM projects, including hydrogen and ammonia technologies, in countries and regions other than the current partner countries. In particular, India is listed as the first region with high business needs for Japanese companies, and the high expectations of Japanese industry toward India have surfaced (Keidanren 2023).

8 Conclusion

This chapter deals with the potential for deeper cooperation between India and Japan in the hydrogen area. In the hydrogen area, Japan started investing in the hydrogen area earlier than India and has accumulated a certain amount of assets, while India is rapidly advancing R&D in the hydrogen area and currently has very high R&D potential. Thanks to the roadmap on R&D prepared by India, the needs of the Indian side have become clearer. In order to bring the assets accumulated by Japan to India and develop technologies that can contribute to the world between the two countries, it is important to appropriately lower the mental hurdle for technology transfer to India and secure a win–win environment as early as possible, so it is effective to secure appropriate intellectual property and to make the information regarding the secured environment known. In addition, early establishment of the JCM will contribute for promoting to establish joint projects between two countries. A system that provides incentives for both countries to invest in hydrogen projects, such as the JCM, is certainly expected to influence the investment behavior of commercial companies in both countries and promote the implementation of active joint projects. From a mid- to long-term perspective, there is much room for cooperation, such as the realization of hydrogen transportation between the two countries and the composition of a hydrogen supply chain that is not confined to the two countries. First, as a short-term goal, it is hoped that joint R&D projects between the two countries will be promoted and many connections that will develop into mid- to long-term projects will be revived.

References

Cabinet Secretariat of Japan (CAS) (2023) Toward the realization of green transformation in our country. https://www.cas.go.jp/jp/seisaku/gx_jikkou_kaigi/dai7/siryou1.pdf (accessed on 14 September 2023)

IEA (2021) India energy outlook 2021, IEA, Paris. https://www.iea.org/reports/india-energy-outlook-2021 (accessed on 21 January 2023)

IEA (2022) Electricity information—data product. IEA. https://www.iea.org/data-and-statistics/data-product/electricity-information#overview (accessed on 13 February 2023)

IHI Corporation (IHI) (2022) IHI and Partners to Explore Ammonia Co-Firing at Indian Thermal Power Plant as Part of Japanese National Project. https://www.ihi.co.jp/en/all_news/2022/resources_energy_environment/1198023_3488.html (accessed on 13 February 2023)

Invest India (2023) Renewable energy in India—Indian power industry investment. In India: Investment Opportunities in the Pow. https://www.investindia.gov.in/sector/renewable-energy

Japan External Trade Organization (JETRO) (2022) Intellectual properties situation in India. WIPO. https://www.wipo.int/edocs/mdocs/mdocs/ja/wipo_webinar_wjo_2022_16/wipo_webinar_wjo_2022_16_pres.pdf

Japan External Trade Organization (JETRO) (2023) Mitsui Kinzoku Mining and Indian Institute of Technology Delhi sign MOU for joint research. https://www.jetro.go.jp/biznews/2023/12/027c0cbc95b72a2a.html

JERA (2022) Development of New catalyst for ammonia production and start of technology demonstration for establishment of fuel ammonia supply chain—adopted as green innovation fund project: press release. https://www.jera.co.jp/information/20220107_824 (accessed on 22 January 2023)

Ministry of Chemicals and Fertilizers (MoCF) (2022). Year- end review of the department of Fertilizers-2022. Press Information Bureau. https://pib.gov.in/PressReleasePage.aspx?PRID=1886054#:~:text=Total%20Subsidy%20Outgo&text=The%20Subsidy%20approved%20by%20Cabinet,(SSP)%20through%20freight%20subsidy (accessed on 12 February 2023)

Ministry of Commerce and Industry of India (MoC) (2019) Procedure guidelines for patent prosecution highway (PPH). https://ipindia.gov.in/writereaddata/portal/news/591_1_pph_procedure_guideline_combined_20191128_final.pdf (accessed on 15 February 2023)
Ministry of Economy, Trade and Industry (METI) (2017) Basic Hydrogen Strategy. https://warp.da.ndl.go.jp/
Ministry of Economy, Trade and Industry (METI) (2021) The Sixth Basic Energy Plan. https://www.meti.go.jp/press/2021/10/20211022005/20211022005-1.pdf (accessed on 14 September 2023)
Ministry of Economy, Trade and Industry of Japan (METI) (2022) Study Group on Structural Changes in Mobility and the Direction of Automobile Policies for 2030 and Beyond. https://www.meti.go.jp/shingikai/mono_info_service/mobility_kozo_henka/index.html (accessed on 23 January 2023)
Ministry of Economy, Trade and Industry (METI) (2023a) Basic Hydrogen Strategy. https://www.meti.go.jp/shingikai/enecho/shoene_shinene/suiso_seisaku/pdf/20230606_5.pdf (accessed on 14 September 2023)
Ministry of Economy, Trade and Industry (METI) (2023b) JCM (Bilateral Crediting Mechanism). https://www.meti.go.jp/policy/energy_environment/global_warming/jcm/index.html
Ministry of the Environment (MoE) (2023) Signing of an Aide-Memoire between Japan and India confirming their intention to establish a Bilateral Crediting Mechanism (JCM). https://www.env.go.jp/press/press_01415.html
Ministry of Environment, Forest and Climate Change (MoEFCC) (2023) India Stands Committed to Reducing Emissions Intensity of GDP by 45% by 2030 and Reach to Net-Zero by 2070, While Developing Sustainably: Shri Bhupender Yadav. Press Information Bureau. https://www.pib.gov.in/PressReleseDetail.aspx?PRID=1897093
Ministry of Foreign Affairs (MoFA) (2023) 15th Japan-India foreign ministers' strategic dialogue. Ministry of Foreign Affairs of Japan. https://www.mofa.go.jp/s_sa/sw/in/page1e_000738.html
Ministry of Land, Infrastructure, Transport and Tourism (MLIT) (2023) Public Call for Additional R&D for Social Implementation of Ammonia Fueled Ships. https://www.mlit.go.jp/report/press/kaiji07_hh_000297.html
Ministry of New and Renewable Energy (MNRE) (2021) India set to achieve 450 GW renewable energy installed capacity by 2030. https://pib.gov.in/Pressreleaseshare.aspx?PRID=1762960
Ministry of New and Renewable Energy of India (MNRE) (2023a) Cabinet approves national green hydrogen mission. Press Information Bureau. 4 January 2023. https://pib.gov.in/PressReleasePage.aspx?PRID=1888547
Ministry of New and Renewable Energy (MNRE) (2023b) R&D roadmap for Green Hydrogen Ecosystem in India. Ministry of New and Renewable Energy | India. https://mnre.gov.in/notice/rd-roadmap-released-by-mnre-on-the-eve-of-world-hydrogen-day/
Ministry of Science & Technology (MST) (2022) Year-end review -2022: department of science & technology. Press Information Bureau. https://pib.gov.in/PressReleseDetail.aspx?PRID=1886528
New Energy and Indsutrial Technology Development Organization (NEDO) (2023a) Hydrogen utilization in the steelmaking process. https://green-innovation.nedo.go.jp/project/utilization-hydrogen-steelmaking/progress/
New Energy and Indsutrial Technology Development Organization (NEDO) (2023b) Development of next-generation vessels. https://green-innovation.nedo.go.jp/project/development-next-generation-vessels/progress/
New Energy and Indsutrial Technology Development Organization (NEDO) (2023c) Development of next-generation aircraft. https://green-innovation.nedo.go.jp/project/development-next-generation-aircraft/progress/
NITI Aayog and RMI (2022) Harnessing Green Hydrogen. https://rmi.org/insight/harnessing-green-Hydrogen/ (accessed on 22 January 2023)
Otaki Shaw (2023) The potential of collaboration between India and Japan in the hydrogen sector. Energies 16:3596. https://doi.org/10.3390/en16083596

Otto M, Chagoya KL, Blair RG, Hick SM, Kapat JS (2022) Optimal hydrogen carrier: holistic evaluation of hydrogen storage and transportation concepts for power generation, aviation, and transportation. J Energy Storage 55:105714. https://doi.org/10.1016/j.est.2022.105714

Sato T (2017) Small and medium enterprises' expansion into India—a study based on the 2013–2014 questionnaire survey. https://shokosoken.or.jp/shokokinyuu/2017/11/201711_2.pdf (accessed on 14 February 2023)

Schrotenboer AH, Veenstra AAT, uit het Broek MAJ, Ursavas EA (2022) Green hydrogen energy system: optimal control strategies for integrated hydrogen storage and power generation with wind energy. Renew Sustain Energy Rev 168:112744. https://doi.org/10.1016/j.rser.2022.112744

The Economic Times (2022) India's first hydrogen fuel cell bus unveiled in Pune—India's First. The economic times. https://economictimes.indiatimes.com/industry/renewables/indias-first-hydrogen-fuel-cell-bus-unveiled-in-pune/indias-first/slideshow/93752486.cms?from=mdr (accessed on 12 February 2023)

The Economic Times (2023) Indian oil unveils India's first green hydrogen-run bus that emits just water. https://economictimes.indiatimes.com/industry/renewables/indianoil-unveils-indias-first-green-hydrogen-run-bus-that-emits-just-water/articleshow/103931084.cms

Toyota Motor Corporation (TMC) (2023) Toyota to accelerate plant carbon neutrality initiatives in fukushima by utilizing hydrogen: corporate: Global Newsroom. https://global.toyota/jp/newsroom/corporate/38917305.html

Volza. India Fertilizer Imports–Volza (2022). https://www.volza.com/p/fertilizer/import/import-in-india/ (ac-cessed on 12 February 2023)

Vaccine Diplomacy: Partnering Global Goods

Srabani Roy Choudhury, Shashank Khandwe, and Megha Sharma

Abstract The pandemic had a profound impact on the world. It showed at large the vulnerabilities of developed and developing nations and expanded the non-traditional security basket to involve the health care system. Moving beyond the pandemic hinged on a vaccine, nations raced to find an answer. India is known as the "pharmacy of the world" as it has the manufacturing capabilities for mass production. Thus, India put its best foot out to develop a vaccine, mass manufacturing it and then disbursing commercially and in the form of aid to many global south nations. Japan's vaccine story was marred in inept policy direction; however, it reached out to provide financial support to its neighbouring nations to buy and disburse vaccination drives. This chapter maps India and Japan's vaccine stories and focus on two critical aspects of vaccine diplomacy. One is how India gained ascendancy in the diplomatic circle through its vaccine diplomacy. Two is how India and Japan collaborated to actively engage in vaccine procurement, distribution, and support throughout the global recovery from the pandemic.

Keywords Vaccine-diplomacy · Vaccine-development · QUAD · COVID-19 · Healthcare

1 Introduction

Diplomacy is all about building goodwill and exerting influence on the global stage. The pandemic, which consumed the world for over two years, initiated new avenues of diplomacy related to the securitization of nations. A critical element of the pandemic was to find a vaccine that would build immunity in the people and enable them to fight the coronavirus. While the solution looked simple, anyone with some knowledge of vaccine development know that the process is riddled with hurdles. As coronavirus-inflicted nations, the attempt to find a cure as much as prevention became the focus.

Present Address:
S. R. Choudhury (✉) · S. Khandwe · M. Sharma
School of International Studies, Jawaharlal Nehru University, New Delhi, India
e-mail: sroyc2@gmail.com

R. Shaw and S. R. Choudhury (eds.), *India, Japan and Beyond*,
https://doi.org/10.1007/978-981-97-3282-1_9

Thrown into the chaos of how to function, most nations followed the principle of isolation as the first round of protection. However, the healthcare system brought nations to their knees. Developed, developing, and less developed nations were all found wanting in their healthcare system. It was realized that all nations must work in unison to fight this virus.

Effective vaccine and healthcare for affected people needed a collective approach. This profoundly affected how to fight the coronavirus as there was a copious need for finance, men, and materials. Vaccine diplomacy grew out of this compulsive need to recalibrate the world and restore order. Vaccine diplomacy is non-traditional security and straddles as a soft power tool. Coined by Joseph Nye, soft power is the ability to shape others' choices and preferences through co-optive tools, which in modern international politics can also take the form of health diplomacy. Vaccine diplomacy has many threads, from simple vaccine donations to countries to help in the production and sharing of technology. The net result was that vaccines became a diplomatic tool to garner goodwill by helping nations and a bargaining chip for negotiation.

As vaccine was the panacea to move beyond the pandemic, nations raced to find an answer. Over a period of time, India's strength of skilled pharmacists and low-cost economy led it to be popularly known as the "pharmacy of the world" as it has the manufacturing capabilities for mass production. Thus, India put its best foot out to develop a vaccine, mass manufacturing it and then disbursing commercially and in the form of aid to many global south nations. India's soft power implementation through vaccine diplomacy has enabled it to become the voice of the global south.

Unforeseen challenges marred Japan's vaccine story. Uncalled for leadership change, at the critical juncture of the pandemic, the decision about the Tokyo Olympics and a constitution that did not give any right of emergency caused uncertainties, resulting in delayed decisions. In many ways, during this period, Japan lost significantly its position in the international community. However, Japan lived by its commitment to being a major global contributor through aid programs. Japan's forte has been using its capital resources to counter power play in its neighboring region. Given China's strong presence in this region and its vaccine diplomacy agenda, Japan reached out to provide financial support to its neighboring nations to buy and disburse vaccination drives.

This chapter maps India and Japan's vaccine stories and focus on two critical aspects of vaccine diplomacy. One is how India gained ascendancy in the diplomatic circle through its vaccine diplomacy. Two is how India and Japan collaborated to actively engage in vaccine procurement, distribution, and support throughout the global recovery from the pandemic.

2 India's Vaccine Development

India is the largest populated nation, with a significant number of its population below the poverty line. Its challenges were multitude to that of many other nations. India's first attempt at coping with the pandemic was to follow a strict shutdown. People were isolated in their homes, and all economic activities were curtailed. A larger number of marginal workers who were immigrants to larger cities and towns lost their earnings and had no option but to try getting back to their native villages or towns. So, while the pandemic was essentially a health issue, this policy impacted society, creating chaos and hardship.

Even within the healthcare system, the coronavirus outbreak came as a long-overdue test of the resilience of Indian healthcare infrastructure, revealing several lacunae and incompetencies. In the gloom that prevailed, one ray of hope was that India had the capability to develop a vaccine, having achieved the distinction of already being the vaccine production hub for many vaccines. The pandemic would encourage the pharma sector to increase its capacity to enable mass vaccine production at low cost with hassle-free distribution.[1] However, the immediate challenge was to develop a vaccine that was safe at minimum cost and potential for easy distribution.

The feat of developing the COVID-19 vaccine within a 10-month period, which formerly used to take 10–15 years, is a culmination of years of steps taken in the right direction. The growth trajectory can be traced back to pre-independence India when public sector entities (PSEs) undertook major research on prevailing diseases of the time (like smallpox, kala-azar, cholera, typhoid, rabies, polio, influenza, malaria, tuberculosis, etc.).[2] Eradication of such epidemics (especially smallpox in the 1970s, polio in the 1980s, and measles in the 2010s) improved the state health system through national immunization programs, which provided not only a structure supported by trained vaccinators, a cold chain equipment system but also a network for surveillance of vaccine-preventable diseases.[3]

Vaccine development history is rooted in India's "self-sufficient economy" policy. Thus, till around the turn of the century, vaccine production was primarily undertaken by the public sector. As the economy opened up for foreign participation, the private sector gradually started taking over vaccines listed in UIP (Universal Immunization Programs) and non-UIP markets. This is attributed to the private sector's comparatively higher capacity utilization status,[4] and greater aptness to newer IHR vaccines, which require technologically more advanced manufacturing processes. However, the public sector did not lose its prominence as it continued to collaborate with international networks like the Coalition for Epidemic Preparedness Innovations (CEPI) and the International Vaccine Institute (IVI) and subsequently emerged successful in the importation of technologies as well as strategies for developing vaccines. Another forum through which India gained ascendency is the "Developing Countries Vaccine

[1] The use of cold chain for distribution can severely impact reaching the hinterland.

[2] Hooda (2023).

[3] Lahariya (2014).

[4] Hooda (2023).

Manufacturers Network" (DCVMN), a public health-driven alliance of corporate vaccine manufacturers to share technical information, best practices, and prospects.[5] India also expanded its vaccine manufacturing capacity categorically through foreign partnerships, for instance, with a South Korean biotech company (EuBiologics). This helped produce an optimized OVC product that was not only more suitable but also enabled higher volume supply and uptake.[6] By harnessing and utilizing such partnerships, Indian pharmaceutical institutes gained research expertise and opportunities to test quality standards. This, coupled with the legacy and experience of administrative machinery with mass immunization programs, resulted in a more systematic healthcare structure in the country. In addition, high education quality and low-cost skilled manpower, along with the dexterity of Indian biotech engineers in leveraging the latest technology, helped expand the domestic vaccine manufacturing base while keeping costs economically viable.

India's road to gaining the "pharmacy of the world" entitlement rests on some critical vaccines developed under rigorous WHO procedures and standards. In 2010, the Serum Institute of India launched the first meningitis vaccine prequalified by WHO.[7] In 2013, Bharat Biotech International launched the world's first typhoid conjugate vaccine (TCV).[8] Looking back, these two vaccines helped the companies to understand the procedures, documentation, and templates to follow to ensure WHO approval. In 2020, as the race to develop a vaccine for coronavirus ensued, these private companies developed and manufactured India's first indigenous COVID-19 vaccine- Covaxin (BBV152), with scientific assistance from public sector research institutions (PSRIs), namely the Indian Council of Medical Research and National Institute of Virology.[9] This vaccine followed the traditional technology of using inactivated coronavirus, similar to the inactivated polio vaccine.

India, today, stands as the only nation in the Developing Countries Vaccine Manufacturers Network (DCVMN) with up to 264 WHO-prequalified vaccines (WHO 2022)[10] This not only instills confidence in the healthcare industry domestically but also garners international acceptance of Indian vaccines. In addition, the low-cost production of Indian vaccines makes them more affordable and thus more preferable among low—and middle-income countries. Therefore, the primary reason why vaccine exports from India between 2011 and 2020 grew at the rate of 13.7%, as opposed to the world export rate of 4.14%, is this.[11] These exports directly contributed

[5] Pathak and Magdalene (2022).

[6] COVID-19 Vaccine Development and Administration in India/Odevall et al. (2018).

[7] Pathak and Magdalene (2022).

[8] Bharat Biotech. Bharat Biotech launches typebar-3TCV™, the world's first clinically proven typhoid conjugate vaccine for infants six months and older and adults. Developing Countries Vaccine Manufacturing Network.

[9] Hooda (2023).

[10] Hooda (2023).

[11] Hooda (2023).

to over half of the country's vaccine industry turnover. India, in total, contributed 60% to the global vaccine supply even before Covid pandemic.[12]

However, it is important to note that even after being the avant-garde of vaccine manufacturing, India had to sign agreements with multinational pharmaceutical companies like AstraZeneca and Novovax to pre-purchase 20% of the global share of vaccines[13] owing to the overwhelming demand from the huge population within the country, which shortens the manufacturing capacity of the indigenous producers.

The success of quick vaccine development, namely Covaxin, rests on the ascendency gained prior to the onset of the pandemic. While WHO took a while to certify the vaccine, India began mass immunization as it attempted to curtail the spread of the coronavirus, enabling a more extensive dataset on which Covaxin could seek WHO's approval. The world acknowledged India's technical capability and production capacity in advance, and when WHO gave its clearance, India was ready to embark on its vaccine diplomacy.

3 India's Vaccine Diplomacy During COVID-19

While credit for vaccine development rests on the scientists, a large army of clinical researchers, and the people who participated in clinical testing, vaccine diplomacy was largely a test of India's astute political and diplomatic overtures.

Covaxin, the vaccine produced at low cost and required a nominal cold chain to that of Pfizer by India, found salience among global south nations. Trust in India's vaccine has grown because of the efficacy of other vaccines already imported from India for other immunization purposes. Further, having the expertise, India offered advice on all facets of COVID-19 vaccine administration, i.e., finances, logistics, storage, distribution, communications, etc.; the government of India established the National Expert Group on Vaccine Administration for COVID-19 (NEGVAC)[14]. In August 2020. Having developed the vaccine to contain the spread of the highly contagious disease, India launched the world's biggest COVID-19 vaccination program in January 2021. The National Digital Health Mission (NDHM) and Co-WIN platform were deployed as innovative measures to ensure rapid immunization.

Based on the philosophy of humanitarianism through the principle of '*Vasudhaiva Kutumbakam*' (The world is one family) and on the practicality of diplomacy through the policy of 'Neighbourhood First,' India sought to harness its vaccine prowess for better relations with immediate neighboring countries.[15] While most of the developed countries of the West were preoccupied with 'Vaccine Nationalism,' an integral element of New Delhi's foreign policy during the pandemic was 'Vaccine Maitri' or 'Vaccine Friendship'. Through this initiative, the Covishield vaccine manufactured

[12] Sharma and Varshney (2021).

[13] Koop (2020).

[14] Ministry of Health and Family Welfare (2020).

[15] Singh et al. (2023).

by 'Serum Institute of India' (SII) was delivered to the South Asian Association for Regional Cooperation (SAARC) countries. To put forth its image as a regional 'net security provider,' a quarter of the country's total COVID-19 vaccine supply was distributed to South Asian countries as grants- Bangladesh (3.3 million doses), Myanmar (1.7 million), Nepal (1.1 million), Sri Lanka (0.5 million), Afghanistan (0.5 million) and Maldives (0.2 million).[16] To control the global health crisis, India dispatched 35.79 million vaccine doses as commercial exports to various countries.[17] Producing 1.5 billion doses per year, India was the heavy lifter out of 190 countries that led the multilateral initiative called COVID-19 Vaccine Global Access Facility or COVAX, which aimed at enabling equitable access to the vaccine worldwide through grants or commercial exports. Although India ranks eight among top 20 exporters in terms of value, however, with a proportion of 11.01%, India ranked fourth globally in terms of the volume of vaccinations supplied.[18] The incongruity between the low export share in value and the large share in volume indicates the affordability of vaccines from India.

The aid and assistance extended to countries were not just through vaccines but also through telemedicine services, testing kits, personal protection equipment, respirators, extended credit lines, etc.[19] India dispatched teams of military medics, essential medicines, and hospital consumables through Operation Sanjeevani to Maldives. Additionally, New Delhi also sent "Rapid Response" teams to nations Kuwait, the Maldives, Nepal, Morocco, Oman, and Seychelles and conducted online training as a capacity-building exercise. To manage China's overwhelming influence on these island countries, New Delhi utilized Mission SAGAR (the Security and Growth for All in the Region) for this diplomatic mission to Mauritius, Madagascar, and Comoros. According to Mashal and Yee (2021), India's vaccination campaign gave a tough challenge to China's long-standing political advances in neighboring countries.[20]

India's altruistic vaccine diplomacy faced a severe test between May and June 2021, when the country was inflicted with the deadly delta variant. Over 300,000 new cases and almost 5,000 deaths per day were registered in India, making it the second-highest number of reported COVID-19 cases after the United States.[21] Struggling to deal with the outbreak of coronavirus and the shortage of medical supplies, the Indian government realized the need to halt vaccine outreach and prioritize immunization drives within the country. While the realist would argue that national needs supersede the altruistic behavior of a nation, in this case, India soon resumed its commitments, and by its actions towards its neighborhood in the first phase, both

[16] Ministry of External Affairs—Government of India. Vaccine Supply. https://www.mea.gov.in/vaccine-supply.htm.

[17] India's role in COVID-19 vaccine diplomacy/Ministry of External Affairs—Government of India. Vaccine Supply. https://www.mea.gov.in/vaccine-supply.htm.

[18] Hooda (2022).

[19] Sharun and Dhama (2021).

[20] Mashal and Yee (2021).

[21] World Health Organization. Coronavirus disease (COVID-19) dashboard.

Nepal and Bangladesh extended assistance through supplies of cryogenic oxygen tankers and critical drugs.

The void created by India's absence in vaccine supply during the brief period proved an excellent opportunity for Beijing to propel its vaccine diplomacy through the 'Health Silk Road' initiative to countries, especially India's neighbors-Bangladesh, Sri Lanka, and Nepal, as to make them reliant on Chinese vaccine exports. For China, it was an attempt to divert the world's attention, which pined China as the perpetrator of the pandemic. Further, it was a prospect for China to create a wedge between India and its neighbors. However, China's inefficacious vaccines caused the surge in export phenomena to fade quickly, whereas Indian exports continued to gain popularity in the latter months of 2022–2023.[22]

4 Japan's Crisis During the Pandemic

Japan's story of the pandemic crisis has evolved with a confluence of both early success and later complications, culminating with a partnered approach with like-minded countries for global comprehensive health security. As with all nations, Japan too had to work at trying to balance between containing the spread of the disease, economic activity, and public anxiety. In addition, Japan faced a unique situation. It did not have a legal mechanism for regulating the movement of people and creating restrictions by shutting down commercial operations, etc. For example, In sharp contrast to India, Japan having no legal framework, the government could only ask people to stay home or businesses to close down, but could not enforce the same through legal measures. Thus, in the beginning, the Japanese government relied mainly on the abiding approach of Japanese citizens, which was visible during many natural disasters. As Tokyo continued functioning normally when most other countries were shutting down, it created public anxiety. The slow decisions during the pandemic exposed political and bureaucratic inefficiency, and as the Tokyo Olympics approached, with the world drawn to how Japan could execute this mega event, Japan's healthcare, regulations, law, and implementation came under scrutiny. What was surprising in the case of the pandemic was that Japan, a country that has overcome many natural disasters successfully, seemed to be radarless for the most part of the pandemic. Cohesiveness between politicians and bureaucracy at the national and local levels was missing.

The Ministry of Health, Labour, and Welfare (MHLW), Government of Japan, registered the country's first official case on January 15, 2020, originating from Wuhan, Hubei Province, People's Republic of China.[23] The COVID-19 administrative measures initially undertaken by the Japanese government remained quite relaxed and unrestrictive compared to the West. The Diamond Princess cruise ship,

[22] Hooda (2022).

[23] Novel Coronavirus (2019-nCoV) Situation Report-1, January 21, 2020, Official Website of the World Health Organisation (WHO).

which docked in Yokohama on February 3, 2020, had cases of coronavirus. However, the Japanese government allowed people to embark and use public transport to reach their destinations. This incident raised serious concerns about infections,[24] following which the government sprang into action. Japanese government came under criticism for its piecemeal approach towards the serious health issue and nonchalant attitude. As an attempt to prove himself, Shinzo Abe's government announced the revision of a 2012 special measure law initially designed to protect Japan from new types of influenza. This change gave Abe's government the power to declare an emergency.

A consolidated Novel Coronavirus disease control policy came up, and entry restrictions were imposed for those traveling from China and South Korea. Nevertheless, the positive case count of COVID-19 remained low, and the fatality rate was stable at 2.8% at the peak in April 2020.[25] These figures were attributed to an effective reduction in contact rates, accustomed use of face masks, and people following government instructions without enforcement measures, a cultural trait of Japanese. The measures included the '3C' approach of avoiding closed spaces, crowded places, and close-contact settings,[26] along with opting for a 'cluster-based' approach to tackle cases. The success in Japan with little enforcement resulted in Abe promoting the 'Japan model' of tackling COVID-19, which became a discourse on effective approaches to plug infections. As of June 2020, with only 0.7% of the Japanese population was affected by the virus.[27] This number was far less than that of most of its neighboring countries. Thus, concern about economic revival measures gained momentum, domestic travel restrictions were lifted, and policies like 'Go to Travel,'[28] campaign was formulated to revive the economy. Despite this, the news around the region, especially the havoc it created in China, was unnerving to the citizens. While the government remained optimistic, critics called it a 'laissez-faire' approach to emergency situations with no priority to health concerns. This was reflected in the low PCR and antigen testing undertaken during this time.

Abe's health issues made him step down in September 2020 amidst these administrative challenges, and his long-term Chief Cabinet Secretary, Yoshihide Suga, took over. The Suga government quickly came under attack from opponents for prioritizing economic revival over health infrastructure build-up as cases began to climb.[29] Suga had to combat the pandemic and also conduct the Tokyo 2020 Olympics. Thus, he imposed the second emergency only around January 2021 as the fourth wave emerged in Japan and continued his efforts for the delayed Tokyo Olympics 2020. In February 2021, Administrative Reform Minister Taro Kono was also given the additional charge of looking after the vaccine roll-out. On his appointment, he promised

---

[24] Ratcliffe and Fonbuena (2020).

[25] Sayeed and Hossain (2020).

[26] Amengual and Tatsuya (2020).

[27] Wingfield-Hayes (2020).

[28] Kazuki et al. (2020).

[29] 'In a trying week, Suga fails to quell disquiet at the Diet over coronavirus response,' *The Japan Times*, 28 h January 2021.

to "tackle the challenge with a complicated logistics plan to deliver an effective blitz of vaccinations quickly."[30]

5 The Vaccine Discourses in Japan

The Olympic aspirants, the International Olympic Committee in particular, and the world at large questioned Japan's delay in vaccination roll-out, as it was imperative for Olympic participants to feel safe in Tokyo. This delay was due to the fundamental problem of lack of cooperation and coordination between governmental agencies and central and local governments. For example, the vaccine roll-out was the responsibility of the Ministry of Health, Labour, and Welfare, and distribution logistics rested with the Ministry of Land, Infrastructure, Transport and Tourism and the Ministry of Internal Affairs and Communications, coordination between the central government and local governments. This is quite similar to other nations, however, Japan faltered in the interagency process.

The vaccine program in Japan was rolled out around February 2021, and the Tokyo Olympics were just months away. Unlike the usual conformity shown by the Japanese, vaccine hesitancy further challenged the government this time. According to a study conducted by the Lancet group prior to the vaccine roll-out, 32.9% of Japanese were unsure about getting vaccinated, and 11% of the Japanese public were unwilling to get vaccinated.[31] Vaccine hesitancy comes with a history of vaccination mishaps that surfaced in Japan throughout all these years. Some of these incidents include the issues with the Diphtheria, Tetanus, and whole-cell Pertussis vaccine (DTwP) in 1974 and the Measles, Mumps, and Rubella (MMR) vaccine in 1993.[32] In the latter case, the vaccine was suspected of leading to considerable chances of aseptic meningitis, which led to its government discontinuation. Similarly, the Human Papillomavirus (HPV) vaccine that rolled out in 2013 allegedly caused severe headaches and seizures, following which the health ministry withdrew and resulted in a subsequent fall in its vaccination rates.[33] This has also made the government gradually ease the restrictions on immunization policy. These factors are reflected in the vaccination figures as around 7.5 million people were vaccinated within the first three months of vaccine roll-out (February–April 2021).

Despite this initial sluggish behavior, the Tokyo Olympics 2020 (conducted in 2021) and the government's dropping approval ratings resulted in a big push to vaccinate people swiftly. Around July 2021, the vaccination drive gained momentum

[30] Tsukimoro and Sugiyama. Japan's new Covid 19 vaccine czar Taro Kano commits to rapid yet safe roll out" https://www.japantimes.co.jp/news/2021/01/19/national/politics-diplomacy/taro-kono-coronavirus-vaccines/

[31] Ghaznavi et al. (2022).

[32] Ly Nguyen and Candice Ruck, 'Japan's vaccine hesitancy has implication for COVID control, and the Summer Olympics,' Asia Pacific Foundation of Canada, November 2023.

[33] Du and Huang (2020).

in the months leading to the Olympic games. It was stimulated by a 160 billion Yen allocation to boost domestic vaccine industries, along with innovative measures like using the "Line" social media platform to spread vaccine awareness, etc.[34] The initial shortage was compensated for by the prompt entry of Moderna and AstraZeneca vaccines despite Pfizer holding the larger share of administered vaccines.

Japan's mismanagement of vaccine roll-out comes as a surprise, considering that the World Health Organisation (WHO), in its health report on Japan, has highlighted the gradual development that the Japanese health infrastructure has undergone through all these decades and has achieved remarkable success, particularly in Information management and surveillance.[35] Despite these, vaccine roll-out brought forth that Japan faces trouble in some structural aspects of its health management system, which was quite visible in how both Abe and Suga administrations struggled with the COVID-19 pandemic. These include the limited powers that governments at the national level possess on these matters. Even though the cabinet and the Ministry of Health, Labour, and Welfare (MHLW) hold the power to designate infectious diseases and draw the basic policies to tackle them, the bulk of the actual pandemic response lies with the prefectural governors.[36] Such drawbacks call for inter-prefectural cooperation that allows the system to effectively manage 'Infection Prevention and Control' (IPC), as emphasized by the WHO report.

Against the backdrop of all these challenges, Japan saw a change in the leadership when Kishida Fumio was elected as Prime Minister in September 2021, who gave new vigor to the national fight against the pandemic as the delta variant continued to impact Japan. His first policy speech stressed Japan's 'spiritual and economic revival' and focused on tackling the COVID-19 pandemic conscientiously. On these lines, he committed to increasing the medical personnel and hospital bed capacity. Secondly, the focus was also on the second and possibly third vaccination dosage, including oral medication support. Third, digital certification of vaccinations (achieved by December 2021) and free-of-charge testing were stressed. The economic assistance to support businesses was targeted through subsidies and aid programs.[37] His efforts found fruition as Japan, in November 2023, finally got the approval for the domestically manufactured vaccine 'Daichirona'.[38]

Prioritizing containment of COVID-19, Kishida embarked on the harshest measures among G7 countries.[39] With the emergence of the Omicron variant, the much-delayed state of 'Quasi-Emergency' was achieved, as foreign entries were banned around November 2021. These border measures remained in place for almost

---

[34] 'Tokyo offers proof of COVID-19 vaccinations through Line app,' *The Asahi Shimbun*, November 2, 2021.

[35] 'Building an effective infection prevention and control system in Japan: a historical overview,' World Health Organization (WHO) 2023 report.

[36] Takenaka (2021).

[37] 'The First Policy Speech by Prime Minister Kishida Fumio,' The Government of Japan, November 18, 2021.

[38] Tomoko (2023).

[39] PM Kishida Fumio speech excerpt, 'Building a Post-COVID-19 New Japan.' www.japan.go.jp.

a year despite Japan's infection rates being many times lower than those of other developed countries and vaccination rate being 79%.[40] The idea of a 'new form of capitalism' became a prominent discourse as the prime minister prepared for the new international order unravelling in the form of the Ukraine crisis, increasing security concerns in its neighborhood, and the domestic economic paralysis due to border closures. Gradually, borders were allowed to open in June 2022 as Kishida faced pressure from business circles and university students.[41]

6 Japan's Vaccine Diplomacy: Hinging on Multilateralism

Despite domestic reproach and criticism, Japan's efforts to resolve the unforeseen global health crisis were supplemented by its gradual multilateral engagements. With its own 'Global Health Strategy,' the Kishida administration has mapped out a cross-sectoral, multi-stakeholder approach to strengthening the global health architecture.[42] This outline was developed as the prime minister continued engaging with the Quadrilateral framework and prepared to host the G7 Summit 2023 in Hiroshima, his home turf. The strategy focused on utilizing existing schemes like the ODA and grant aids to support its vision of "Universal Health Coverage".

Japan has previously used the G7/G8 forum to draw global attention to common health concerns and led initiatives in an attempt to achieve convergences. As a step a commitment was made at the G8 Summit 2000 at Nago, Okinawa (Japan). It played a prominent role in establishing the Global Fund to Fight AIDS, Tuberculosis, and Malaria (GFTAM). Similarly, the G8 Summit, 2008 at Toyako, Hokkaido (Japan), became the platform to raise efforts for neglected tropical diseases (NTDs), and the G7 Ise Shima Summit 2016 stressed research and development on innovative medication initiatives, among other goals.[43] As the G20 chair in 2019, Japan reiterated its idea of 'Universal Health Coverage' and hosted the first-ever joint session of G20 Finance and Health Ministers, stressing the need for health financing in developing countries as a major agenda.[44] Japan's institutional efforts in the global healthcare framework have been remarkable. These efforts have ranged from research and development to training programs, data collection, and support for local health infrastructure.

The health financing aspect of their efforts was facilitated by the Japan International Cooperation Agency (JICA), which became a critical vehicle in carrying

[40] Soma and Tiberghien (2022).

[41] https://asia.nikkei.com/Spotlight/Coronavirus/Kishida-announces-limited-reopening-of-Japan-s-borders/.

[42] Japan's Global Health Strategy Outline, www.kantei.go.jp.

[43] https://www.isdp.eu/japans-global-health-strategy-a-diplomatic-foresight-on-the-g7-hiroshima-summit/.

[44] 'Japan's International Cooperation for Global Health,' Public Relations Office, Government of Japan.

out health diplomacy by extending necessary support to vulnerable countries worldwide. Japan stands as the largest contributor to the World Health Organisation's immunization program. As soon as WHO declared COVID-19 as a 'Public Health Emergency of International Concern,' JICA jumped into action by supporting the low and middle-income countries. The assistance began with providing necessities such as health supplies like masks, which are estimated to be around 3.9 billion yen (between January 2020 and March 2022). During the same period, they supported technical cooperation and resources of roughly 3.5 billion yen in 54 countries for COVID-19 control. This has been followed by necessary financial assistance in the form of Overseas Development Assistance (ODA) loans worth 652.9 billion yen to 23 countries. JICA's central pillars in these efforts have been around strengthening treatment and diagnostic systems, enhancing infectious disease research and alert systems, and promoting infectious disease prevention through awareness campaigns.[45]

Japan's health diplomacy has persisted with the stand taken by PM Abe in his government's involvement with the Sustainable Development Goals 2030,[46] framed by the United Nations General Assembly in 2015. The idea of 'Universal Health Coverage' has been showcased with the principle of "leaving no one's health behind" as the world fights against a common threat to human security. Some of the many efforts that Japan has made on this front include the supply of vaccines individually and through the COVAX facility and providing 'last one mile support' with the assistance of cold chain system development. It further helped expand the coverage of diagnostics and therapeutics, strengthen the surveillance system, and develop the capacity of frontline health workers.[47] In no uncertain terms, Japan has been a trailblazer in global health discourse, which is part of the bigger umbrella of human security concern.

7 India and Japan: The Asian Health Guards

Japan's traditional outreach to achieve health security for developing countries has been fruitful. India has been one of the beneficiaries of this outreach, being the first country to receive Overseas Development Assistance (ODA) in 1958. Health and welfare have been critical pillars of support for Japan in this assistance, and it saw a dramatic increase with the coming of the COVID-19 pandemic. Japan aided India with an assistance loan worth JPY 50 billion for crisis response support in August 2020 and a grant worth JPY 1 billion for critical medical equipment provision.[48] This

---

[45] 'JICA's Initiative for Global Health and Medicine-Interim Report: The Two-year Response to COVID-19 and Future Prospects,' Japan International Cooperation Agency (JICA), Human Development Department, September 2022.

[46] Shinzo Abe (2015).

[47] 'Japan's COVID-19 related cooperation towards achieving Universal Health Coverage,' Ministry of Foreign Affairs, June 2022.

[48] https://indbiz.gov.in/japans-commits-rs-35bn-to-indias-covid-19-efforts/.

was followed by JICA extending support to the Government of India in its social protection scheme, 'Pradhan Mantri Garib Kalyan Yojana', by parallel financing of a loan of JPY 30 billion. This was supplemented by emergency assistance with oxygen concentrators, ventilators, testing equipment, etc.[49] The Government of Japan, along with UNICEF, extended further help in 2021 by providing Cold Chain Equipment (CCE) support to India, which included walk-in Coolers, Freezers, Solar Direct Drives, Vaccine Carriers, and Voltage Stabilizers.[50] Japan agreed to support regional countries through an aid of $3.3 billion in the 'COVID-19 Crisis Response Emergency Support Loan Program'.[51] During the 2022 QUAD leaders' summit, it was further announced that the Bank of Japan would invest $100 billion to facilitate vaccine manufacturing in India. These efforts align with Japan's long history of financial and technical support under its aid program.

In the wake of the pandemic, Indian Prime Minister Narendra Modi also proposed a US$10 million joint emergency fund based on international and regional health cooperation in the face of the global health crisis. Across Asia, Africa, the Caribbean, and Europe, India's vaccine outreach was appreciated so much that UN Secretary-General Antonio Guterres applauded India's vaccine capacity as the "best asset" to the world during the crisis.[52] India emerged as a natural leader globally and proved its firmly committed to equity and equality for vaccine disbursement.

The pandemic showed Japan's unpreparedness in domestic vaccine manufacturing, whereas India proved to be a leader. This gives a natural thrust to their partnership, both bilaterally and within the framework of the QUAD partnership. With this realization, Japan supported the Indian vaccine manufacturing facilities with financial and technical assistance. Further, these two nations push forth the QUAD arrangement of delivering vaccines to Southeast Asia, Indo-Pacific, and Africa. This was done as a attempt of partnering in the global commons and to counter China's overtures in this region.

7.1 *QUAD: Confluence for Addressing Vaccine Diplomacy*

The COVID-19 pandemic strengthened Japan's engagement with like-minded partners on several multilateral forums to comprehend the common challenges human society faces. As the Asian member of the G7 grouping, Japan pledged to coordinate

---

[49] JICA brochure on 'JICA Assisted Health Sector Projects In India'. www.jica.go.jp.

[50] https://www.unicef.org/india/press-releases/japanese-assistance-support-government-india-strengthen-covid-19-vaccines-drive.

[51] 'Fact Sheet: Quad Leaders' Summit,' statements and releases, www.whitehouse.gov, September 24, 2021.

[52] India TV. (2021, January 29). India's vaccine production capacity is the best asset the world has today, says UNChief. https://www.indiatvnews.com/news/world/india-vaccine-production-capacity-best-asset-world-un-chief-antonio-guterres-681182.

necessary public health care measures in the March 2020 summit to protect people at risk from COVID-19.[53]

The Quadrilateral framework of the United States, Japan, Australia, and India emerged as a platform for these countries to leverage their shared norms for the Indo-Pacific region and answer the challenges from growing Chinese ambitions. The tragedy of the 2004 tsunami became an occasion to initiate a positive vision for their joint efforts. The COVID-19 pandemic, along with other concerns of climate change and security challenges, have given a renewed purpose to the institution.[54] QUAD agenda has gradually raised a discourse of being a 'force of global good'.[55] By its shared belief in a rules-based world order.

On these lines, the leaders from the QUAD framework met for their first summit virtually in March 2021. They agreed to integrate their efforts to expand safe, affordable, and effective vaccine production and make its access equitable, with the belief that 'no one is safe until everyone is safe'. The grouping agreed to combine their individual medical, scientific, manufacturing, financing and delivery, and development capabilities for these goals. Along with this, a joint expert working group was established to become a vehicle to implement these goals.[56] The expert group would investigate the vaccination program's planning, financing, production, and distribution functions and support and strengthen other international organizations, such as WHO, UNICEF, COVAX, GAVI, CEPI, G7, and ASEAN, in their health security efforts.

By the time the QUAD leaders' summit took place, India was the only partner actively supplying vaccines to other countries under its 'Vaccine Maitri' campaign. The quadrilateral partners realized that by mid-2021, China had supplied more than a billion vaccine doses to low and middle-income countries. Meanwhile, Japan had no domestically produced vaccine and no record of exporting COVID-19 vaccines. Australia was in a similar situation. But, both had rendered financial help to their neighbors to procure vaccines. The United States had provided some support to its neighboring countries. In contrast, India stood as an exception and had provided over 66 million vaccine doses to countries worldwide, but the second COVID-19 wave at home became an impediment[57] to this global outreach.

The QUAD's pledge to support the Indo-Pacific region comes in the wake of the realization that global commons have strong altruistic components, and previous experience of converging on global commons showed that it pays good dividends. In their first in-person meeting in September 2021, the grouping affirmed its support by providing 1.2 billion vaccine doses globally in addition to those facilitated through the COVAX initiative. In the joint statement, entitled "The Spirit of QUAD," the

---

[53] Joint Statement of the G7 leaders March 2020, Ministry of Foreign Affairs, Government of Japan.

[54] Abhijnan Rej (2021).

[55] Jyotsna Mehra (2022).

[56] Fact sheet titled 'Quad Leaders' Joint Statement: The Spirit of the Quad,' statements and releases from www.whitehouse.gov.

[57] Maiko Ichihara and Atsushi Yamada (2021).

member countries envisioned an effective vaccine distribution, making it the central goal of QUAD.

The arrangement arrived in the quadrilateral framework where India would carry out production, the United States and Japan would handle the finance, and Australia would help with the logistics of vaccine distribution. The details worked out in following format. India's Hyderabad-based drug maker Biological E would manufacture 1 billion doses of COVID-19 vaccines, including the one developed by Johnson & Johnson, by 2022 under the QUAD commitment. It was further decided that the US would help Biological E expand its COVID-19 vaccine output capacity, while Tokyo was to provide concessional loans to India to expand the production of vaccines for export. Australia would assist in vaccine supply while focusing on last-mile delivery support in Southeast Asia.

Despite the positive ambitions for health security, the targets could not be met as the grouping delivered only 79 million doses by December 2021.[58] In continuation of their commitment, the QUAD Joint Leaders' Statement of May 24, 2022, stated that Prime Minister Anthony Albanese of Australia, Prime Minister Narendra Modi of India, Prime Minister Fumio Kishida of Japan, and President Joe Biden of the United States celebrated the delivery of more than 670 million doses, including at least 265 million doses to the Indo-Pacific, including Thailand and Cambodia.[59] They also communicated their commitment to strengthening the global health architecture and pandemic prevention, preparedness, and response (PPR) and supporting economic and social revitalization in the Indo-Pacific region.[60] This was seconded, with their future aims of donating around $2 billion to 115 countries as aid and coordinating efforts through initiatives like the 'COVID-19 Prioritized Global Action Plan for Enhanced Engagement' (GAP) and the COVAX vaccine delivery partnership.[61] QUAD further pledged $524 million to the Coalition for Epidemic Preparedness Innovations (CEPI) to support efforts for the development of new vaccines and prevent and contain infectious diseases.

7.2 *Vaccine Diplomacy as a Strategic Tool*

While it is appreciated that the QUAD is focused on supporting the socio-economic revitalization of the Indo-Pacific, it falls short of China's efforts on these fronts. By the time the QUAD leaders met in Tokyo in May 2022, Beijing had donated around 1.9 billion doses to around 118 countries globally and an additional donation of around 240 million vaccine doses.[62] Chinese efforts have been followed by their discourse of labelling QUAD as an 'anti-China' reactionary grouping with a 'cold-war mentality'.

---

[58] Anuar and Hussain (2022).

[59] https://www.pm.gov.au/media/quad-joint-leaders-statement.

[60] Ibid.

[61] Quad Joint Leaders' Statement, May 24, 2022.

[62] Liu et al. (2022).

China's vaccine diplomacy as strategic reach can be understood by examining Cambodia, an ASEAN country that is central to the QUAD vaccine diplomacy discourse. In April 2022, Cambodia became the first country to receive the consignment of 325,000 doses of Made in India Covishield vaccines.[63] Nevertheless, these donations dwarf the Chinese delivery of around 18.7 million doses within seven months of the vaccine partnership agreed in January 2021. China's efforts made Cambodia one of the first countries in Southeast Asia to achieve a high vaccination rate.[64] Realizing other factors that affected this partnership, the Cambodian example highlighted the problems QUAD countries faced regarding repeated delays and hurdles in production issues that were not addressed effectively. The Chinese ambitions to achieve diplomatic victory through its 'Health Silk Road' were only interrupted by the growing concerns of several researchers about the effectiveness of Chinese vaccines.[65] These concerns helped the QUAD grouping push for its vaccines, such as Moderna and AstraZeneca, the production of which was carried out through a single facility at the Biological E firm in Hyderabad, India.

There has been a shift in the discourse of vaccine diplomacy. From being an "extending a hand" to combat the virus, vaccine diplomacy is increasingly perceived as a strategic counterweight to China's growing assertiveness in Asia. At the beginning of the pandemic, QUAD got an impetus to raise its stature above the complex regional security conundrums and regain its collective action identity. As vaccine diplomacy gained strategic currency, what followed was a diplomatic approach to facilitate vaccine supply chain resilience as a countermeasure to China's tactics.

The shortcomings of this minilateral must be realized by the member states as newer initiatives and goals are mapped in subsequent deliberations. The 2023 leaders' summit became a benchmark as the countries agreed to grow their vaccine partnership to a 'QUAD Health Security Partnership,' which aims to strengthen their coordination and cooperation to support health security in the Indo-Pacific region.[66]QUAD plans to broaden its agenda to include rapid response to pandemic-potential diseases, support for healthcare workers, surveillance and electronic health information systems, and coordination of outbreak responses. This would primarily involve initiatives like the QUAD pandemic preparedness exercise. Other areas of cooperation would include strengthening the vaccine data system and treatment capacity of core hospitals.[67] All this gives both India and Japan the opportunity to find more areas of alignment to liaise, collaborate, and cooperate within the QUAD and draw Australia and the US to deepen their commitment to the Indo-Pacific.

---

[63] https://www.mea.gov.in/pressreleases.htm?dtl/35188/Inaugural_delivery_of_Covid_Vaccines_under_the_Quads_Vaccine_Partnership.

[64] Wanyi Zhao and Mingjiang Li (2022).

[65] https://www.economist.com/graphic-detail/2022/04/19/how-chinas-sinovac-compares-with-biontechs-mrna-vaccine.

[66] Quad Leaders' Joint Statement, May 20 2023. www.whitehouse.gov.

[67] https://www.whitehouse.gov/briefing-room/statements-releases/2023/05/20/quad-leaders-summit-fact-sheet/.

8 Conclusion

The COVID-19 pandemic, which began in 2020, completely upended the global community and made the world reflect on major policy programs, both at domestic and international level. On the one hand, it brought out the inequalities amongst countries through issues like 'vaccine nationalism' while also opening newer avenues of cooperation for them to readjust the global governance framework. At the forefront of these efforts was the global health infrastructure, which was under scrutiny for even the most developed countries as their defaults became apparent.

Vaccine development and diplomacy raised the stature of India in the region and the world. India proved the vital link as it developed a low-cost vaccine and followed the policy of extending vaccine to its neighbors in the region, providing global visibility to the country's manufacturing capabilities. India's commitment to the international community, particularly the global south, was evident in its vaccine outreach. It has significantly helped India's aspiration to be a global power. Japan, on the other hand, struggled with vaccine development. Plagued with political uncertainties, bureaucratic dysfunction, and erroneous policy decisions, it faltered in vaccine roll-out and vaccination drive. However, as has been its policy, it came up with financial aid to support its neighboring nations.

The pandemic has brought focus on a new set of non-traditional security concerns, one of which is vaccination. Vaccine diplomacy provides a new arena of coordination and data sharing among pharmaceutical and biotechnology regulatory authorities across the Indo-Pacific region, thus helping to institutionalize the forum better. India and Japan, with their strength, have found yet another convergence. The pharmaceutical industries of both countries have an opportunity to converge on their strengths and seek out opportunities for vaccines and drugs to cure many diseases. The ultimate aim should be to go beyond their bilateral to provide effective solutions to the region. As the world rebalances itself, the approach of the QUAD in fostering the appeal of small states, island states, and ASEAN to target global commons works in tandem with India and Japan's vision for the region. Vaccine diplomacy stands as an illustration for future negotiations on other global commons.

References

Abhijnan Rej (2021) In historic summit quad commits to meeting key Indo-pacific challenges. The Diplomat

Amengual O, Tatsuya A (2020) COVID-19 pandemic in Japan. PMC-Springer Nature publications

Anuar A, Hussain N (2022) The quad and regional health security: implications and prospects for the Indo-Pacific. Policy Report, RSIS-NTU

Du L, Huang G (2020) Japan's bitter vaccine history creates Hurdles in COVID-19 fight. The Japan Times

Ghaznavi C, Daisuke Y, Takayuki K, others (2022) Factors associated with reversals of COVID-19 vaccination willingness: Results from two longitudinal national surveys in Japan 2021–2022, vol 27. www.thelancet.com

Hooda SK (2022) India in the global vaccine market prior to and during COVID-19: some structural issues. ISID Working Paper 250, August, https://isid.org.in/wp-content/uploads/2022/10/WP250.pdf

Hooda SK (2023) State of Vaccine manufacturing in india: structure, size, and competitiveness. Econ Polit Week 8(31)

Jyotsna Mehra (2022) An evolving Agenda for the Quad. The Henry L. Stimson Centre

Kazuki S, Taro K, Yasuharu T, Kenji S (2020) An open letter to Japan's new Prime Minister, vol 396. www.thelancet.com

Koop A (2020) Chart: tracking COVID-19 vaccines around the world. Visual Capitalist

Lahariya C (2014) A brief history of vaccines & vaccination in India. INDIAN J MED RES

Liu L, Huang Y, Jin J (2022) China's vaccine diplomacy, and its implications for global health governance. Healthcare (basel) 10(7):1276. https://doi.org/10.3390/healthcare10071276.PMID: 35885803;PMCID:PMC9317735

Maiko Ichihara, Atsushi Yamada (2021) Japan as an agenda Setter for Quad's Vaccine Diplomacy. The Diplomat

Mashal M, Yee V (2021) The newest diplomatic currency: COVID-19 vaccines. https://www.nytimes.com/2021/02/11/world/asia/vaccine-diplomacy-india-china.html

Ministry of Health and Family Welfare (2020) COVID-19 Vaccine Operational Guidelines. https://main.mohfw.gov.in/sites/default/files/COVID19VaccineOG111Chapter16.pdf

Odevall L, Hong D, Digilio L, Sahastrabuddhe S, Mogasale V, Baik Y, Choi S, Kim JH, Lunch J (2018) The Euvichol story—development and licensure of a safe, effective, and affordable oral cholera vaccine through global public-private partnerships. Vaccine 36(45):6606–6614

Pathak D, Magdalene AP (2022) COVID-19 Vaccine Development and Administration in India Health Dimensions of COVID-19 in India and Beyond (pp 129–154) https://doi.org/10.1007/978-981-16-7385-6_7

Ratcliffe R, Fonbuena C (2020) Inside the cruise ship that became a coronavirus breeding ground. The Guardian

Sayeed U, Hossain A (2020) How Japan managed to curb the pandemic early on: Lessons learned from the first eight months of COVID-19. J Global Health (JoGH) 10(2)

Sharma J, Varshney SK (2021) India's vaccine diplomacy aids global access to COVID-19 jabs. Nat India. https://doi.org/10.1038/nindia.2021.31

Sharun K, Dhama K (2021) India's role in COVID-19 vaccine diplomacy. J Travel Med 28(7):taab064

Shinzo Abe (2015) Japan's vision for a peaceful and healthier world. PMC-Lancet

Singh B, Singh S, Singh B, Chattu VK (2023) India's neighbourhood vaccine diplomacy during COVID-19 pandemic: humanitarian and geopolitical perspectives. J Asian Afr Stud 58(6):1021–1037. https://doi.org/10.1177/00219096221079310

Soma S, Tiberghien Y (2022) Japan Slams the Borders Shut on Omicron. East Asia Forum

Takenaka H (2021) Japan's fragmented COVID response: a systemic failure of national leadership. www.Nippon.com

Tomoko O (2023) Japan approves first domestically developed COVID-19 XBB vaccine. The Japan Times

Wanyi Zhao, Mingjiang Li (2022) China's COVID-19 Vaccine Cooperation with Southeast Asia: A Boost for the Health Silk Road Initiative. East Asian Affairs 2(2). https://doi.org/10.1142/S2737557922500097

Wingfield-Hayes R (2020) Coronavirus: Japan's mysteriously low virus death rate,' BBC News

Indian IT Industry and Innovation

Yukio Takeyari

Abstract The Indian IT industry is undergoing significant changes. India's leading IT services companies serve clients around the world and have grown into giant companies. Global MNCs are setting up R&D centers in India and accelerating the pace of developing new technologies. The surge in the number of start-ups is remarkable, with unicorns emerging one after the other. As global IT technology innovation and business shifts to emerging markets, there's a proactive effort by US and European companies to increase collaboration with the Indian IT industry. On the other hand, efforts by Japanese companies have been limited. This chapter looks at India's IT industry, which has continued to grow despite the pandemic, and how India is becoming a center of innovation as digitalization progresses. The chapter also examines the challenges Japan faces in the IoT/AI era and the potential for Japan and India collaboration to address and solve these challenges.

Keywords Indian IT industry · Bengaluru · R&D center · Start-ups · Innovation

1 Introduction

In October 2008, I was appointed as the head of Sony's software development center in India (Sony India Software Centre Pvt. Ltd.) and move to Bengaluru, which is known as India's Silicon Valley. Since joining Sony, I have been engaged in the development of computer products such as workstation and personal computer and has frequently visited Silicon Valley in the U.S. for work. I thought I understood very well that the center of the world's IT industry was the U.S. Speaking of India, I had an image of undertaking software development and support work in developed countries at a low price. After that, I was stationed there for seven years, and Sony's development center has nearly tripled to approximately 2,000 engineers, making it the largest development center for Sony outside of Japan. Although it is the larger

Present Address:
Y. Takeyari (✉)
India Japan Laboratory, Keio Research Institute at SFC, Keio University, Fujisawa, Japan
e-mail: takeyari@sfc.keio.ac.jp

R. Shaw and S. R. Choudhury (eds.), *India, Japan and Beyond*,
https://doi.org/10.1007/978-981-97-3282-1_10

among Japanese companies, many of the U.S., European, and even Korean companies are even larger. The Indian IT industry has grown and changed drastically more than I could have imagined. After coming back to Japan, I really wanted many people to know about the potential and importance of Indian IT industry, and in March 2018, I published a book in Japan titled "India Shift: Why are the world's top companies establishing R&D centers in Bengaluru?" (Takeyari 2018). Based on the latest industry data, this chapter introduces the drastically changing Indian IT industry and examines the possibility of collaboration between Japan and India.

2 Indian IT Industry Undergoing Drastic Change

Amid the acceleration of IT innovation worldwide, the Indian IT industry continues to expand steadily. According to NASSCOM (The trade association of Indian IT BPM industry), Indian IT-BPM industry's revenues have increased from $8 billion in 2000 to $245 billion in 2023, a 30-fold increase (see Fig. 1). India's nominal GDP has grown about 8 times in 23 years, but it has grown much faster than that. About 80 percent of that, or $194 billion, comes from overseas. In other words, it is a global business that deals with the world. Initially, Indian IT company took on the downstream processes (coding, testing, maintenance) of software development in developed countries, but now also responsible for the upstream processes (requirements analysis, design, etc.) and further whole system, service and products. In addition, the emergence of disruptive new technologies such as big data, AI, IoT, and blockchain is accelerating its growth. In 2022 (April 2021 to March 2022), during the coronavirus pandemic, there was a rapid increase of more than 15%. While many companies around the world have suffered business damage due to the pandemic, India's IT industry is accelerating its growth due to the growing need for global digitalization. It is projected to grow to $500 billion by 2030.

This is because the Indian IT industry, which works with Silicon Valley companies and has an unusually large number of young high-level IT talents, is overwhelmingly quick to catch up with these new technologies. Its IT technology is about to approach Silicon Valley in the United State.

The center of this upheaval is the South Indian city of Bengaluru, which is now spreading rapidly to other Indian cities.

As of 2023, there are approximately 5.4 million direct employments in the Indian IT-BMP industry. This is overwhelming the scale of about 1 million IT talents in Japan (Ministry of Economy, Trade and Industry 2019). The reason for such a large scale is mainly to undertake large-scale work for the U.S. and European companies. Looking at India's IT industry exports by region in FY2019, the U.S. is by far the largest at 62%. If we add the UK (17%) and Europe (11%), these regions account for about 90%. Japan is included in the "other" category of less than 2%, but if you look at Japan alone, it is less than 1%, and the amount is less than $1 billion. At present, it can be said that the collaboration of the Indian IT industry by Japanese companies is extremely limited (see Fig. 2).

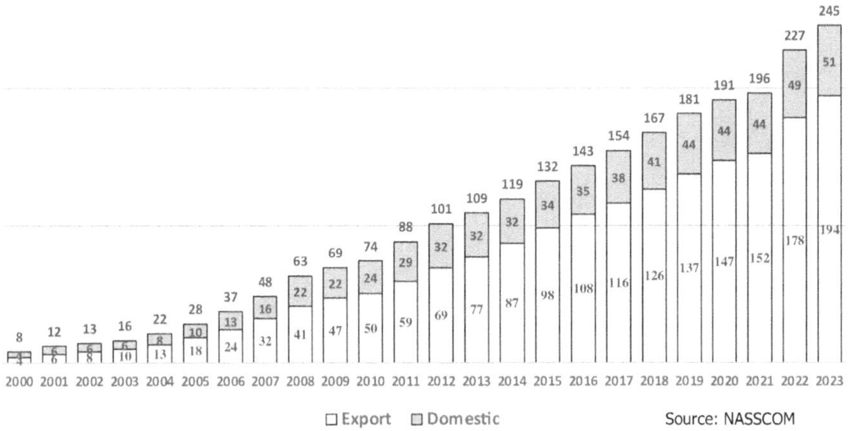

Note: Figures for each year are from April of the previous year to March of that year.

Fig. 1 Indian IT-BPM industry's revenues

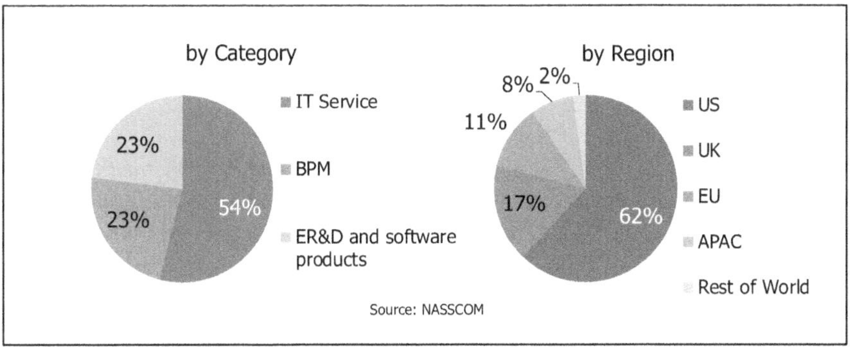

Fig. 2 Exports of Indian IT-BPM industry's revenues (FY2019)

3 Indian IT Services Companies Growing Giant

The growth of the Indian IT-BPM industry is supported by many Indian IT service companies. It continues to expand in scale and become huge. Figure 3 compares the number of employees of Indian IT services companies and multinational IT services companies worldwide. There are five Indian IT services companies employing more than 100,000 people, with Tata Consultancy Services (TCS), the largest, employing around 610,000 employees. Bengaluru is home to the headquarters of Infosys, the second largest company in the industry, with more than 310,000 employees, and Wipro, the third largest company with about 250,000 employees. These Indian IT services companies employ an estimated 80–90% of their employees in India. On the other hand, Accenture, a multinational global IT services company, employs

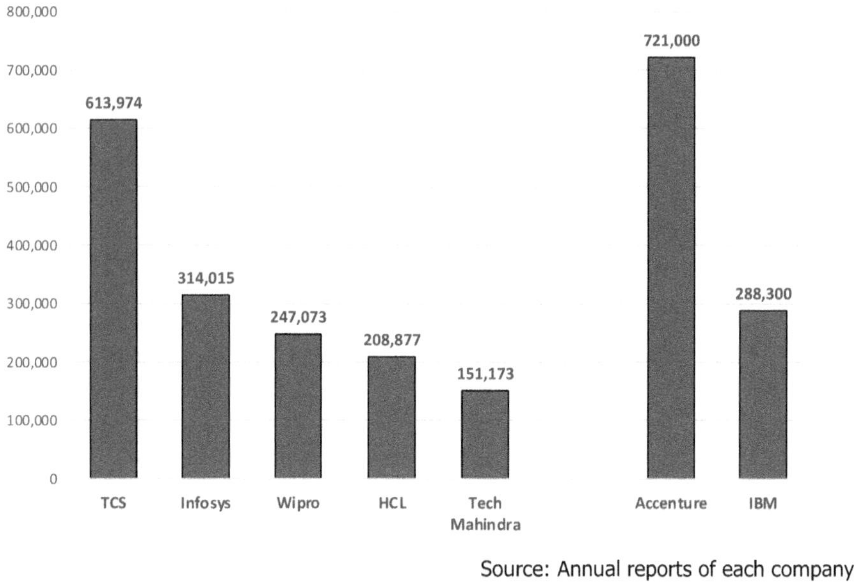

Fig. 3 Number of employees in Indian IT services companies and multinational IT services companies at the end of 2022

about 40% of its 720,000 employees in India, and IBM employs about 30% of its 290,000 employees in India, and the ratio is increasing year by year. Indian talents are making a significant contribution to the world's IT service business. In 2019, TCS surpassed IBM in market capitalization to become the world's second-largest IT services company after Accenture.

4 MNC's Development Center in India Transforming into Strategic Center

The number of multinational companies with their own development centers in India is increasing year by year, and the scale of these companies is also expanding. It is safe to say that most major companies in the U.S. and Europe have set up centers. Many of them are the largest center outside of their home country. These in-house development centers in India are called GCC (Global Capability Centers). Their role varies from company to company, but they often relate to strategic areas such as R&D and product development.

In FY2015, there were about 1,000 companies, but in FY2023, the number has increased to 1,580, employing 1.66 million people. In FY2021-FY2022 alone, 150 companies have set up new centers. It is worth about 24% of India's IT industry's exports, or $46 billion (NASSCOM (2023-1) 2023). The industry is not limited to the world's leading IT companies, but also covers a wide range of industries, including telecommunication, semiconductor, electronics, automobile, industrial machinery, retailer, and BFSI. By region, 1,025 (65%) are in the U.S., 434 (27%) are in EMEA, and 121 (7.7%) are in APAC. 49 Japan companies (3%) are included in APAC. Many companies have multiple locations, but 30% of them are in Bengaluru. Table 1 is part of a list of global companies with development centers in Bengaluru.

Among European company, the presence of the German company's GCC is significant. In Germany, the public and private sectors are working together to promote Industry 4.0, and the major companies such as Bosch, Siemens, Mercedes-Benz, and SAP have a large GCC in Bangalore, which plays a strategic role and conducts various research and development for Industry 4.0.

Many Japan companies conduct R&D and product development in Japan and consider India as a market. However, Western companies are already developing global products at their development centers in India. In addition, it is being used as a strategic center for business in emerging countries including India.

During the coronavirus pandemic, several Japan companies have begun to establish GCCs. These include Mitsubishi UFJ Bank (established in 2020, location: Bengaluru), Raksul (2020, Bengaluru), Mercari (2022, Bengaluru), Suzuki (2022, Delhi), and PayPay (2022, Delhi).

Apart from the existing GCC, Sony (Bengaluru, 2022) and Fujitsu (Bengaluru, 2022) have established new research laboratories.

In addition, Hitachi and Panasonic acquired GlobalLogic and Blue Yonder, respectively, and about half of the employees of the acquired companies were employed at their centers in India, so the number of Indian talents employed by each company has increased rapidly.

Table 1 MNC's GCC in Bengaluru

Software, Internet, IT equipment Microsoft, Google, Amazon, Oracle, SAP, Adobe, HP, Dell EMC, NetApp, Akamai, Ebay, Intuit, Cadence, Synopsys, Citrix Systems, VMware, Verisign, McAfee, LinkedIn, Rakuten **IT services, consultant** IBM, Accenture, Capgemini, Cognizant, NTT Data **Semiconductor** Texas Instruments, Intel, Qualcomm, AMD, Nvidia, Arm, ST Micro, Broadcom, Freescale, Cypress **Telecomm/network equipment, telecomm carriers** Cisco Systems, Juniper Networks, Nokia, Ericsson, Huawei, Avaya, AT&T, British Telecom	**Electronics, automobile, industrial machinery** Sony, Toshiba, Hitachi, Yokogawa, Samsung, LG, Philips, Bosch, Mercedes-Benz, Honeywell, GE, Siemens, Boeing, Airbus, Rolls-Royce **Retailer** Walmart, Target, Tesco, Victoria Secret, Loews, JC Penny **BFSI** Goldman Sachs, Morgan Stanley, JP Morgan Chase, Fidelity, Wells Fargo, Bank of Australia-New Zealand, Visa, CME Group **Others** 3 M, Exxon Mobil, Cargill **Unicorn (outside of India)** Uber, Airbnb, Grab Taxi, GO-JEK

5 Rapidly Growing Start-Ups in India

According to a report by NASSCOM, between 25,000 and 27,000 start-ups were founded in the decade from 2012 to 2022. Among them, more than 3,100 companies (12% of the total) are using the latest technologies called deep tech: AI (64%), IoT (19%), big data (18%), blockchain (9%), AR/VR (7%), and 3D printing (4%).

Not only are there many start-ups, but as of the end of 2022, the number of unicorns (a privately held start-up company valued at over US$1 billion) was 89, behinds the U.S. (619) and China (312) (NASSCOM (2023–2) 2023). It has already overtaken the UK (57), Israel (44) and Germany (30). There are only six unicorns in Japan. India is already the world's third-largest start-up nation.

After 2020, the coronavirus pandemic seems to have taken a toll on many start-ups globally, but the number of unicorns in India has skyrocketed, with a record growth rate in 2021(see Fig. 4).

A breakdown of Indian unicorns and potential unicorns (start-ups founded between 2012 and 2022 and with total funding more than $50Mn +) shows that 50% are B2B businesses and 52% are global expansion, and many of them are growing not only for the huge Indian domestic market, but also for the global market.

The background to the increase in start-ups is the existence of many venture capitalists, angels, and individual investors, as well as the establishment of an ecosystem that supports entrepreneurs. More than 520 (FY2020) start-up support organizations have been established. There are various forms such as universities, companies, and governments. Most of premium universities have incubation centers. Many GCCs have also started accelerator programs for the purpose of open innovation.

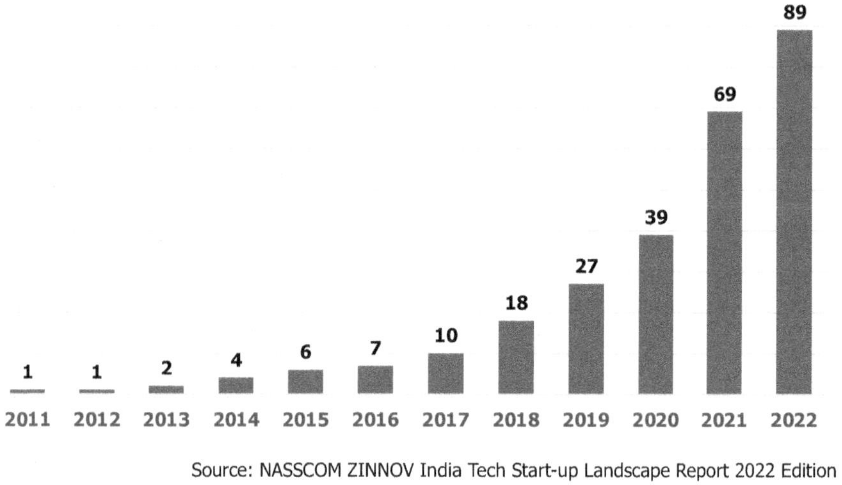

Fig. 4 Number of unicorns in India

6 India's Accelerating Digitalization

With the rapid spread of smartphones, India's digitalization is accelerating. This is due to the presence of the Indian IT industry, which is growing by incorporating the world's digital technology and the emergence of start-up companies that provide various smartphone-based services. In addition, the impact of the Indian government's "Digital India" policy is significant. Among them, the "Digital ID System–Aadhaar" and the "India Stack" deserve attention.

Aadhaar, is a project to assign 12-digit ID numbers to 1.3 billion people living in India. Originally, India did not have a personal identification number, such as a resident card or social security number, and there was no way to prove one's identity. As a result, the 400 million poor people were unable to access government services such as welfare. In fact, they couldn't even open a bank account. Therefore, the government started a project with the idea that by disseminating this ID, it would solve the problem and provide various services.

Aadhaar uses biometric technology to authenticate individuals. When you register, you will receive a 12-digit number by providing a photo of your face, 10 fingerprints on your hands, and information about the iris of your eyes. And if you need to authenticate your ID at a bank or government office, you can prove your identity by simply presenting a 12-digit number and placing your finger on the fingerprint machine at the counter.

Initially, registration was not compulsory, but a voluntary system was adopted, and registration centers were set up in various parts of India. In particular, the registration was started in areas where there were many poor people, and at first, it seems that many people did not know much about the meaning of it and registered.

In September 2010, this project began issuing ID numbers, steadily increasing the number of registrants, and finally exceeded 1 billion in April 2016. It is said that the number of registrants has exceeded 1 billion in five and a half years since the start of registration, which is faster than the spread of Facebook and Android OS as a digital platform. Currently, it is mandatory to present a number for tax filing, but as of the end of 2022, more than 94% of all Indians were registered.

On the other hand, Japan's national ID system, "My Number Card", started in October 2015, and as of April 2021, five and a half years later, about 36 million cards were issued, and the issuance rate was 28.3%. As of June 2023, it is about 70%.

The rapid spread of the Indian Aadhaar was because there was no way to prove the individual. In the case of Japan, there are official documents that can prove an individual, such as a copy of the family register, a certificate of residence, and a driver's license, so there is no problem even if you do not have a My Number Card. On the other hand, in the case of India, there is no other way than Aadhaar, and it can be said that obtaining it has led to various benefits and accelerated its spread. This is exactly the case where mobile phones and smartphones have spread all at once because fixed-line phones are not as widespread as in Japan, but because there is nothing, dynamic changes can be made.

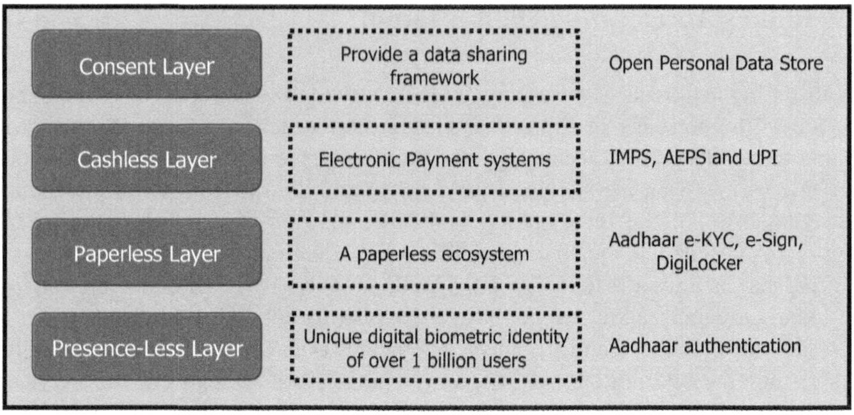

Fig. 5 India stack

Based on Aadhaar, a software architecture called "India Stack" has been developed on the cloud. The India Stack consists of four independent layers. It is called the "Presenceless Layer", "Paperless Layer", "Cacheless Layer", and "Consent layer", and all the APIs (Application Programming Interfaces) of each layer are exposed. Not only governments, but also companies and start-ups can develop various services (see Fig. 5).

Already, services such as "e-KYC" for electronic identity verification, "e-Sign" for signatures, and "DigiLocker" for storing documents have already been launched. In addition, NCPI (National Payments Corporation of India) has released the Unified Payment Interface (UPI), an interface for unified communication between banks. This UPI is already used in many smartphone applications and is used for payments. It has now exceeded the total volume of credit and debit card transactions. "Consent Layer" provides a mechanism in which personal data is held by individuals and individuals provide permission to use it as necessary.

The goal is to realize a "presenceless, paperless, and cashless society". There is no need to go to government offices or banks, all necessary documents are digitized, and cash is not used. A country that did not have a personal ID a few years ago is now transforming into a country that provides the world's leading and advanced services. In fact, the lives of the people of India are changing dramatically.

In addition, the MOSIP (Modular Open Source Identity Platform) project has been launched to open-source this technology and provide it to the world free of charge. Countries in Southeast Asia, Africa, South America, and elsewhere are already interested, and 11 countries are introducing or considering introducing it. Countries that have been slow to digitalize as emerging countries will accelerate digitalization at once. It is highly likely that it will be the creation of global innovation originating in India.

India is a country rich in diversity beyond imagination in terms of culture, religion, language, food, rich and poor. On the other hand, there are many social issues and problems, and there are many infrastructure and cost constraints to solve them. On the other hand, there is an IT industry that have acquired cutting-edge IT and digital technologies, and there are many high-level IT talents who support it, and many of them have an entrepreneurial spirit, and the number of start-up companies is increasing rapidly. Solving problems in India is likely to contribute to solving problems in the world, especially in other emerging countries, and it can be said that India is transforming into a global innovation powerhouse.

7 Possibility of Japan-India Collaboration for the IoT/AI Era

The need to accelerate Japan's digitalization has been pointed out in several reports.

In the "International Comparison of Labor Productivity 2022", Japan ranks 27th out of 37 OECD member countries, and continues to be at the bottom of the list among the seven major industrialized countries (Japan Productivity Center 2022).

In the "World Digital Competitiveness Ranking 2022", Japan ranked 29th out of 63 countries, the lowest ranking ever (IMD 2022).

According to the Ministry of Economy, Trade and Industry's report "The Cliff of 2025" (September 7, 2018), companies need to promote digital transformation (DX) to grow and strengthen their competitiveness in the future, but if DX cannot be realized by renewing existing systems, it is pointed out that a large amount of economic loss may occur after 2025 (Ministry of Economy, Trade and Industry 2018).

On the other hand, it is also predicted that there will be a shortage of 450,000 IT personnel required for digitalization by 2030 (Ministry of Economy, Trade and Industry 2019).

To drastically change this situation, I believe that collaboration between Japan and India in the IT field is important, but at present, there are few Japan companies that utilize Indian IT service companies or have their own development centers in India, and collaboration between Japan and India in the IT field is limited.

The reason for this is a lack of understanding of India, especially the drastically changing Indian IT industry. It is also true that there is a problem with the language of English in collaboration with India. I also think that there is a unique way of working in Japan. The Indian IT industry has actively adopted Western-style development methods, and in a sense, it can be said that it is close to global standards. To keep up with the speed of IT innovation, Japan needs to be aware and prepared to change itself.

With the advent of the IoT/AI era, there is great potential for collaboration between Japan and India. Japan still has an overwhelming strength in design and manufacturing. On the other hand, the shortage of IT personnel required is a major problem. In addition, because of the advancement of social infrastructure, it is becoming difficult to expand globally with products and services for Japan because they are too special. It will be further accelerated by IoT/AI technology. On the other hand, India has acquired cutting-edge IT technology through collaboration with the U.S., which is already an advanced IT country, and has created abundant IT talents. There is also a huge Indian market.

Until now, Japan's collaboration with the Indian IT industry has been limited to small-scale offshore projects aimed at cost reduction. Unfortunately, this model doesn't seem to have produced much success. Nowadays, it is necessary for all companies to put IT technology at the center and innovate. To that end, strategic collaboration with India has great potential.

Partnering with Indian IT services companies to develop solutions for the Indian and global markets. Collaborating Indian start-up companies to aim for open innovation. There are various possibilities, such as setting up a company's GCC, hiring top IT talents in India, and creating new products and services that combine the company's own technology with the latest IT technology.

India is one of the few countries in the world that is an advanced IT country despite being an emerging country. Various innovations originating in India are also beginning to occur. I believe that the combination of Japan's experience and wisdom and the Indian IT industry will be a powerful strategy for Japanese companies to survive in the era of global competition.

8 Conclusion

In 2018, the Japan-India Digital Partnership Agreement was signed between the governments of Japan and India, and various movements are taking place, such as the establishment of a Japan-India Startup Hub in Bengaluru.

To further promote Japan-India cooperation in the future, I believe it is important to increase people-to-people exchanges and deepen mutual understanding.

According to the Ministry of Foreign Affairs' Survey of the Number of Japanese Residents Abroad (2022), the number of Japanese living abroad (about 1.31 million) is the largest in the U.S. (about 420,000), China (about 100,000), Australia (about 95,000), and India is only 8,145. On the other hand, the nationality of foreign residents (about 3.08 million) living in Japan is Chinese (about 760,000), South Korea (about 410,000), Vietnam (about 490,000), and India is about 44,000 (1.4%), so it can be said that there are very few people-to-people exchanges between Japan and India (Ministry of Justice 2022).

In addition, the number of foreign students in Japan from abroad is approximately 231,000 (as of May 1, 2022), with China (104,000), Vietnam (37,000), and Nepal (24,000) in that order, and India with 1,532 (0.7%) only (Japan Student Services Organization 2022). Meanwhile, about 269,000 people from India are studying in the U.S. (Open Doors Report 2022–2023). International students from India tend to prefer English-speaking countries, but even in non-English-speaking Germany, 43,000 students (Federal Statistical Office of Germany, 2022–2023) and about 10,000 students study in China (August 2023), and the number of students studying in Japan is extremely small.

In view of this situation, I believe that it is necessary for Japan to actively make efforts to increase people-to-people exchanges, including those involving international students. To increase exchanges in the field of technology with India, mutual exchanges between students, researchers, and engineers are extremely important. India-Japan laboratory of Keio Research Institute at SFC also must play a major role.

References

IMD (2022) World digital competitiveness ranking 2022 (28th September 2022). https://www.imd.org/centers/wcc/world-competitiveness-center/rankings/world-digital-competitiveness-ranking/. Accessed 30 Nov 2023

Japan Productivity Center (2022) International comparison of labor productivity 2022 (December 19, 2022). https://www.jpc-net.jp/research/assets/pdf/report_2022.pdf. Accessed 30 Nov 2023

Japan Student Services Organization (2022) International student enrollment survey results. https://www.studyinjapan.go.jp/ja/_mt/2023/03/date2022z.pdf. Accessed 30 Nov 2023

Ministry of Economy, Trade and Industry (2018) D X report~overcoming the IT system "2025 Cliff" and full-scale development of DX~(September 7, 30). https://www.meti.go.jp/shingikai/mono_info_service/digital_transformation/20180907_report.html. Accessed 30 Nov 2023

Ministry of Economy, Trade and Industry (2019) Survey on the supply and demand of IT human resources (overview) (April 2019). https://www.meti.go.jp/policy/it_policy/jinzai/gaiyou.pdf. Accessed 30 Nov 2023

Ministry of Foreign Affairs (2022) Survey statistics on the number of japanese residents abroad reiwa 4th year edition (as of October 1, 2022). https://www.mofa.go.jp/mofaj/toko/tokei/hojin/index.html. Accessed 30 Nov 2023

Ministry of Justice (2022) Statistics on foreign residents December 2022. https://www.e-stat.go.jp/stat-search/files?page=1&layout=datalist&toukei=00250012&tstat=000001018034&cycle=1&year=20220&month=24101212&tclass1=000001060399&stat_infid=000040068461&tclass2val=0. Accessed 30 Nov 2023

NASSCOM (2023–1) (2023) GCC 4.0 India redefining the globalization blueprint. https://nasscom.in/knowledge-center/publications/gcc-40-india-redefining-globalization-blueprint. Accessed 30 Nov 2023

NASSCOM (2023–2) (2023) India tech start-up landscape report 2022 Edition. https://nasscom.in/knowledge-center/publications/nasscom-tech-start-report-2022-rising-above-uncertainty-2022-saga. Accessed 30 Nov 2023

Takeyari Y (2018) India Shift: Why are the world's top companies establishing R&D centers in Bengaluru? PHP Research Institute

Rural and Social Innovation: India Japan Collaboration

Kotaro Kataoka

Abstract This chapter examines rural innovation in India through the Suzuki Innovation Centre (SIC), a collaborative platform initiated by Indian Institute of Technology (IIT) Hyderabad and Suzuki Motor Corporation in 2022. Despite the rural population ratio having constantly decreased in the world, the actual number of rural populations has surged, indicating a need for innovations to enhance rural life. SIC's activities, symbolized "Engage, Exchange, and Innovate", involve short-term projects to accumulate diverse experiences and address social issues in an agile manner. This approach positions SIC as a complementary force in the broader open innovation landscape. SIC uniquely views India as a source of untapped inspiration rather than merely a vast market. While rigorous evaluations are the future tasks, SIC has gained insights into rural society dynamics through real-world projects, establishing a broad network. The chapter presents SIC as a promising initiative contributing to the evolving perspectives of rural innovation and Indo-Japan collaboration.

Keywords Rural and social innovation · Indo-Japan collaboration · Academia-Industry collaboration · Agile development · Cross-cultural exchange

1 Introduction

While the ratio of the rural population to the total population has constantly decreased from 82.1% (in 1960) to 64.3% (in 2022) over several decades, the rural population itself has increased from 366,021,680 (in 1960) to 908,804,812 (in 2022) (World Bank 2018). These statistics indicate that innovations in rural India have a lot of opportunities to improve daily life, and technologies and social entrepreneurship play an important role. Suzuki Innovation Centre (SIC) is a unique initiative jointly founded by Academia (IIT Hyderabad) and Industry (Suzuki Motor Corporation)

K. Kataoka (✉)
Department of Computer Science and Engineering, Indian Institute of Technology Hyderabad, Kandi, Sangareddy 502284, Telangana, India
e-mail: kotaro@cse.iith.ac.in

R. Shaw and S. R. Choudhury (eds.), *India, Japan and Beyond*,
https://doi.org/10.1007/978-981-97-3282-1_11

aiming for establishing an India-Japan collaboration platform. Rural Innovation is one of the important topics for SIC. This chapter takes SIC as a case study for attempting to trigger rural innovation and reports its multiple activities from the ground-level explorations of social issues and grassroots innovation in rural India, and agile solution development.

2 Suzuki Innovation Centre as an Open Innovation Platform

IIT Hyderabad and Suzuki Motor Corporation have had a long relationship since 2009 as a part of "IITH Consortium" which was established in the Ministry of Foreign Affairs (MOFA), Government of Japan (MOFA 2008). While IIT Hyderabad and Suzuki have initiated R&D collaboration projects, both organizations agreed to enable an initiative Suzuki Innovation Centre (SIC) in 2022 that focuses on open innovation through the collaboration between India and Japan (SIC 2022). The activities of SIC are solely for the social good and the facilitation of Indo-Japan collaboration with various stakeholders beyond IIT Hyderabad and Suzuki including industries, universities, and startups from both India and Japan.

SIC has three categories of activities: Engage (involving future collaborators), Exchange (sharing values and synergy between India and Japan), and Innovate (bringing balanced growth in rural India and society). All the projects in SIC fall in any one or a combination of these categories and initiate with a short duration of up to 1 year or even shorter. The motivation for keeping each project short is to accumulate the unique know-how and thought process as SIC through executing as many projects as possible in an agile fashion and acquiring both (1) depth of experience including failures and best practices in rural Innovation and (2) breath to address broad social issues and futuristic topics in India and Japan. This way of executing the projects helps SIC to be complementary to other broad stakeholders on open innovation.

3 Grassroots Innovations Pilot Project

Grassroots Innovations Pilot Project (GRIPP) is a trial for accelerating and upscaling innovation in rural areas through interconnecting grassroots entrepreneurs, technology and business experts, and markets. GRIPP also serves as a trial for SIC itself in terms of learning and accumulating the best practices of rural innovation through real-world experience.

One of the difficulties in supporting rural innovation is identifying entrepreneurs to be a part of the project. SIC collaborates with Gujarat Grassroots Innovation Augmentation Network (GIAN) (GIAN 2023) for sourcing the potential entrepreneurs

Fig. 1 SIC team visiting the GIAN office in Ahmedabad, Gujarat, and attending the exhibition of grassroots innovation products held at the GIAN office

to collaborate (Fig. 1). GIAN has a database of more than 10,000 grassroots entrepreneurs. SIC selected 6 potential rural entrepreneurs working in the agricultural domain through the rigorous process out of 70 high-impact candidates found in the GIAN's database.

Through the collaboration and interaction with the selected entrepreneurs (Figs. 2 and 3), GRIPP has reconfirmed the several difficulties that are likely typical, and well-known, but hard to practically realize without directly working with them. These difficulties were found in the following domains but not limited to: quality and functionality of product, scalability, and sustainability in product manufacturing, finance for a business in the premature stage, sales and marketing, after-sales support, the geographical disadvantage of "rural", language barrier due to the many diverse languages in India.

One key uniqueness of GRIPP is the involvement of technology experts from IIT Hyderabad and industry to practically improve the innovative solutions developed by grassroots entrepreneurs rather than simply operating GRIPP as a networking platform. For example, the Design department of IIT Hyderabad is providing technical support to improve the product design of a rice transplanter in terms of its competitive features, ergonomics, and efficiency. In further stages, GRIPP can also seek industrial involvement by Suzuki to leverage its mechanical and production expertise.

As of today, the quantum and quality of GRIPP's social impact as well as its contribution to rural innovation have not been rigorously evaluated and consolidated. However, GRIPP expanded SIC's reachability to the wide range of rural

Fig. 2 A grassroots entrepreneur demonstrating his solution, Groundnut Digger, at Rajkot in Gujarat, India

entrepreneurs, and villages in various regions in India. In addition, the know-how of operating GRIPP can be used by SIC in day-by-day and future activities.

4 Agile Projects

Agile Projects are collaborative programs where Japanese young software developers and Indian students form a team, explore social issues in rural villages, identify the problem, and solve it together through software development and real-world deployment. As "Agile" tells in the project name, each project is concluded within 4 months of a reasonably short duration. This section shares two programs: Water Quality Assessment and Career Guidance and Mentorship for Rural Kids.

Fig. 3 A grassroots entrepreneur demonstrating his solution, a Mini tractor at Amreli in Gujarat, India

4.1 Water Quality Assessment Project

Project members from Japan and India visited villages near IIT Hyderabad and found out the fact that the aquaculture using natural ponds in rural villages is suffering from uncertain and low yield. The team identified the work items in the program to be an IoT-based Water Quality Assessment System, a Knowledge Center, and developed the IoT Device by integrating various sensors to monitor water temperature, acidity, etc. This program has also collaborated with the local fisheries society so that the program can be conducted with proper approvals by the local authority and awareness by the local community (Fig. 4).

This project developed an IoT Device, and a Web Application to monitor the sensor data about the water quality and share it with the local fisheries society in real time. The sensor nodes were connected through the mobile data connection, and the sensing data was periodically uploaded to the cloud server (Fig. 5). The visualized data were provided to the local fisherman community to understand the water quality in the target pond (Fig. 6).

4.2 Kimochi Project

Kimochi Project aims to construct a career guidance and mentorship platform for rural kids ('Kimochi' refers to 'Feeling, Emotion' in Japanese). The project members visited schools in rural villages and realized that career guidance is hardly available

Fig. 4 Project members visited a pond at an aquaculture site and discussed the issues and difficulties there with a local fisherman community

Fig. 5 An actual sensor node that was deployed at an aquaculture site

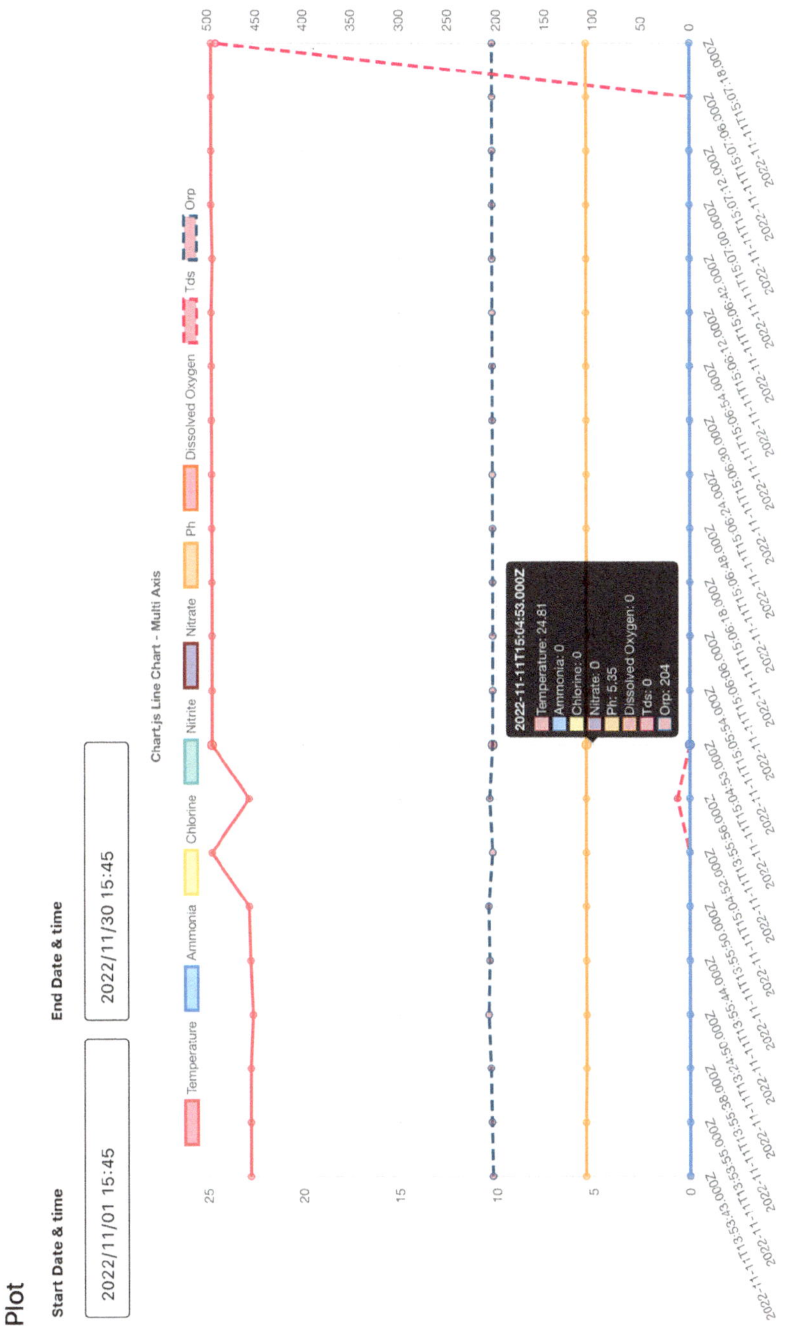

*Right y-axis is for TDS and ORP and left y-axis is for all other.

Fig. 6 The data visualization of the sensor data

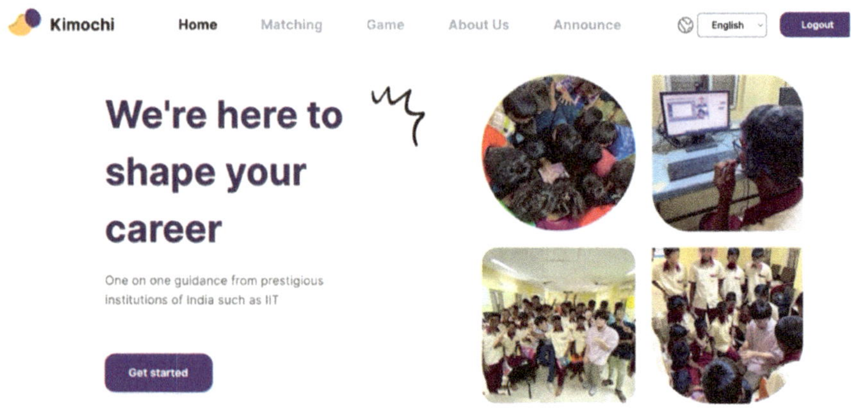

Fig. 7 The web UI of Kimochi platform

due to the lack of career education and as well as role models that school kids can refer to. Therefore, even though school kids can tell their dreams, they have difficulties figuring out or imagining their career path.

"Kimochi" is an online platform that facilitates career guidance and mentorship among university students as mentors and public junior high to high school students in Indian villages (Fig. 7). This platform has been developed as a solution for the limitation of interaction between mentors and mentees that are long geographical distance between universities and schools, language barrier, lack of continuity of mentoring program, lack of role models.

Kimochi project collaborates with the National Service Scheme (NSS) at IIT Hyderabad. The objective of the project is alleviating the broad difficulties of in-person career guidance, and enabling online interaction can effectively reinforce career guidance among university students and rural school students. The UI/UX is fine-tuned so that rural students can intuitively navigate themselves on the Kimochi application and enjoy using it without needing much technical assistance to operate it (Fig. 8).

In case that personal guidance is not available due to insufficient equipment, Kimochi platform also supports "Home Room" interaction where multiple students from the university and rural schools can (1) share dreams, (2) know cultures, and (3) learn subjects with customized contents for Home Room. The online platform has already been deployed among IIT Hyderabad, NIT Delhi, and several rural schools in 2023 August (Fig. 9), and the career guidance has continued weekly basis with active participation from both mentor and mentee groups.

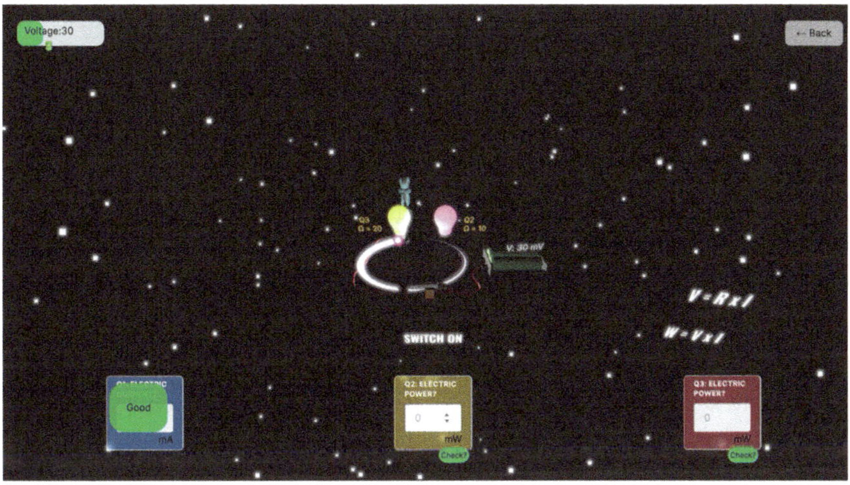

Fig. 8 The gamified UI of a science quiz

Fig. 9 Kimochi project members demonstrated the first release of Kimochi app in a rural school

5 Learn in Bharat

Awareness is one of the most important factors to initiate rural innovation. At the same time, sharing the awareness among broad stakeholders is equally important especially because any rural innovation needs diverse perspectives and inspirations to look into the social issues, collective efforts that can happen through the collaboration among multiple stakeholders, many potential stakeholders have not been aware of the issues and opportunities through rural innovations.

Learn in Bharat (LIB) is an intensive study tour in India for Japanese university students and working professionals. LIB provides an exact answer to "where to begin with for social innovation?" to the participants. Particularly, LIB aims to (1)

expose the Japanese participants to the front line and reality of Indian society and various ecosystems, (2) encourage entrepreneurship in India as well as (3) explore a new business of very high social impact. In the first program in August 2023, the LIB program visited the four major cities in India, Delhi (Fig. 10), Ahmedabad (Fig. 11), Hyderabad and Bangalore, and conducted the various modes of interaction with broad stakeholders. These stakeholders are from Academia (IIT Hyderabad and IIM Ahmedabad), Industry (Suzuki, Maruti Suzuki, KAI India, and Rakuten), Government Organization (JETRO New Delhi Office), Rural Villagers in Ahmedabad, Startup Ecosystem (NASSCOM Bangalore), and many more those are connected through Suzuki Innovation Centre.

Unlike an ordinary study tour, in LIB the participants stayed one night in a rural village to conduct the workshop there after exposing themselves to the actual rural life and interactions with the rural people (Fig. 12). The selection of the village was made so that the village itself is not touristy, and foreigners are safe and welcomed to visit and stay.

The important milestone for all the participants was to develop an entrepreneurial idea throughout the program, present it in the final business plan pitch with the presence of the examiners from IIT Hyderabad and Suzuki Motor Corporation. The overall feedback of LIB was positive, and SIC formed an alumni community for the LIB participants so that they could always stay in touch with SIC and be a part of any relevant event or activities in SIC.

Fig. 10 The Delhi session to explore "success stories of Japanese companies based in India". This session included the address by Mr. Kawazu (Deputy chief of mission, Embassy of Japan in India), Mr. Takashi Suzuki (Chief Director General, JETRO New Delhi), and Mr. Hisashi Takeuchi (Managing Director, Maruti Suzuki)

Fig. 11 The IIM Ahmedabad session covered the lectures on "Innovation from India". The LIB participants also interacted with startup founders in the Centre for Innovation Incubation & Entrepreneurship (CIIE) at IIM Ahmedabad

Fig. 12 LIB participants visited a village in Ahmedabad, and interacted with people in the village to understand the real rural life

6 Seed Innovation Bootcamp

Seed Innovation Bootcamp (SIB) is a trial program to intensively nurture young entrepreneurs by ideating a zero-to-one (0 to 1) solution for social entrepreneurship from scratch to its prototype. SIB was conducted from July to September 2023 as

Fig. 13 The SIB kick-off meeting facilitated by the faculty mentors from IIM Ahmedabad

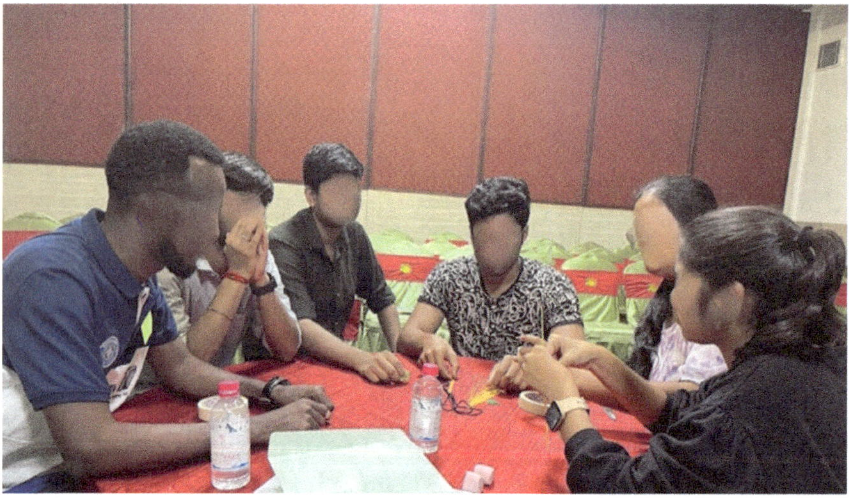

Fig. 14 The group discussion among the young entrepreneurs in the SIB kick-off meeting

a collaboration with two professors from the Indian Institute of Management (IIM) Ahmedabad as well as nine startup founders who are already successful in India (Figs. 13 and 14). 15 young enthusiastic entrepreneurs were selected to participate in SIB out of a total of 74 applications. The selected entrepreneurs included two international members from Japan and Afghanistan.

SIB conducted various activities during the three-month program so that the participants could acquire the theoretical and practical understanding of entrepreneurship:

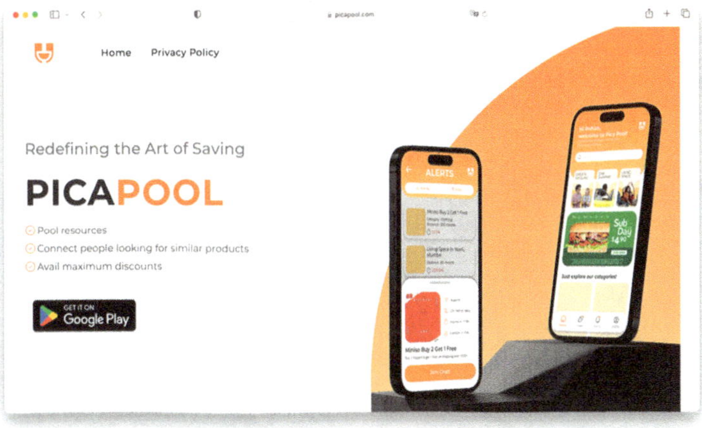

Fig. 15 The prototype of "PICAPOOL" mobile application is available online

"Immersion" program at IIM Ahmedabad, Pitch and Network Event at T-Hub, Mid-term and Final Demo of the actual business plan. Also, the participants continuously worked on the process of developing a business model for a startup: Pre-Customer Discovery, Problem Statement Validation, Customer Discovery, Idea Redefinition, Solution & On-paper Prototyping, Customer Validation, Product Market Fit, etc. In addition, SIB organized Weekly Founder's Talk to cover various entrepreneurial topics throughout the program period so that the participants are fully focused on learning as well as their own projects.

As a result of the first SIB, "PICAPOOL" (Fig. 15), an online platform connecting people looking for similar products and allowing them to pool their orders and access exclusive discounts, was selected as the best business idea with a PoC implementation among various ideas in health care (Fig. 16), waste management, community development, etc. The post-program feedback by the participants helped the SIB program to evaluate the effectiveness of each activity.

7 Supply Chain Project

In order to seek rural innovation, understanding the real situations and issues in the rural areas is very important. Focusing on the supply chain in rural villages, Supply Chain project conducted a survey in five villages where approximately 200 people participated in a form of interview, and analyzed the time to access Fast-Moving Consumer Goods (FMCG) in rural areas. For example, dairy products, soaps and detergents, breads and biscuits, stationary goods (pens, notebooks, etc.), and other basic groceries (sugar, tea, etc.) are available within a 5 km range (Fig. 17). However, people in the village regularly travel 5 to 10 km to purchase vegetables,

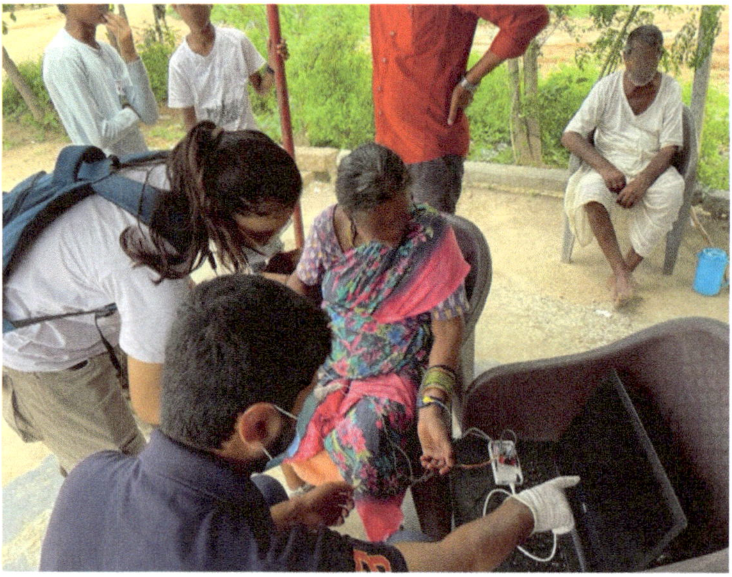

Fig. 16 Field trial of a low-cost portable sphygmomanometer prototyped by a participant

cooking oils, spices, fresh meat and fish, pulses (daal, etc.), and sanitary goods. For clothes, utensils, fertilizers, bangles, and jewelry, they need to travel more than 10 km distance.

In typical rural villages, mobile application-based cab service (Uber, Ola, etc.) are not available. According to our survey, the two most popular modes of travel to a market are shared auto-rickshaws and motorbikes. Usually, in the case of motorbikes, people travel with family (2 people) to buy goods. In addition to the distance and mode of travel to access FMCG, the Supply Chain project also summarized the major and minor challenges in rural villages.

1. Counterfeit and/or low-quality products
2. Unreasonable pricing (higher than Maximum Retail Price)
3. Limited availability of international/regional brand products
4. Dependency on unreliable transportation
5. Mindset of "see-before-buy" leads to a long travel to the city rather than buying online
6. Limited availability of goods for childcare

Fig. 17 A glossary shop in one of the villages where the project visited

Given these challenges, the Supply Chain project installed vending machines (Figs. 18 and 19) in seven villages as a part of the trial to further understand the purchase behavior in the villages and how the vending machines may change the lifestyle in the rural villages.

The installation of a vending machine can also lead to the increase of litter as a downside. The Supply Chain project has displayed an educational poster for garbage segregation together with dust bins (Fig. 20). The garbage segregation made a reasonable effect in mitigating littering. Also, the project facilitated the exposure to digital literacy through the Unified Payments Interface (UPI) to enable purchases resulting in the adoption of protocols by majority.

According to the preliminary survey results after the installation of vending machines, the consumers accepted the vending machines and used them on a daily basis. As a natural consequence, the burden of traveling to the market and the city was significantly reduced. Currently, the project has also initiated a pilot program to enable bulk purchases by a collection of orders, collaborating with a women's self-help group. The activities in the Supply Chain project have involved six local female entrepreneurs as resident service providers in villages.

Fig. 18 The outlook of a vending machine installed in a village

8 Learning so Far and Way Forward

SIC looks into India as the place to explore unfound inspirations, to harvest seeds of innovative ideas, and to sow them and see how they grow. This is totally different from a typical Japanese way of looking into India: a potential market having 1.4 billion population.

The common aspect among the various programs running in SIC is the involvement of local people who are the potential end users of innovative ideas and solutions that each program explores. Being in the second year of SIC, most activities are still ongoing and their concrete evaluation has not been conducted yet. However, SIC has already accumulated the awareness of how rural society works through the real-world project execution, and constructed a broader network in rural society in India. Together with Learn in Bharat and other human exchange programs, SIC is in the process of constructing the platform where people from India, Japan and broad countries can figure out what could be done for rural and social innovation in India. And this is going well.

Fig. 19 Fast-moving consumer goods made available in a vending machine. The choice of these goods was adjusted according to the purchase trends in each village

Fig. 20 Educational posters for garbage segregation

References

GIAN (2023) Gujarat grassroots innovations augmentation network (GIAN). https://gian.org/. Accessed 27 Dec 2023

MOFA (2008) Report of the Japan-India working group on the cooperation toward establishment of a new IIT. Ministry of Foreign Affairs. https://www.mofa.go.jp/region/asia-paci/india/pmv0810/report.pdf. Accessed 27 Dec 2023

SIC (2022) Suzuki Innovation Centre (SIC). https://sicglobal.org/. Accessed 27 Dec 2023

World Bank (2018) Rural population (% of total population)–India. World Bank. https://data.worldbank.org/indicator/SP.RUR.TOTL.ZS?end=2022&locations=IN&start=1960&view=chart. Accessed 6 Sep 2023

Future of Agritech Startups: Perspectives from India and Japan

Divya Suresh, Abhishek Choudhury, and Rajib Shaw

Abstract Global climate change poses many threats, significantly impacting crop productivity and food security. Digitalization in agriculture is envisioned to address the multi-faceted problems the agricultural sector faces, such as eliminating information asymmetry amongst diverse stakeholders, lowering operational and transaction costs, improving market access, enabling finance access, and promoting sustainable farming practices. The readiness of digital technologies is a critical factor in the agricultural sector transformation. The landscape and categories of technologies used in agriculture significantly vary across countries. This chapter delves into the case of India and Japan, their unique agritech categories, potential, and policies enabling the sprouting of agritech startups that develop them.

Keywords Agritech · India-Japan relationship · Digitalisation · Sustainability · Data-driven farming

1 Introduction

Globally, agriculture faces multifaceted problems such as climate change, labor shortages, evolving consumer demand, supply chain issues, etc. The agricultural sector remains underdeveloped mainly due to the numerous difficulties smallholder farmers face, such as lack of access to modern equipment, agricultural information, limited access to secure financing, lack of access to data to validate credit risk, demotivated young people to pursue farming as a career because of the non-lucrative side of the sector, weather vagaries affecting crop yield. It is the least digitized and innovated sector (Arvanitis and Symeonaki 2020; 10 Digital Technologies That Are Transforming Agriculture 2022).

Digitalization in agriculture is envisioned to bridge information and knowledge gaps in agriculture, promote food security, increase climate change/environmental sustainability, future automation of agricultural activities, provide employment and

D. Suresh (✉) · A. Choudhury · R. Shaw
Graduate School of Media and Governance, Keio University, Fujisawa, Japan
e-mail: divi.info@gmail.com

R. Shaw and S. R. Choudhury (eds.), *India, Japan and Beyond*,
https://doi.org/10.1007/978-981-97-3282-1_12

empower the youth, promote gender and women empowerment, and enhance livelihood resilience in rural areas. It also addresses various Sustainability Development Goals such as No Poverty, Zero Hunger, Good health and well-being, Responsible Consumption and production, and climate action. Notably, the different key pain points across the agricultural value chain can be addressed by applying technology in agriculture. Agricultural technology plays a vital role in transforming the agricultural production process, which is the heart of agriculture. The readiness of digital technologies is a critical factor in facilitating transformation (Goh 2021). With the introduction of data and technology in agriculture, data-driven farming techniques are changing the status quo and challenging traditional farming techniques, which are labor-intensive and time-consuming. The application of agtech in agriculture is multi-faceted: solving farm input needs, agricultural production, crop monitoring, harvesting, and produce management (From Farm to Table: AgTech Innovations That Drive Agricultural Efficiency Across the Value Chain 2023).

The agility and dynamism of agricultural transformation are being enabled through the emergence of agricultural technology or agritech start-ups. These small and innovative companies are the origins of providing unique business models and agricultural services with value propositions to farmers leveraging old and emerging technologies.

The agri-tech solution is classified under the following 14 categories: Biotechnology, Agribusiness marketplaces, Bioenergy and biomaterials, Cloud retail infrastructure, eGrocery, Farm management software, sensing & IoT, Farm Robotics, mechanization and equipment, Home & cooking tech, Innovative food, In-store retail & restaurant tech, Midstream Technologies, Novel farming Systems, Online Restaurants & Meal kits, and restaurant marketplaces. (What Is Agri-FoodTech?–AgFunderNews, n.d.) The agility and dynamism of agricultural transformation are being enabled through the emergence of agricultural technology or agritech start-ups. These small and innovative companies are the origins of providing unique business models and agricultural services with value propositions to farmers leveraging old and emerging technologies.

The framework for classifying agrifood tech companies and startups provided by the Agfunders Investing Report 2017 is under upstream and downstream categories. The "upstream" agrifood tech is listed under eight types: Ag Biotechnology; Farm Management Software, Sensing, and IoT; Farm Robotics, Mechanization, and Equipment; Bioenergy and Biomaterials; Novel Farming Systems; SupplyChain Technologies; Agribusiness Marketplaces; and Innovative Food. On the other hand, the "downstream" agritech companies can be classified under six types: In-store retail & restaurant tech, restaurant marketplaces, eGrocery, Home & cooking tech, Online Restaurants and meal kits, and Cloud retail infrastructure (Steup et al. 2019).

In this chapter, we highlight agritech categories and potential areas from India and Japan, highlighting government startup policies shaping agritech companies' emergence in the respective geographies.

2 Current Agritech Landscape in India

Agriculture forms a significant part of the Indian economy, with a sizeable rural population dependent on farming as a primary occupation. And so, agritech startups often aim at large-scale impact solutions. The primary focus is often on increasing productivity, improving market access for farmers, and addressing post-harvest losses. Indian startups are primarily geared towards providing digital solutions, supply chain management, and affordable technology to increase yield and profitability for small-scale farmers.

The table below illustrates nine major tech categories in India, their definitions, and some illustrative players (Table 1).

3 Current Agritech Landscape in Japan

The agriculture sector in Japan forms a minor part of the economy. The technological solution is geared towards precision agriculture, quality enhancement, and sustainable practices, considering the country's position as a developed economy. The solutions focus on addressing labor reduction because of the aging society. The table below illustrates six major agritech categories in Japan, their definitions, and some illustrative players (Table 2).

4 Potential Significant Areas for India and Japan

This table explains key agri-tech areas poised to transform India's and Japan's agricultural landscapes. From precision agriculture harnessing cutting-edge technologies to intelligent farming methodologies driven by Artificial Intelligence (AI) and the Internet of Things (IoT), these innovations offer tailored solutions addressing distinct needs and contexts in each country.

In India, where agriculture remains a primary livelihood for a significant portion of the population, adopting Agri-Tech opens avenues to enhance resource efficiency, crop yield, and resilience against environmental challenges. Conversely, in Japan, renowned for its technological expertise and limited fertile land, Agri-Tech is pivotal in maximizing production within confined spaces, ensuring high-quality yields, and addressing labor shortages through automation (Medvediev 2020).

This table delves into specific aspects of Agri-Tech, explaining how India and Japan can leverage these technologies to strengthen their agricultural capabilities. By exploring precision agriculture, AI-driven crop monitoring, sustainable water management, and other innovative facets, it aims to highlight the potential impact and unique strategies each country can employ to propel their agricultural sectors into a new era of efficiency, sustainability, and productivity.

Table 1 Agritech categories in India

Category	Description	Example of Start-up
Agriculture fintech	Enabling sustainable financing options for agricultural stakeholders in rural areas	I support farming
Agriculture mechanization/ automation	Providing machineries and automation technologies to use on farm as a service	Tractor junction
Upstream agtech	Enables input linkages through a B2C platform for inputs such as seeds, nutrition, agrochemicals, etc	Agrostar, Big Haat
Farm-to-fork brands	Connects end consumers directly with farmers through a platform thereby help improving farming practices	Eggoz nutrition, Akshayakalpa
Agriculture biotech	Leverage biotechnology to create sustainable food products. This includes biotechnology along with materials such as bio inputs, alternative ingredients, etc	String bio, Sea6 energy
Mid-stream agtech	Provision of supply chain that helps with proper management and storage of agriproduce	Arya, Ecozen
Digital solutions and precision agtech	Providers of mobile or sensor-based solutions to farmers. Ex: digital advisory services,	Aquaconnect, MoooFarm, Cropin
End to end ecosystem	Platforms span across agricultural value chain	DeHaat
Downstream agtech	Platforms enable connecting farmers with businesses or consumers	Ninjacar, absolute, waycool

Source McKinsey (2023)

This table outlines noteworthy Agricultural Technology (Agri-Tech) domains in India and Japan. It emphasizes pivotal sectors for technological advancement in agriculture, including precision farming, automation, climate-resilient agriculture, agricultural biotechnology, optimization of supply chains, data-centric decision-making, technology in aquaculture, Agri-fintech, and urban farming technology. These sectors offer possibilities for utilizing technology to amplify agricultural efficiency, sustainability, and market reach in India and Japan (Table 3).

Agriculture is a vital sector in India, intertwining with the livelihood of numerous small farming households. Supporting the employment of approximately one billion individuals globally directly impacts multiple industries. Notably, over 608 million

Table 2 Agritech categories in Japan

Category	Description	Example of start-up
Precision farming	Offers GPS guidance systems for tractors and vehicle robot systems for unmanned devices and vehicles	ListenField
Agricultural robot systems	Ranging across agricultural robots such as grafting robots to manipulator robots like harvesting robotics until assisting robots like exoskeletons	Agrist
Drone solutions	Broadly enables farm monitoring services and agrichemical spraying services	NileWorks
Cultivation support solutions	Production support solutions for dairy farming. Platforms for control devices and sensors such as light, heat, humidity etc. Also farm work management cloud based application	Inaho, GRA Inc., SenSprout
Management support solutions	Systems such as accounting software for farming and accounting services for agricultural corporations	Farm OS Co., Ltd
Sales support solutions	Leveraging ICT to connect various stakeholders and actors to optimize the sales process. Also offer services towards sales support	Pocket Marche

Source Schilling (2021)

smallholder farmers, responsible for producing around 35% of the world's food, are crucial in achieving the Sustainable Development Goal (SDG) of eliminating poverty and hunger by 2030. However, the agriculture sector faces diverse challenges, including production, market linkages, supply chain dynamics, and climate change effects, leading to significant losses for smallholders. India's burgeoning Agri-Tech sector actively addresses these challenges through innovative solutions, necessitating substantial support and investment to enhance smallholder productivity and achieve SDGs (The Rise of Agri-Tech in India 2023). With the government of India backing initiatives such as 'Made in India' and 'Digital India,' the country's agri-tech sector holds immense potential for value creation and growth opportunities (Table 4).

As of 2020, the average age of individuals primarily involved in Japanese agriculture is around 68 years, and there is a declining trend in the number of people participating in farming. Given these demographic changes and the imperative for sustainable cultivation methods amid challenges like climate change, the significance of Agricultural Technology (Agri-Tech) in Japan is progressively gaining recognition (Prepared by the New Zealand Embassy in Tokyo 2022). Agri-tech, the convergence of agriculture and technology, encompasses a broad spectrum of technologies designed to simplify and enhance farming practices. The utilization of technology, including advancements in hydroponics, drone technology, artificial intelligence (AI), and data analytics for crop monitoring, is anticipated to witness global growth in scale and significance. Considering this context, the integration and advancement of Agri-Tech solutions in Japan have the potential to play a crucial role in reshaping the agriculture sector, fostering improvements in productivity, sustainability, and resilience (Hive Life Magazine 2021) (Table 5).

Table 3 Common potential areas for India and Japan

Potential significant areas for Agri-Tech	India	Japan
Precision agriculture	Adopting precision agriculture entails leveraging technology to enhance resource efficiency, like water and fertilizers, in farming practices	Utilizing advanced farming techniques for maximizing resource efficiency and improving agricultural output
Automation and robotics	Incorporating automated systems and robotics into agricultural procedures to enhance efficiency while diminishing reliance on manual labor	Advancing automation and robotics to streamline farming operations and address labour shortages
Climate-smart farming	Creating methods and advancing technology aimed at adjusting to climate shifts, endorsing eco-friendly farming, and diminishing ecological footprint	Concentrating on innovative agricultural techniques designed to address climate challenges, thereby promoting sustainable farming methods amidst changing environmental conditions
Agri-biotechnology	Leveraging biotechnological methods to enhance crop quality, manage pests, and boost agricultural output	Investing in Agri-biotechnology for crop enhancement, disease resistance, and sustainable agricultural development
Supply chain optimization	Implementing technology-driven supply chain solutions to reduce wastage, improve logistical efficiency, and ensure better market access for farmers	Optimizing the agricultural supply chain through advanced technologies to enhance efficiency and minimize waste
Data-driven decision making	Leveraging data analytics and technology for informed decision-making in farming practices, crop management, and market insights	Emphasizing data-driven decision-making to enhance agricultural productivity, optimize resource allocation, and foster market intelligence
Aquaculture technology	Advancing technologies for sustainable and efficient aquaculture practices to meet the growing demand for fish and seafood	Developing innovative aquaculture technologies to support sustainable fish and seafood production
Agri-Fintech	Integrating financial technology to improve access to credit, insurance, and financial services for farmers	Enhancing access to financial services through Agri-fintech solutions, supporting farmers with credit and risk management tools

(continued)

Table 3 (continued)

Potential significant areas for Agri-Tech	India	Japan
Urban farming technology	Leveraging technology for urban farming practices, including vertical farming and hydroponics, to address food security challenges in urban areas	Expanding urban farming technology to address urban food security issues and promote sustainable food production in urban environments
Farmer-centric applications and tools	Developing user-friendly applications and tools to empower farmers with information, market access, and advisory services	Focusing on farmer-centric applications and tools to provide access to information, markets, and advisory services tailored to their specific needs

Leveraging its expansive and varied agricultural terrain, India prioritizes adaptable solutions for smallholder farmers. Employing strategies like precision agriculture, drone-enabled crop monitoring, and comprehensive data analysis, the focus remains on empowering farmers, elevating output, and addressing issues linked to limited resources. In contrast, Japan, constrained by little arable land yet equipped with advanced infrastructure, directs its efforts toward sophisticated precision farming. Harnessing technologies like cutting-edge robotics, satellite-based imaging, and refined biotech applications, the goal is to maximize efficiency, ensure top-tier produce, and sustainably manage agricultural resources. While both countries share common ground in advocating sustainable farming methods, innovating supply chains, and utilizing data-driven decision-making, their methodologies are tailored to fit their distinct agricultural landscapes, technological capacities, and the specific needs of their farming communities.

5 Agri-Tech Startup Policy in India and Japan

Policies surrounding Agricultural Technology (Agri-tech) serve as governmental blueprints directed at propelling agricultural advancements through technological innovations. These policies encompass diverse strategies, including financial incentives, infrastructure development, regulatory frameworks, and educational campaigns. Their purpose is to champion research, progression, and integration of state-of-the-art technologies in agriculture, such as artificial intelligence, data analytics, IoT, and robotics.

In India, agri-tech policies have considerable influence in reshaping agricultural landscapes. Prioritizing heightened productivity, these policies introduce technological interventions like precision farming, IoT-infused monitoring systems, and intelligent irrigation approaches. These innovations yield amplified crop yields, optimized resource utilization, and refined agricultural techniques. Furthermore,

Table 4 Potential agritech categories in India

Potential significant areas	How it will help the agricultural scenario in India
Precision agriculture	The unique agricultural landscape of India necessitates customized strategies to maximize resource efficiency, minimize waste, and enhance crop productivity. Precision farming, incorporating GPS, sensors, and data analysis, enables precise resource allocation tailored to specific local conditions, offering advantages to farmers of all scales
IoT-based farm management systems	Agriculture in India relies significantly on weather elements and soil quality. Utilizing IoT-driven technology, these systems offer up-to-the-minute information on climate trends, soil moisture levels, and crop well-being. This empowers agricultural workers to make well-informed choices regarding watering, pest management, and overall farm administration, pivotal in enhancing effectiveness and output
AI-driven crop monitoring	AI-enabled monitoring of crops across diverse geographical areas assists in early identification of diseases, forecast of yields, and efficient irrigation management. This technological approach provides customized strategies for India's wide array of crops, facilitating prompt actions and augmenting the overall agricultural output (Agritech in India: Industry Overview and Scope for Investments 2022).
Agri-drones and UAVs	Monitoring India's expansive agricultural regions presents significant difficulties. Drones provide a cost-efficient solution for overseeing, evaluating crop health, and precisely applying necessary resources. They assist in promptly managing pests, irrigation, and delivering precise mapping, particularly in areas that are remote or hard to reach (Rise of Agri-Tech in India, 2023).
Sustainable water management	Water shortages in various parts of India are a pressing concern. The adoption of advanced techniques enhancing irrigation efficiency, recycling water, and preserving resources is imperative for maintaining sustainable agriculture. By integrating these methods, it becomes possible to maximize water utilization, tackle scarcity problems, and uphold the long-lasting viability of agricultural practices
Agri-robotics for harvesting	During specific periods when labor is scarce and efficient harvesting becomes essential, robotics emerges as a crucial domain. Implementing automated systems for crop harvesting and processing not only eases the strain on labor but also guarantees timely and effective crop administration
Crop health imaging technologies	Cutting-edge imaging tools such as satellite imagery and unmanned aerial vehicles play a crucial role in identifying diseases, pest invasions, and insufficient nutrients at an early stage. These innovative tools offer valuable guidance to farmers, facilitating swift actions and enhancing the overall well-being and productivity of crops

(continued)

Table 4 (continued)

Potential significant areas	How it will help the agricultural scenario in India
Mobile-based agri-advisory services	In India, the substantial agricultural population frequently faces a dearth of immediate guidance and expert support. Addressing this issue, mobile-centric advisory solutions serve as a conduit, furnishing prompt updates on weather patterns, market rates, and optimal farming methodologies. This empowers farmers with indispensable knowledge and guidance
Blockchain in supply chain management	The agricultural distribution system in India commonly faces issues like inefficiency, opacity, and exploitation by intermediaries. Blockchain offers a solution by establishing transparent, traceable, and equitable transactions, fostering trust and responsibility through a secure and unalterable transaction log, benefiting farmers with fair returns
Agri-waste management innovations	Challenges exist within the realm of agricultural residue. Solutions involving waste handling innovations hold the potential to transform these remnants into sought-after assets such as biofuels or natural fertilizers. This approach fosters sustainability and curtails the ecological footprint of agricultural methods
Climate-smart farming	The agricultural sector in India faces vulnerabilities because f climate change, such as unpredictable precipitation patterns and fluctuating temperatures. Implementing agricultural strategies that are sensitive to climate variations involves utilizing resilient crop breeds, conservation approaches for water usage, and specialized soil management methodologies. These practices are designed to accommodate and adjust to evolving climate circumstances, safeguarding both food production stability and environmental sustainability
Agri-biotechnology	Biotechnology presents vast possibilities in enhancing crop quality, fortifying against illnesses, and boosting nutritional value. Leveraging genetic alterations and advanced biotech techniques facilitates the creation of resilient crops with amplified productivity, resilience against pests and diseases, and adaptability to environmental pressures, crucial in tackling issues surrounding food security
Aquaculture technology	India, endowed with extensive coastlines and inland water resources, stands poised to harness the potential of aquaculture. Advances in aquaculture technology drive enhancements in fish yield, streamline feed administration, and promote sustainable aquaculture methods, fulfilling the escalating need for seafood

(continued)

Table 4 (continued)

Potential significant areas	How it will help the agricultural scenario in India
Agri-fintech	Access to financial services and credit stands as a critical need for farmers. Agricultural financial technology (Agri-Fintech) solutions provide customized digital lending avenues, crop insurance programs, and payment structures designed specifically for the agricultural community, nurturing financial stability and advancement within this sector
Urban farming technology	Urban farming addresses food security challenges in rapidly growing urban areas. Technology-enabled vertical farming, hydroponics, and rooftop gardening allow cultivation in limited spaces, reducing food miles and ensuring access to fresh produce (Rise of Agri-Tech in India 2023).

endeavors promoting digital marketplaces and mobile applications empower farmers by granting access to markets, valuable insights, and advisory services, fostering informed decision-making and potentially elevating incomes.

In addition, Agri-tech policies in India nurture economic expansion by fostering innovation and entrepreneurial endeavors. The backing provided to startups and widespread technology adoption generates employment prospects and significantly contributes to the country's GDP growth. These policies catalyze technological strides within the agricultural sector and position India as a prominent player in global agricultural innovation (Fig. 1).

Similarly, Japan's Agri-tech policies concentrate on modernizing age-old farming practices by integrating advanced technologies. These initiatives target heightened efficiency, reduced labor reliance, and augmented agricultural produce quality. The integration of cutting-edge technologies facilitates precision agriculture, optimizes resource utilization, and curbs environmental impact by minimizing chemical usage and promoting efficient water management (Hive Life Magazine 2021).

Moreover, Japan's Agri-tech policies harmonize to revitalize rural areas. By championing innovation in agriculture, these policies bolster the economic sustainability of rural regions and preserve traditional farming communities. This emphasis on sustainable practices ensures environmental guardianship and fortifies the long-term viability of rural locales.

5.1 India's Agri-Tech Startup Policies:

Globally, there has been a significant rise in the culture of entrepreneurship in recent times. The shift from economies centered around controlled and license-regulated public companies to those rooted in knowledge has spurred individuals to embark on entrepreneurial ventures. The term "startup" refers to the initial phase of a business that incorporates science-based innovations capable of substantial growth. Essentially, all founders of startups are entrepreneurs, yet not all entrepreneurs are startup

Table 5 Potential agritech categories in Japan

Potential significant areas	How it will help the agricultural scenario in Japan
Precision agriculture	In Japan, the scarcity of cultivable land calls for precise agricultural methodologies. The integration of drones, sensors, and Internet of Things (IoT) devices enables meticulous supervision of crop health, soil quality, and water consumption. This method aids in fine-tuning resource distribution and enhancing agricultural output (Tokyoesque 2022).
Vertical farming	Due to Japan's high urban population density, vertical farming has become increasingly popular. Cutting-edge hydroponic and aeroponic methods, combined with LED lighting and precise climate regulation, allow for effective farming in controlled settings, optimizing space usage and maximizing yield
Robotics and automation	Japan leads the way in agricultural robotics innovation. Robotic systems are employed for activities such as planting, harvesting, and crop sorting. Given an aging workforce and scarcity in labor, these technological advancements are vital for sustaining productivity levels in agriculture
Smart greenhouses	State-of-the-art greenhouses, outfitted with AI-managed climate setups, automated watering mechanisms, and surveillance devices, create ideal environments for cultivating crops. Such facilities ensure a steady output of premium quality harvests consistently throughout the year (Washizu and Nakano 2022).
Biotechnology and genomics	Studies in biotechnology and genomics concentrate on cultivating crops that can withstand Japan's specific climate conditions, pests, and diseases. Through genetic modification and selective breeding methods, the goal is to produce stronger and more nourishing crops tailored to these challenges
Blockchain in agriculture	In Japan, a strong focus is placed on ensuring the safety and traceability of food. Employing blockchain technology enables the meticulous tracking and verification of agricultural product origins, quality, and distribution, guaranteeing transparency throughout the supply chain to benefit consumers
Agri-fintech	Agricultural-specific financial technology aids farmers by providing avenues for capital, insurance, and risk mitigation. These platforms furnish micro-financing, crop insurance, and assorted financial offerings, fostering sustainable agricultural methodologies
Waste reduction and recycling	In Japan, a strong emphasis lies on sustainability. Agricultural technology endeavours prioritize the reduction of food wastage by employing inventive packaging, preservation techniques, and the transformation of organic waste into either fertilizers or energy sources (OECD 2019).
Agri-education and training	In the context of agriculture's technological evolution, the education of farmers regarding Agri-Tech tools and methods emerges as a pivotal factor. Specialized programs and workshops are designed with the intent to furnish farmers with the necessary expertise for embracing and customizing novel technologies

Fig. 1 A general framework of typical startups. https://naas.org.in/Policy%20Papers/policy%20108.pdf

founders. Startups are emerging as platforms for achieving societal objectives, channels for realizing technological prowess, the wings of imaginative creators transporting them to the realms of their visions, and favored paths toward financial success. These startups are poised to become the driving forces propelling the Indian economy toward the envisioned US$5 trillion mark and the realization of the Atmanirbhar Bharat dream.

India boasts over 100 Agri-focused incubators, primarily within academic and research institutions, notably at the ICAR and agricultural universities. These incubators receive support from initiatives like Startup India, Atal Innovation Mission, NSTEDB, RKVY-RAFTAAR, and ICAR. Within India, 36 agriculture-centered incubation centers are promoted and supported through the NIDHI scheme by the Department of Science & Technology (DST). Among the National Agricultural Research and Education System (NARES), only seven Technology Business Incubators (TBIs) exist, predominantly housed at ICAR-IARI, New Delhi; ICAR-NDRI, Karnal; ICAR-NAARM, Hyderabad; ICAR-IIHR, Bangalore; TNAU, Coimbatore; and ICRISAT, Patancheru. Correspondingly, roughly 45 NIDHI PRAYAS centers operate in India, with only one within NARES (located at NAARM). ICAR has established 50 Agribusiness Incubators (ABIs) across various research institutes with an annual budget of Rs10 crores (Rs 20 lakhs each for ABI). Similarly, 29 ABIs have been instituted at state agricultural universities (SAUs) under the RKVY-RAFTAAR scheme of the Ministry of Agriculture & Farmers' Welfare, receiving grants of Rs 2–3 crores each for R-ABI. Presently, these ABIs support over 750 startups and Agri-entrepreneurs in the agricultural domain, including farmer-entrepreneurs and producers' companies (FPOs). India, a farming powerhouse, recognizes the pivotal role of technology in transforming the sector. (Srinivasrao and Kumar 2022).

The nation's policies underscore a multi-faceted approach aimed at empowering farmers and fostering innovation:

1. **National Policy on Electronics (NPE)**: Envisioning growth in domestic manufacturing, this policy extends its purview to encompass agricultural technology. It aims to bolster the sector's technological capabilities, encouraging local production and innovation in Agri-tech solutions.
2. **National Agriculture Market (NAM)**: Addressing the fragmented nature of agricultural markets, NAM establishes an online trading platform, leveraging technology to create a unified national market for agricultural commodities. The goal is to ensure fairer prices and improved market access for farmers.
3. **National Policy for Farmers (PM-AASHA)**: Designed to support farmers, PM-AASHA aims to provide farmers with remunerative prices for their produce. The

policy aims to enhance farmers' incomes through a combination of mechanisms, including price support, procurement, and market interventions.

4. **Atal Innovation Mission (AIM)**: AIM's Agricultural Grand Challenge is a beacon for agri-tech startups, encouraging them to develop innovative solutions to address agricultural challenges. This initiative fosters a culture of innovation by providing platforms and support for startups to thrive.
5. **NITI Aayog Initiatives**: NITI Aayog is pivotal in nurturing Agri-tech startups. Through various schemes and Atal Incubation Centres, it provides financial and infrastructural support to startups, fostering an ecosystem conducive to innovation and growth in the agricultural technology domain (Press Information Bureau, n.d.).

5.2 Japan's Agri-Tech Startup Policies:

In Japan, a confluence of geographical, economic, policy, and cultural elements has contributed to the current challenges confronting its agricultural sector. These include the increasing abandonment of farmlands, exacerbating the country's already limited available land, the prevalence of small farm sizes that hinder scalability, an imbalanced array of cultivated products (where rice cultivation, for instance, is protected, leading to setbacks for wheat and soy), and a substantial dependency on food imports. Among these challenges, the most pressing concern is the aging demographic within the farming community. The average age of Japanese farmers now stands at 67, and projections indicate a shortfall of 200,000 laborers in agriculture by 2040, further compounded by the loss of extensive agricultural expertise developed over many years.

The projected growth of Agri-Tech in Japan, anticipated to rise from US$103 million in 2017 to an estimated US$322 million by 2023, signals a proactive stance toward addressing pertinent challenges. With labor shortages looming, a pivotal emphasis rests on technological advancements supporting and substituting human labor. Since a significant portion of Japan's farmers operate part-time, contributing to 60% of total rice production, adopting labor-saving technology is essential in enabling them to sustain this arrangement. Despite limited land availability, Japan leverages its abundant resources in technology and capital. The government's Strategic Innovation Promotion Program prioritizes WAGRI (next-generation agriculture) and smart farming. Concurrently, the spotlight is on drone-based automated equipment alongside the development of the highly precise QZ Satellite System. Relaxed regulations on drone usage since 2019 have paved the way for expanding drone-approved farmland to encompass 1 million hectares by 2022. Another focal area of research and development pertains to automated greenhouse systems.

Japan, with its technological expertise, prioritizes innovation in agriculture through policies emphasizing integration and modernization:

1. **Support for Agri-Tech Startups**: The government collaborates with the private sector to provide Agri-tech startups funding, resources, and mentorship. Japan

aims to drive advancements in agricultural practices by incentivizing innovation and technology adoption (Iwasaki 2022).
2. **Japan Agricultural Cooperatives (JA Group)**: Collaborations between the JA Group and startups enable the infusion of technology into farming. These partnerships facilitate adopting innovative practices, modernizing agriculture, and ensuring sustainable growth (ZEN-NOH n.d.).
3. **Regional Revitalization Policy**: Recognizing the significance of agriculture in rural areas, Japan's policy focuses on revitalizing regional economies through agricultural innovation. By promoting the adoption of technology and modern farming practices, this initiative aims to bolster rural communities (Regional Revitalization of JapanGov 2021).
4. **WAGRI**: WAGRI is a publicly accessible cloud service offering a range of agriculture-centric data and applications, including weather forecasts, farmland information, and yield projections. This platform seamlessly allows ICT vendors and agricultural machinery manufacturers to amalgamate diverse agricultural data and programs. This facilitates the effortless creation of web services and applications to enhance farmers' efficiency and financial gains. These solutions can be crafted and made accessible through websites or smartphone applications (Aboutwagri 2019).

India's approach directly focuses on supporting farmers while fostering innovation through targeted initiatives like AIM and NITI Aayog's schemes. In contrast, Japan emphasizes the comprehensive integration of technology into agricultural practices, leveraging collaborations between established cooperatives and startups. Both India and Japan stand as exemplars in promoting Agri-tech startups through tailored policies. Their efforts signify a commitment to harnessing technology for agricultural advancement, ultimately steering towards sustainable farming practices, increased productivity, and improved livelihoods for farming communities.

These policies support startups and pave the way for a progressive and resilient agricultural sector poised to address contemporary challenges while ensuring global food security and sustainability.

6 Existing India-Japan Agritech Collaboration

India and Japan's collaboration on a broader framework has been on digital services, clean energy, healthcare, mobility and electric vehicles, and aerospace. India-Japan Digital Partnership, clean energy partnership, India-Japan Cooperative Science Program, etc., are a few examples of technology redefining the shared partnership between the two countries (Department of Science & Technology 2023). In a recent cooperation, India and Japan signed a MoU to expand their partnership on semiconductor design, manufacturing, and building resilience in the semiconductor supply chain. Many seminars are being conducted, such as Indo-Japan deep tech Innovation and clean energy seminars, India-Japan Emerging Tech initiative, etc., to explore the cooperation in technology and innovation between two geographies leveraging India's cost-effective and innovative ecosystem and high technology industrial base of Japan. Generally, in the collaboration, it has been emphasized that the role of startups is essential for collaboration between the two countries (Asia Pathways 2023).

Below is a list of a few cases of Agritech collaborations between India and Japan:

1. **Sagri**, a Japanese startup in association with the Japan International Cooperation Agency (JICA), was adopted as the first India-Japan startup Hub In Bengaluru, India, promoted by the Japan External Trade organization. Through this project, they have been instrumental in improving farmland productivity by visualizing farmland with satellite data, raising farmers' awareness through educational activities, and expanding future possibilities for local farmers (JICA 2023).
2. **Makuta Amenity Co., Ltd**: In association with JICA, has piloted projects in Uttarakhand leveraging AI analysis technology for image analysis for visualizing the taste of fruits and vegetables.
3. **Casley Consulting, Inc**: In association with JICA, the initiative is about improving the quality of contract farmers and building a supply chain based on its customer data through the Agri SCM platform. The related services include deciphering the traceability of crop information and planning based on market demand and marketing.
4. **MoU for irrigation and water resource management**: This MoU facilitates bilateral cooperation towards sustainable water resources for agriculture and strengthens water security. Efforts have been undertaken to improve knowledge management and technology infusion and assess the domains for potential bilateral projects in delta management and related activities.
5. **Japan International Cooperation Agency (JICA)**: JICA has facilitated development in the agriculture sector through the Jharkhand Horticulture Intensification by Micro Drip Irrigation (MDI) project, the Rengali Irrigation project, maximizing soybean production in Madhya Pradesh, improving skills of sericulture farmers in the southern state of India.
6. **Act East Forum**: development of the North-Eastern region of India (2017): The scope of initiative under this MoU includes the capacity enhancement for sustainable agriculture and Irrigation development in Mizoram. It also involves the development of agro-industries, such as strengthening the bamboo value chain in the northeast, food processing, organic farming, and horticulture, primarily through small and medium entrepreneurs (Ministry of External Affairs Government of India 2022).
7. **Japanese Ministry of Agriculture, Forestry, and Fisheries (MAFF)**: An initiative under MAFF leverages technologies, products, and services of some sixteen Japanese companies and support of the Self-Employed Women's Association (SEWA) to implement J-methods at Anand in Gujarat. They have grown delicious tomatoes and cabbage (Foodtechbiz 2022).
8. **An alliance between original equipment manufacturers, the Japanese subsidiary of Mahindra & Mahindra, Mitsubishi Mahindra Agricultural**

Machinery Co., Japan, and Kubota Co., Japan: The collaboration addresses diversifying customer needs and contributing to the sustainable development of Japanese agriculture by expanding mutual OEM supply and utilizing products compatible with Kubota's IoT platform "KSAS," which is the basis of smart agriculture.

7 Conclusion

Agricultural technology has the unique potential to transform traditional manpower-intensive agricultural setups by creating value for the farmer through efficient and cost-effective services. Increasing trends in agritech start-ups have been realized because of the increasing numbers of farmers facing them and the growth of the agricultural food tech industry. Despite this, as illustrated in Fig. 2 below, some significant problems with adopting agritech are the high cost of technology, unclear return on investment, high expectation on ROI, and low willingness to pay. Despite the various sustainability goals addressed by agricultural technologies, there are significant barriers to adopting agritech, which varies across geographies. The more agritech companies get to know the farmer, the better products or services they can build for the farmer.

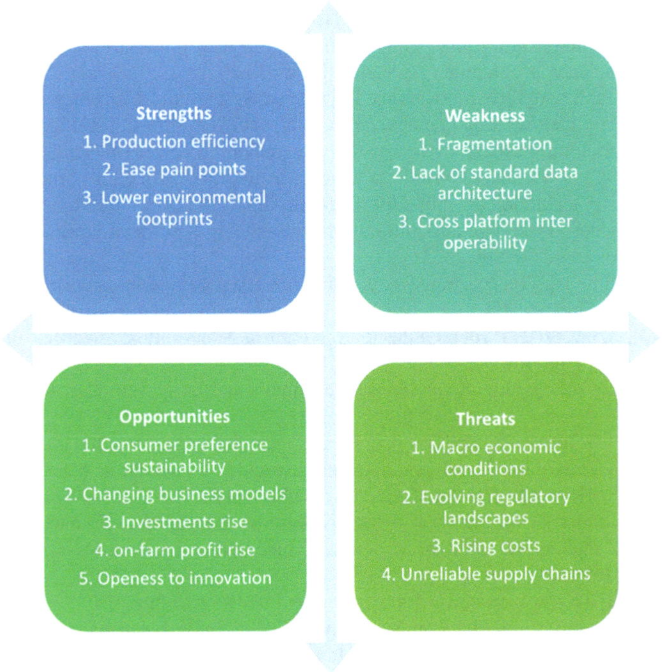

Fig. 2 SWOT analysis of agritech. *Source* By authors

Acknowledgements The first author is thankful to Japan Science and Technology (JST) for the scholarship supporting Ph.D. research.

References

10 Digital Technologies That Are Transforming Agriculture (2022) https://www.forbes.com/sites/forbestechcouncil/2022/04/26/10-digital-technologies-that-are-transforming-agriculture/?sh=1d242cd7bafc. Accessed 30 Nov 2023

Aboutwagri (2019) Agricalture data plarform-WAGRI. https://wagri.net/en-us/aboutwagri. Accessed 30 Nov 2023

Agritech in India: Industry Overview and Scope for Investments (2022) https://www.india-briefing.com/news/agritech-in-india-overview-and-scope-investments-23628.html/. Accessed 30 Nov 2023

Arvanitis KG, Symeonaki EG (2020) Agriculture 4.0: the role of innovative smart technologies towards sustainable farm management. Open Agric J 14(1):130–135. https://doi.org/10.2174/1874331502014010130

Asia Pathways (2023) Japan guides the way on smart farming technology adoption. https://www.asiapathways-adbi.org/2023/01/japan-guides-the-way-on-smart-farming-technology-adoption/. Accessed 30 Nov 2023

Department Of Science Technology (2023) Call for proposal-2022 under "India-Japan Cooperative Science Programme".https://dst.gov.in/callforproposals/call-proposal-2022-under-india-japan-cooperative-science-programme. Accessed 30 Nov 2023
From Farm to Table: AgTech Innovations That Drive Agricultural Efficiency Across the Value Chain (2023) https://www.plugandplayapac.com/post/from-farm-to-table-agtech-innovations-that-drive-agricultural-efficiency-across-the-value-chain. Accessed 30 Nov 2023
Foodtechbiz (2022) J-Methods Farming, a concept to improve agriculture productivity in India with Japanese technologies. https://www.foodtechbiz.com/agriculture/j-methods-farming-a-win-win-concept-to-improve-agricultural-productivities-in-india-with-japans-agricultural-technologies. Accessed 30 Nov 2023
Goh, L (2021) How AgriTech is transforming traditional agriculture in emerging markets "Think Big, Act Fast, Start Small"
Hive Life Magazine (2021) Your Guide to AgriTech in Japan. https://hivelife.com/agritech-in-japan/. Accessed 30 Nov 2023
Iwasaki K (2022) Policy support for Japan to become a startup powerhouse. https://www.jri.co.jp/en
JICA (2023) A revolution for a billion people–agri-tech challenges in india. https://www.jica.go.jp/english/about/dx/jicadx/dxlab/goodpractice/casestudy_1/index.html. Accessed 30 Nov 2023
McKinsey (2023) How agtech is transforming agriculture in India. https://www.mckinsey.com/industries/agriculture/our-insights/how-agtech-is-poised-to-transform-india-into-a-farming-powerhouse?utm_medium=DSMN8&utm_source=LinkedIn&utm_user=1441923777391313 2/. Accessed19 July 2023
Medvediev A (2020) Agritech–towards transforming Indian agriculture EY
Ministry of External Affairs Government of India (2022) India-Japan sustainable development initiative for the North Eastern Region of India. https://www.mea.gov.in/bilateral-documents.htm?dtl/34993/IndiaJapan+Sustainable+Development+Initiative++for+the+North+Eastern+Region+of+India. Accessed 30 Nov 2023
OECD (2019) Innovation Agricultural Productivity and Sustainability in Japan. OECD Food and Agricultural Reviews, Paris. https://doi.org/10.1787/92b8dff7-en
Prepared by the New Zealand Embassy in Tokyo (2022)
Press Information Bureau (2023) https://pib.gov.in/PressReleasePage.aspx?PRID=1868937. Accessed 30 Nov 2023
Regional Revitalization of JapanGov (2021) https://www.japan.go.jp/key_policies_of_the_suga_cabinet/regional_revitalization.html. Accessed 30 Nov 2023
Rise of Agri-tech in India (2023) https://www.drishtiias.com/daily-updates/daily-news-editorials/rise-of-agri-tech-in-india. Accessed 30 Nov 2023
Schilling (2021) Smart farming technology in Japan and opportunities for EU companies. http://www.EUbusinessinJapan.eu
Srinivasrao C, Kumar R (2022) Agri-startups in India: opportunities, challenges and way forward National Academy of Agricultural Sciences New Delhi
The Economic Times (2023) India-Japan script an expanding partnership for a shared future. https://economictimes.indiatimes.com/tech/startups/india-japan-script-an-expanding-partnership-for-a-shared-future/articleshow/102139039.cms. Accessed 30 Nov 2023
The Rise of Agri-tech in India (2023) https://www.investindia.gov.in/team-india-blogs/rise-agri-tech-india. Accessed 30 Nov 2023
Tokyoesque (2022) Growth And Innovation: Overview And Opportunities Of Agritech In Japan 2022. https://tokyoesque.com/growth-and-innovation-opportunities-agritech-japan/. Accessed 30 Nov 2023
Washizu A, Nakano S (2022) Exploring the characteristics of smart agricultural development in Japan: analysis using a smart agricultural kaizen level technology map. Comput Electron Agric 198. https://doi.org/10.1016/j.compag.2022.107001
ZEN-NOH (n.d.) The JA Group. Our Cooperatives. https://www.zennoh.or.jp/english/cooperatives/jagroup.html. Accessed 30 Nov 2023

Culture and People

Role of Yoga in Post-COVID Stress Management in Urban Areas: Analysis of Japan with Reference to India

Akiko Kato, Kazuaki Ohhashi, and Rajib Shaw

Abstract Human stress is naturally influenced by the environment. Changes in the environment in which people live will directly lead to changes in the causes of stress. Stress has a strong link of urbanization. Due to urbanization, and lack of green spaces, the sports or outdoor activities are often restricted. This is same both in Japan and India. During COVID-19, people had to stay in home and this caused significant increase of stress level among different age groups. The youth groups are possibly the worst sufferer. Through two specific surveys (one among university students, and one adult population) in Japan, this chapter argues the importance of yoga in stress management. The Yoga Association of Japan has been promoting yoga in different groups (including sports people, diet members, different levels of schools and universities, maritime force etc.). It is argued that regular practice of yoga can reduce both physical and mental issues. Yoga can also be considered as a communication platform.

Keywords Covid-19 · Urbanization · Stress · Yoga · Communication platform

1 Introduction

Human stress is naturally influenced by the environment. Changes in the environment in which people live will directly lead to changes in the causes of stress. For example, there is a relationship between stress and temperature, with rising temperatures affecting stress (Zhang 2014), especially at night, when high temperatures cause autonomic imbalance. Environmental changes, such as urbanization and related phenomena become another cause of stress. Urbanization has reduced the amount

---

A. Kato · R. Shaw (✉)
Graduate School of Media and Governance, Keio University, Fujisawa, Japan
e-mail: shaw@sfc.keio.ac.jp

K. Ohhashi
Yoga Organization of Japan, Tokyo, Japan

R. Shaw and S. R. Choudhury (eds.), *India, Japan and Beyond*,
https://doi.org/10.1007/978-981-97-3282-1_13

of greenery in urban areas, and there are fewer places to do exercise. The concentration of population due to urbanization, on the other hand, causes traffic congestion and stress builds up during the commute to and from work. Coronavirus infections (COVID-19) had a significant impact on urbanized societies.

Since the outbreak of novel coronavirus infection (COVID-19) was reported in Wuhan, Hubei Province, People's Republic of China, in December 2019 and beyond (TMG 2021), the number of infected people worldwide has exceeded exponentially through several phases of virus infection over a period of 2+ years. In response to the large-scale infectious disease, the Japanese government, like other countries had not only tried to save patients at medical institutions and by medical personnel, but it also managed the community health to combat this new virus in every aspect, including finance, economy, and daily life of the people.

On April 7, 2020, the Japanese government declared a state of emergency under Article 32 of the revised Law Concerning Special Measures against Influenza Pandemic and Other New Strains of Influenza, stating that "a nationwide and rapid epidemic has occurred that may have a serious impact on the lives of the people and the national economy" (CAO 2020). The emergency declaration is specifically a request to the public to refrain from leaving their homes, a measure to reduce economic activities and keep people indoors (unlike the lockdown in other countries, including India). This first attempt to have the entire population refrain from leaving the house is creating problems for society as a whole other than direct medical treatment. Of course, in addition to the economic stagnation, more than 100 million people without coronary infections or other diseases have remained in their homes for more than a month, and citizens of all ages, including infants, schoolchildren, working people, and the elderly, have been forced to refrain from going out. In other words, the unprecedented fight against the virus has placed a burden of inactivity and stress on a healthy population.

The stress caused by this refraining from going out has resulted in social and family problems such as domestic violence, and there are already concerns about an increase in suicides due to economic insecurity (Mainichi Shimbun 2020). The stress and lack of exercise associated with this voluntary curfew, known as "Corona Blue," has become a problem not only for adults who work from home, but also for the entire population, from children who cannot commute to school to the elderly who cannot attend day care services. On April 22, 2020, two weeks after the state of emergency was declared, a meeting of government experts to discuss measures to combat the new coronavirus, gave a "situation analysis" and "recommendations" for a 10-point plan to reduce human contact by 80%, online activities were proposed (MHLW 2020). The result was in the form of a recommendation to relieve physical and mental stress and maintain wellness through yoga, which can be done online and in the home.

Keio University had also decided to incorporate video yoga lessons as a means for students to maintain mental and physical wellness at home when they are not allowed to leave the house. In this way, yoga is being recognized again as a useful tool for relieving stress in the body and mind in a society where people are refraining from going out or are prohibited from going out not only in Japan but also in the whole world in order to fight an unprecedented large-scale infectious disease disaster caused by the new coronavirus infection (COVID-19). At that point, it was uncertain whether the new coronavirus infection (COVID-19) would be completely eradicated in near future, and it was highly likely that intermittent and partial curfews and bans will continue for couple years to come, creating a similarly stressful situation around the world.

With this context, this chapter will first examine how people were stressed by the coronavirus infection, the rarest, deepest, and most widespread disaster in human history, which affected people's minds and bodies in India and Japan. The chapter will also analyze the urbanization and stress and growth and evolution of yoga in the two countries. Finally, some analysis results of the practice of yoga among the university students in Japan are presented.

2 Effects of Urbanization and Stress in India and Japan

As already mentioned above, the stresses associated with the 2020 COVID-19 Crisis have brought significant stress to the society, but even before this unprecedented large-scale infectious disease, societies in developed and developing countries, including Japan, were under tremendous stress even in normal times. The annual number of suicides in Japan is 20,840 (2018 data), while in India it is around 230,314 people (2016), who attempted suicide. Japan and India have extremely high suicide rates. As these figures of Japan: 18.5 (out of 100,000) and India: 16.3 (out of 100,000) suicides (WHO 2016) show, it is clear that both countries are stressful societies. In addition to normal times, in 2020, the curfew due to the fight against COVID-19 and similar viruses is expected to disrupt the global economy, leading to an increase in the number of bankruptcies in many microeconomic sectors, such as food service, retail, and tourism Asahi Shimbun (2020), and a corresponding increase in suicides. The large-scale infectious disease disaster will spread from medical and economic problems to the human mind and stress problems as well.

There is a relationship between stress and temperature, with rising temperatures affecting stress (Zhang 2014), especially high nighttime temperatures causing autonomic imbalance. The concentration of population due to urbanization, on the other hand, causes traffic congestion and stress builds up during the commute to and from work. Indian cities are facing the same problems of urbanization that Japan has been facing, creating a stressful society in the post world war development stages (Kono 1988). In Japan, the method of refraining from going out and the stress caused by such refraining differ depending on the situation of each city, such as highly urbanized large cities and their surrounding cities, isolation of sparsely populated regional cities, and isolated islands, especially Okinawa (problems with medical facilities and the number of workers). In India, the economic and medical disparities caused by urbanization are even more pronounced, and the Indian government is exploring how yoga can be used to address the immunity, health, and stress issues of people living in slums with poor sanitary environments (AFP BB 2020). Also, with the lockdown and even inter-city transportation being stopped by the government, there is a need for yoga to provide a way to address the physical and mental wellness issues that accompany the return home on foot of low-income people who have been flowing to the big cities as a result of urbanization (Jiji 2020). Following the Ministry of Yoga, Government of India, the Ministry of Science and Technology, Government of India, has initiated a study on "Appropriate Scientific and Technological Interventions of Yoga and Meditation in the Fight against COVID-19 and similar viruses" (Hitadva 2020).

3 History, Benefits and Yoga Practices in India and Japan

Yoga, which was also practiced by the Indus civilization 5,000 years ago, is said to be "the oldest exercise science in the world". Even in ancient India, yoga was studied and practiced not as an exercise for the body, but as a meditation method for the liberation of the mind (Ayush 2020). In the United States, which is said to be the leading yoga country along with India, the population of yoga is said to be 25 million, 6 million of which are children, and it is used for emotional education (Seven and I 2017). In the U.S. and India, the elucidation of the scientific basis for the effects of yoga is attracting attention. Not only have the NIH (National Institute of Health) and the Morarji Desai National Institute of Yoga demonstrated that yoga can inhibit all possible diseases of the body, such as stroke, heart disease, and high blood pressure, but yoga's ability to manage stress is also attracting attention (Murthy 2009). In India, there is a Ministry of Yoga (AYUSH) that oversees the aforementioned Morarji Desai National Institute of Yoga and the Central Council for Research in Yoga & Naturopathy (National Yoga Research Institute), and the Minister of Yoga has been appointed to this ministry. The nation is working on those studies.

In India, the economic and medical disparities caused by urbanization are more pronounced, and the Indian government is exploring how yoga can be used to address the immunity, health, and stress issues of its citizens living in slums with poor sanitation (AFP BB News 2020). Also, in India, with the city locked down by COVID-19 and even inter-city transportation stopped by the government, yoga is also being sought as a way to address the physical and mental wellness issues associated with the problem of low-income people returning to the big cities on foot, who had been flowing to the cities as a result of urbanization (Jiji 2020). Following the Ministry of Yoga, Government of India, the Ministry of Science and Technology, Government of India, has initiated a study on "Appropriate Scientific and Technological Interventions of Yoga and Meditation in the Fight against COVID-19 and similar viruses" (The Hitavada 2020).

Yoga is currently playing an increasing role in health and industry in Japan. Japan's yoga population is approximately 5.9 million people who do yoga at least once a month, 7.7 million people who do yoga at least once a year, 8.8 million people who would like to do yoga within the next two to three years, and the future yoga population including potentials is approximately 16 million people (Seven and

I 2017). Several companies and municipalities are beginning to utilize yoga for the health of their employees. The city of Nishinoomote in Kagoshima Prefecture is recognized as a yoga hotspot and is working to create a local community through yoga, and elderly people on a remote island can relieve the stress of loneliness by practicing yoga. Kagoshima Prefecture, which aims to be a prefecture of wellness, is actively seeking ways to use yoga, and Kanagawa Prefecture, under Governor Kuroiwa, is already implementing measures to improve unwellness through yoga. In addition to these local governments, Kudaka Island, Nanjyo city in Okinawa Prefecture, Yoshino Town in Nara Prefecture and Izumo City in Shimane Prefecture are also aiming to improve physical and mental health and wellness through yoga. In addition, an increasing number of private companies are also aiming to improve health and wellness through yoga, including Kabaya Ohayo Holdings Japan, Inc. The Skytree has achieved its goal of being recognized as a "sacred yoga site," linking the Skytree to yoga as a symbol of health in Japan.

4 Analysis of Stree Management and Yoga in Japanese Students

In this section, the survey results of the analysis of stress management and yoga is presented. The survey was conducted during December 25, 2020 to January 15, 2021, the peak time of COVID-19, and before the vaccine was used for the youth in Japan. Total number of students were 27, 10 males and 17 females of age group between 18 and 22 from different parts of Japan.

4.1 Causes of Stress

The most common cause of stress was the threat of COVID-19 and the fear that they may have the virus. Some felt excessive nervousness about cleanliness. The stress of not being able to go out or refraining from going out due to fear of infection in the family was also stressful because it restricts one's activities, and young people, who were usually more active, seemed to be more restricted in their activities and more stressed. Not being able to go out also meant not being able to move around much, which limited sports and recreational activities, and physical and physical inactivity contributed to stress.

The stress was not only caused by individuals; relationships with others, inability to share time with others, inability to see others, and lack of communication also contribute to stress. New students in particular felt lonely and anxious about not being able to make friends, and although they were able to use various new tools such as SNS to communicate in new ways, it was rather difficult for them to communicate with others in a simple way. Communication was not possible. In other words, they were stressed by the fact that they had to "go to the trouble" of communicating. Unlike in cities, in rural areas, there is stress in relationships due to suspicion of the disease, and stress due to the lack of walking due to rural life (living by car).

The negative effects of social networking also create new stressors, such as the way people present themselves on social networking services and the watchful eye of the police. Not being able to go out, which not only disrupts the rhythm of life, but also causes increased fights as the result of increased contact with family members, which was also considered stressful for the young people. Social stress was also a new stress. These included the failure to find a job, the cancellation of study abroad programs, and facing vague fears about the future. There was also the mental stress of adapting to a series of various schedule changes due to self-restraint and consensuses and changing schedules. The new normal lifestyle also created new physical stresses such as increased PC time, fatigue from looking at screens all the time, and VDT (visual display terminal) syndrome caused by online activities. Many young people suffered from mild depression or mild mental illness (mental crisis), and their anxiety accelerated the negative bias, leading some to become even more depressed.

4.2 Effects of Yoga as Perceived by the Students

Although many of the students in this study were yoga beginners, many of them seemed to have a new awareness of yoga. Many of them mentioned that "yoga allowed me to move my body anywhere and anytime without having to go outside, and it reduced my physical stress". The physical benefits of yoga were mentioned as reducing back pain and stiff shoulders due to improved blood circulation, elimination of intestinal problems, and relief of the physical aspects of VDT syndrome due to the New Normal. Some mentioned that constant practice of yoga helped them to notice small changes and irregularities in their own bodies, which they used to help them in the hospital. Many simply believed that yoga improves metabolism, increases basal body temperature, and strengthens the immune system.

On the mental side, many thought that meditation and breathing exercises and improved mental health were beneficial, and they were willing to accept yoga as a lifestyle if it not only reduced stress, but also made them more tolerant of stress afterwards. Students who wish to improve their sleep, for example, experience the importance of breathing exercises and postures at the same time, and that the improvement of stress, both physical and mental, leads to good sleep. Many students are also interested in the benefits of being able to listen to voices (introspection), and one student stated that introspection through meditation (focusing only on oneself) can stop mild panic symptoms. They feel experientially that yoga can help alleviate symptoms of depression. Some students responded that the positive effects of yoga are not negative to zero, but more positive to zero: improved blood flow improves concentration, stress management improves life skills, and they are able to manage their own stress and learn how to control their own bodies.

Finally, the sense of connecting with others through yoga not only reduced stress for men and women of all ages, but some students also discovered the benefits of yoga in a larger group, meeting new people (beyond age) in local yoga classes, and using yoga not only to reduce friction between family members but also to create new communities. Some thought it was a good idea.

5 Stress Management and Yoga in Adult Population

Dr. Richard Brown of Columbia University, in an interview on NHK, stated that mindfulness meditation and reducing the frequency of breathing can calm the ectopia and help with stress management. Since the respiratory center is close to the pituitary gland, breath control through yoga calms the pituitary gland and reduces stress by suppressing cortisol secretion from the adrenal cortex. Pranayama (yogic breathing exercises) can reduce blood flow to the pituitary gland, thereby reducing stress (NHK 2019). Also, with regard to the body, the Ministry of Health, Labor and Welfare states that "mortality rates decrease with increased physical activity" (MHLW 2021). People who are physically active have lower rates of hypertension, diabetes, obesity, osteoporosis, colon cancer, and other lifestyle-related diseases, as well as lower rates of mortality. Physical activity (exercise and lifestyle activities) improves mental and lifestyle aspects.

This section will describe how yoga has been used as a tool for stress management, based on the "Questionnaire on Yoga Practice at the Corona Disaster" conducted with the All-Japan Yoga Association. The Indian Ministry of Science and Technology invited proposals to study "Appropriate Scientific and Technological Interventions of Yoga and Meditation in the Fight Against COVID-19 and Similar Viruses". The call was in the form of a broad call for proposals from scientists, clinicians, and experienced yoga and meditation practitioners with proven experience in the field of yoga and meditation research on the following topics: (1) Immunity enhancement, (2) Improve the function of the respiratory system and other organs to overcome respiratory diseases such as COPD, (3) Resolving problems related to stress, anxiety, and depression due to loneliness, various social uncertainties such as school and work, and disruption due to the inability to lead a normal life.

In response to this, the Yoga Organization of Japan conducted the questionnaires between May 15 to June 1, 2020, where 336 people responded, who were interested in yoga, India, health and beauty. The call was given through Keio University India Japan Lab's India class, Yoga class, network of Yoga organizations and social media etc. The age group of the respondents were between 20s and 70s, coming from different nationalities residing in Japan, including Japanese and Indian. Overall, 77.8% of the respondents answered that they were feeling stress of some type, with less motivation, getting irritated easily, less appetite, out of breath, lack of sleep etc. The causes were: (1) fear of infection, (2) isolation, (3) insecurity and (4) relationship with family. 25% used to do yoga to weight loss, physical fitness as well as stress release. Some did other sports, talking to family members etc.

There were several suggestions provided by the respondents: (1) all generations (elderly, mature adults, elementary, middle, high school, college, and young children) can practice yoga, (2) Online yoga needs to be standardized, there is a need to develop a standardized disaster prevention yoga program for general disaster shelters, visualization of the degree of stress is necessary.

6 Examples of Yoga for Stress Management in Different Parts of Japan

Following the COVID-19 experiences, Yoga Association of Japan promoted yoga as a stress management tool in wider section of the Japanese society. Following are some of the examples:

Sports (American football, soccer, sumo): Tom Brady of the American football superstar Buccaneers is a yoga enthusiast. A reporter for The Sportster in Quebec, Canada wrote, "Tom Brady's health care is not complete without embracing the ancient Eastern wisdom of health". He is a serious believer in yoga, and he recognizes the tremendous benefits it has for both the mind and the body. Tom Brady also said of yoga, "Yoga has a positive effect on flexibility, on healing, and on one's mental attitude" (Feel guide 2017). Yoga has also been actively adopted by American Major League Baseball players, the British Rugby Union, and the Australian women's national soccer team. Japanese professional soccer players also use yoga to prevent injury as well as for mental unification. The same is true for Japanese sumo wrestlers. Terunofujizeki of the Isegahama group, Atamifujizeki, Terutsuyoshizekii, Takarafujizeki, and Midorifujizeki, as well as Terunofujizeki of the Isegahama group, all of whom have made miraculous comebacks after falling to the makushita division due to injury after their promotion to Yokozuna (the highest rank), utilize yoga for mental and physical management.

Stress management for infants, elementary schools, junior high schools, and universities: There is a growing trend to incorporate stress management into school education as a form of mental management, including stress management. A yoga program was introduced at Souwa Minami Junior High School in Koga City, Ibaraki Prefecture, to reduce stress and improve concentration. In particular, yoga is being used to help third-year junior high school students, who have retired from extracurricular activities and are less physically active, to regain physical and mental balance. Yoga is also offered in physical education classes at Keio University. Yoga was the most popular remote physical activity class at the Corona Disaster.

Examples of yoga in the Maritime Self-Defense Force, Indian Armed Forces, and U.S. Armed Forces: Military forces of various countries whose activities are themselves subject to stress, as well as Japan's Maritime Self-Defense Force, have introduced yoga programs. In Japan, the Maritime Self-Defense Force Yoga Project (Maritime Self-Defense Yoga and Submarine Yoga) is working with the All-Japan Yoga Federation on a stress management yoga project. Stress management through yoga is being practiced in the Himalayan Army and Indian Navy in the extreme cold regions of India. The U.S. Navy SEALs have been more active in stress management through breathing techniques, and a breathing technique called Box Breathing, which incorporates the Kumbhaka breathing technique of yoga, is used to calm the body and mind in extremely stressful situations.

7 Yoga as a Communication Platform

Finally, yoga's role as a communication platform is also effective for stress management due to urbanization. In addition to the mental and physical benefits of yoga, the benefits of yoga may include the moral values cultivated through yoga and a network of yoga people (yoga friends) from around the world. Yoga is not only about exercise and direct physical benefits, which many people misunderstand; Baltimore, Maryland, USA, has introduced yoga meditation in primary and secondary education and has significantly reduced violence and violence-related mortality (Holistic Life Foundation 2021). The students were able to take control of their minds by returning the school's punishment room to the meditation room.

It is also possible that yoga communities may prevent loneliness and suicide due to loneliness. In a time when many people feel lonely due to both urbanization and depopulation, the existence of a yoga community has the potential to help those who suffer from loneliness, young and old. This is because yoga originally has the meaning of connecting. Yoga connects the mind and body of an individual, connects people within a community (bonding), and finally connects the yoga community to the community (bridging).

8 Conclusion

In the massive disaster of the Corona disaster, the causes of people's stress due to urbanization became prominent. Almost all people felt stress because their living environment had been changed by the corona, and many were forced to confront their stress. For example, there were people who felt lonely because of urbanization, but thanks to Corona's self-restraint, everyone clearly felt lonely. Even those who did not realize that they were stressed more or less realized that they were stressed. Urbanization stress does not necessary equal to corona stress. However, the places where the voluntary curfew was more firmly in place were more urbanized areas, and it is in more urbanized areas that coronavirus stress afflicts more people. Environmental changes such as urbanization and viral infections have created various sources of stress, including fear of viral infections themselves, decreased physical activity due to self-restraint, anxiety about relationships and communication, and anxiety about the burden of adapting to a new social environment and the future. And yoga is effective for a variety of reasons in relieving that stress and managing stress. Many people are already using yoga to combat their stress and create better mental and physical wellness and a healthier body. The yoga community can also "connect" people and save individuals from isolation. We will continue to search for more effective ways to manage stress through yoga.

References

AFP BB News (2020) New corona spreads to India's biggest slum, both infected and dead. https://www.afpbb.com/articles/-/3278408. Accessed 5 May 2021

Asahi Shimbun (2020) Corona-related bankruptcies exceed 100, funds drying up, increase after holidays?. https://www.asahi.com/articles/ASN4Z6Q8PN4ZULFA031.html. Accessed 5 May 2021

CAO (2020) Prime Minister's Office (2020) Headquarters for countermeasures to combat new coronavirus infections (27th meeting). https://www.kantei.go.jp/jp/98_abe/actions/202004/07corona.html (May 2021). Accessed 5 May 2010

FEEL guide (2017) The Tao of Tom: how Tom Brady uses an 80-20 diet, meditation, yoga & one book to age backwards. https://www.feelguide.com/2017/01/15/the-tao-of-tom-how-tom-brady-uses-an-80-20-diet-meditation-yoga-one-book-to-age-backwards/. Accessed 28 June 2021

Holistics Life Foundation (2021) Holistic me after school program. https://hlfinc.org/programs-services/after-school-programs/. Accessed 19 June 2021

Jiji.com News (2020) Corona hits India's poor, girl walks 150 km, runs out of life—a wrinkle in 'total blockade'. https://www.jiji.com/jc/article?k=2020042800778&g=int. Accessed 5 May 2021

Kono T (1988) I-B-32 urban stress and health disorders (stress II - social mental health - 1-). Psychosom Med 28(Abs):48

Mainichi Shimbun (2020) 'Loneliness of not being able to see you' and 'anxiety and stress': a rapid increase in consultations, various backgrounds, and detailed care. https://mainichi.jp/articles/20200424/k00/00m/040/178000c. Accessed 5 May 2021

MHLW (2020) Ministry of Health, Labour and Welfare "Analysis of the situation and recommendations on measures against new coronavirus infection." https://www.mhlw.go.jp/stf/seisakunitsuite/bunya/0000121431_00093.html. Accessed 5 May 2021

MHLW (2021) Ministry of Health, Labor, and Welfare website, "Physical activity and exercise." https://www.mhlw.go.jp/www1/topics/kenko21_11/b2.html. Accessed 19 June 2021

Ministry of Ayush, Government of India (2020) Certification of yoga professionals-official guidebook, pp 6–8

Murthy BTC (2009) Effect of asanas and pranayamas on neurological, neuromoscular & cardio-respiratory function in healthy human volunteers. Central Council for Research in Yoga & Naturopathy

NHK (2019) NHK 3rd Production Unit (Science) (2019) NHK Gatten! Lower blood pressure easily! Saikyou Waza [Strongest Waza]. Shufu to Seikatsu-sha, p 85

Seven & i Publishing Co. (2017) Yoga journal Japan edition: survey of the yoga market in Japan 2017. https://www.7andi-pub.co.jp/pdf/2017/20170307_sevenandi_yoga.pdf. Accessed 5 May 2021

The Hitavada (2020) Govt invites proposals to study effects of yoga, meditation in fighting COVID-19, p 4

TMG (2021) Tokyo Metropolitan Information Center for Infectious Diseases (2021) Information on new coronavirus infection (COVID-19). http://idsc.tokyo-eiken.go.jp/diseases/2019-ncov/. Accessed 5 May 2021

WHO (2016) World Health Organization, Suicide rate. https://www.who.int/data/gho/data/themes/mental-health/suicide-rates. Accessed 4 Jan 2024

Zhang J (2014) Basic research on heat stress inflicted by environmental temperature and humidity on the human body: from physiological, psychological, autonomic nervous system and salivary amylase response. SENKEI J 55(10)

Edu-Tech and Its Implications in Promoting Japanese Language Education in India

Yuka Kataoka, Shraddha Namjoshi, and Mahati Chittem

Abstract With the growing closeness of the India-Japan political and economic relationship over the past decade, Japanese language and cultural education has become important in India as these foster and bolster this growing relationship. However, Japanese remains an uncommon foreign language, and the Japanese language education and learning environment in India for even the basic level of Japanese language learning at universities is not yet equally or well developed. Thus, this somewhat under-representation of Japanese lends itself to limited opportunities for learners to practice and experience the language in India. This chapter focuses on the large-scale Japanese-language education at the Indian Institute of Technology Hyderabad (IITH), which has laid the foundation for an ongoing Japan-India academic, research, and employment collaboration. The chapter introduces the initial phase of Japanese language educational collaborations for IITH students and reviews 11 Japanese language courses at IITH for more than 1687 students in the past seven years on how Information and Communication Technology (ICT) has been effectively utilized to increase learners enthusiasm, knowledge, and skills in Japanese. Using an interactive methodology and ICT techniques, interest in and retention of Japanese culture were sustained among the Indian students, thus enhancing learning effectiveness and enabling a greater desire to engage in a multitude of activities within the Japanese context.

Keywords Japanese language education · Information and communication technology · Learning management system · Oral repetition practice · Electronic feedback

---

The affiliation of the first author is the institution at which the author conducted the research.

Y. Kataoka (✉)
Graduate School of Media and Governance, Keio University, Fujisawa, Japan
e-mail: shori@sfc.wide.ad.jp

S. Namjoshi · M. Chittem
Liberal Arts Department, Indian Institute of Technology Hyderabad, Hyderabad, India

R. Shaw and S. R. Choudhury (eds.), *India, Japan and Beyond*,
https://doi.org/10.1007/978-981-97-3282-1_14

1 Background

Expectations for the use of ICT (Information and Communication Technology) in education are increasing in the international community to complement and enrich education. SDG4, one of the 'SDGs (Sustainable Development Goals),' which is a global goal to be achieved by 2030, sets a goal of high-quality education for all. To improve the quality of education, to decrease educational inequality, to promote lifelong learning opportunities, and to develop teachers' skills, further educational practices utilizing ICT are required. In addition, as the direction of education policy in Japan, the Ministry of Education, Culture, Sports, Science and Technology (MEXT) has been promoting the introduction of ICT-based education and its infrastructure from 2018 to 2022 in the areas of (1) learning guidance in primary and secondary education, (2) improvement of education quality and development toward active learning in higher education, and (3) providing learning opportunities throughout one's life. However, there are still many difficulties and challenges in introducing ICT in education, such as human resource development and budget constraints. In addition, educational methods and models differ in each field of education, and it is unlikely that all problems can be solved simply by introducing ICT, and there are still some educational fields where problems have not yet been clarified. Language education is also no exception. The effective use of ICT and the sharing of its findings in language education are also essential to tackling various problems and further improvement for SDG4.

According to the Japan Foundation's survey (Japan Foundation 2023), the number of Japanese language learners has rapidly expanded approximately 30 times during the two decades (Fig. 1). There might be various factors to cause this increased number of learners: close international relationships, interactions, and corporations between Japan and other countries in many domains such as science and technology, arts and culture, business and commerce, policy and governance, etc. However, the number of Japanese language teachers is still very small compared to the number of Japanese language learners. Therefore, the Japanese language remains underrepresented as a minor foreign language (FL) in many regions outside Japan (Kato 2020; Kataoka et al. 2023b) and it has limited the learning opportunities to be exposed to the Japanese language and culture.

While effective language acquisition requires practical engagement with native speakers, Japanese as a foreign language (JFL) learners face challenges in obtaining opportunities for meaningful speaking practice due to an unbalanced student-teacher ratio in their learning environment. The guidelines outlined by the American Council on the Teaching of Foreign Languages (ACTFL) emphasize the necessity of the provision of regular and substantial interactions between students and instructors, monitored practice and exercises, and individualized feedback during classroom hours to foster students' communicative proficiency (ACTFL 2012). Ideally, the classroom size should be limited to up to 15 students to implement meaningful speaking practice and exercises as ACTFL advocates. In situations where class sizes exceed this limitation, it is necessary to provide instructors with supplementary support to ensure

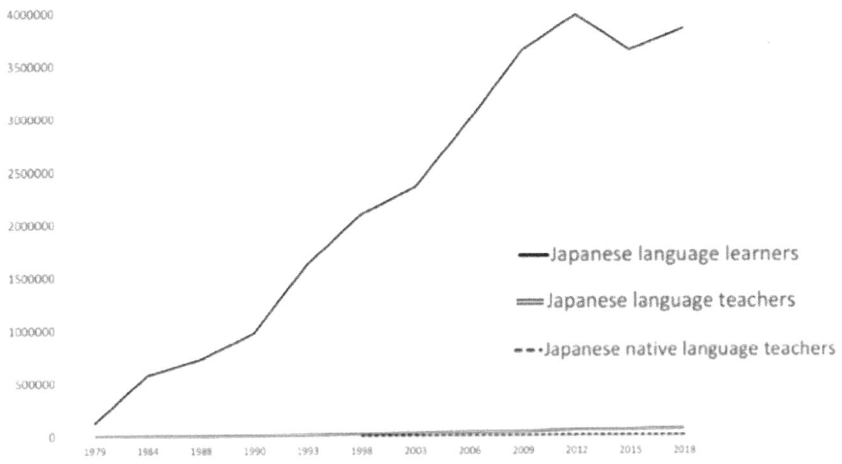

Fig. 1 Situation of Japanese language education abroad (Japan Foundation 2023)

pedagogical best practices. However, the support is deficient for Japanese language instructors working abroad, hindering their capacity to address the multifaceted demands encountered in diverse countries and regions. When the target language is minor FL, there are many constraints to implementing language teaching methods in terms of sociocultural gap, limited number of teachers, and limited opportunities to use the language (Butler 2011).

According to the Japan Foundation (JF)'s latest survey (Japan Foundation 2023) and the past test data on the Japanese Language Proficiency Test (JLPT) website (JLPT 2023a), the number of Japanese language learners in India and the number of JLPT examinees in India have increased during the past two decades (Fig. 2). JLPT has been offered by JF and Japan Educational Exchanges and Services (JEES) since 1984 to evaluate and certify the Japanese proficiency of non-native speakers (JLPT 2023b). JLPT is provided two times a year, in July and December. The test centers in India are in Delhi, Pune, Kolkata, Chennai, Bangalore, Mumbai, Salem, Santiniketan, and Karur. (Some new centers such as Salem, Santiniketan, and Karur provided JLPT once a year.) In general, most reasons why people overseas take JLPT are to measure their level of proficiency and to make use of it for their work such as finding jobs and securing salary increments and promotions, etc.

Specifically, the number of Japanese language learners in India has expanded 6.6 times over 19 years and the number of JLPT examinees in India continued to rise from 2009 to 2019 before the COVID-19 pandemic in 2020 (Fig. 2). Although the examination was suspended due to the COVID-19 pandemic in 2020, the number of examinees rebounded more than 20,000 in 2022. A half number of learners take JLPT every year except during the pandemic periods. These data indicate a higher interest in the Japanese language as well as its qualification in India due to the active international exchanges and collaboration between India and Japan.

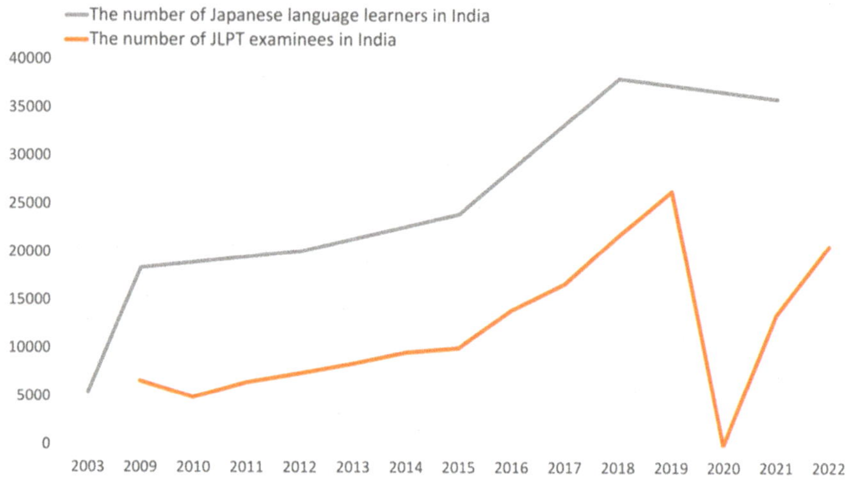

Fig. 2 The number of Japanese language learners and JLPT examinees in India (Japan Foundation 2023; JLPT 2023a)

Although the number of Japanese learners has increased, its population is still small among the total population of India. Therefore, Japanese remains an uncommon foreign language, with not many Indian people frequently viewing it as the foreign language of choice (as opposed to French and German, for example). Indeed, although there are some educational institutions and universities that offer Japanese as a foreign language, Japanese is not routinely considered a part of the curriculum in primary and secondary education for Indian people. The opportunities to be exposed to the Japanese language and culture are tightly limited in their society. In the case of higher education, some universities whose disciplines cover foreign language studies implemented annual Japanese courses. However, science and engineering universities have severe constraints to implementing Japanese courses even as short-term elective courses since there is a specific curriculum for their students.

2 Case of Japanese Language Courses in IITH

The Indian Institute of Technology Hyderabad (IITH) was established in 2008 as the institute of science and technology. IITH offers undergraduate, master's, and Ph.D. programs in classical engineering disciplines, applied science, and some interdisciplinary departments such as liberal arts, design, and entrepreneurship and management (Indian Institute of Technology Hyderabad n.d.). IITH was initially established with Japanese government funding and has laid the foundation for an ongoing Japan-India academic, research and development, and employment collaboration. For example, the DISANET project, which is an Indo-Japan research collaboration on establishing the infrastructure of an information network for natural disaster mitigation and recovery, was conducted from 2009 to 2015. DISANET was conducted on the framework of Science and Technology Research Partnership for Sustainable Development (SATREPS) which is the joint program of Japan Science and Technology (JST) and Japan International Cooperation Agency (JICA). In addition, JICA FRIENDSHIP Scholarship program offers a scholarship for IITH students to study for Ph.D. degrees (including Master's degrees) in the field of science, technology, and engineering at Japanese university (FRIENDSHIP Project Phase 1 continued from 2013 to 2020 (Japan - IIT Hyderabad collaboration 2020) and Phase 2 is ongoing from 2021 till the present (FRIENDSHIP Project 2023)).

Because there have been active human exchanges between India and Japan from the early years of IITH's founding, Japanese language and culture also became fundamental to people in IITH for their smooth communication with people in Japanese universities and the advanced preparation for their trip or stay in Japan. For these reasons, short-term Japanese language training was provided to the IITH students who were selected as FRIENDSHIP scholarship students in 2015. After that, the Japanese course in IITH commenced as an elective course in the Liberal Arts department in 2016.

2.1 The Short-Term Japanese Language Training for FRIENDSHIP Scholarship Students

At the time of 2015, there were not so many, but some Japanese language or culture courses or training were offered to IITH students within the framework of a working group, aiming to investigate and cultivate potential collaboration towards the establishment of IITH as a new IIT (Ministry of Foreign Affairs of Japan 2008). Although Japanese language education was included as a possible means of cooperation, there were not so many customized options for them to learn Japanese depending on various needs. Therefore, we assumed that the scholarship students have worries and anxieties about their new lives in Japan due to the language and cultural differences. The short-term Japanese language training aimed to help the scholarship students adjust to Japanese language and culture more easily, reduce their anxiety, and enhance their interest and motivation for Japanese language and culture. The language training should be effective and not be a burden for them, it was designed as a three-day intensive language course that contains a lot of interactive practice and exercises with Japanese native and non-native language instructors. The course was repeated two times for a different group of students: three students in July and eight students in August 2015 before they shifted to Japan.

As Fig. 3 shows, the course was conducted with a collaboration of native and non-native (local) Japanese language instructors. The native Japanese language instructor was a native Japanese speaker and three-year experience of residing in India but just started working as a Japanese language instructor in India after completing the

420-hour Advanced Japanese Teacher Training Course. The non-native language instructor was a local Japanese language instructor who commands English, Telugu, Hindi, and Japanese. She had three-year English teaching experience at Japanese elementary schools in Japan and had Japanese teaching experience for Indian students since 2014. Native Japanese and non-native instructors have different roles in facilitating language practice and exercises and sharing different perspectives on learning Japanese language and culture. Specifically, students can be exposed to (observe, get used to, and try to imitate) the native instructors' way of speaking and manner of communicating with other people. On the other hand, students can ask some community-based questions related to their local languages and culture, daily lives, and customs to the local language instructors. For example, the students asked some specific questions: "What kind of over-the-counter (OTC) medicine did you prepare when you shifted to Japan?", "Are Indian food products and seasonings available in Japan?", and "Please tell us simple Japanese when we need to tell our food restrictions to others."

As a learning outcome of the course, the students acquired not only grammatical knowledge and some useful expressions but also practical experiences with instructors and peers through communicative and role-play exercises. This helped them to build a sense of confidence to communicate with local Japanese people and adjust to different cultures by easing their worries and anxiety. In addition to this, the course provided cultural experiences and business manners in Japan such as Japanese traditional dress "Yukata" (one of casual Kimono) experience,

Fig. 3 Language instruction for the FRIENDSHIP scholarship students with a collaboration of native and non-native (local) Japanese language instructor

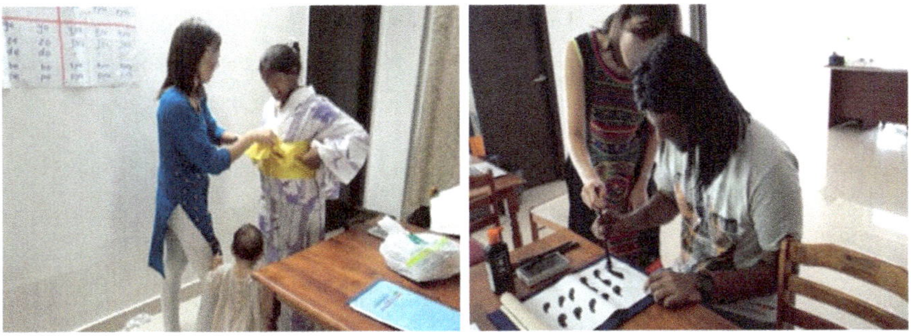

Fig. 4 Japanese cultural sessions for FRIENDSHIP scholarship students (left: Japanese 'Yukata' dressing session, right: Japanese calligraphy session)

Japanese calligraphy "Shodoo" (way of writing), Japanese tea culture, Japanese art of paper folding "Origami", and traditional card game "Karuta" (Fig. 4) in between the sessions or after the session. These cultural activities mainly aimed to build a relaxed and comfortable atmosphere in a classroom, enhance each student's active participation and concentration in language instruction, and help them to find an attitude of politeness, humbleness, and consideration of others in Japanese language and culture. This course clearly distinguished "Completion (Full Attendance)" and "Participation (Partial Attendance)" for the certification. After the three-day course, we conducted a brief ceremony of the Completion and Participation.

2.2 Japanese Language Courses in IITH

As we introduced one of our Japanese language courses as an example in Sect. 2.1, language learning activities are conventionally conducted based on frequent and meaningful one-on-one interactions as the ACTFL guideline states (ACTFL 2012). The language courses that we provided for FRIENDSHIP scholarship students were relatively small size classes (Two instructors for below 10 students), and we believed that the courses provided students with adequate meaningful interactions, monitoring, and personalized feedback they needed. However, when we implemented elective Japanese courses at the IITH, the course needed to be implemented in the short term for a large number of students who were beginner-level (Table 1). As a majority of students are absolute beginners, the teaching language is English and the teaching method is communicative language teaching (CLT) (Richards 2006), which is one of the widely-used language teaching methods for teachers to facilitate students' use of the target language for meaningful communication. In particular, CLT is widely adopted and is used in task-based learning (a method that focuses on completing meaningful tasks) in the Asia Pacific region (Butler 2011).

Table 1 The overview of Japanese language courses in IITH (Kataoka et al. 2023b)

Courses	Spoken Japanese basics course	Japanese reading and writing basics course
Number of teachers	1 certified Japanese language teacher and a native speaker, 1–2 teaching assistants (Indian Master or Ph.D. student from IIT-H)	
Number of students	50–250 students from IIT-H (undergraduate science and engineering students)	
Requirement	No prerequisite Japanese language skills of reading, writing, listening, or speaking	
Student language	English as a first or second language (multilingual students who speak Hindi, Telugu, and other Indian languages as well)	
Teaching language	The lesson topic is explained in English	
Teaching method	Communicative language teaching (Richards 2006)	
Credit	1 Credit (15 h)	
References	'Spoken Japanese beginners 1' (The World Japanese Language Center 2014), 'Japan: then and now' (Eguchi and Dumas 2020)	'Nihongo Shoho' (The Japan Foundation Japanese Language Institute 1981)
Learning outcomes	6–8 lessons in the 'Spoken Japanese beginners 1', Japanese language overview (characters, phonics, pronunciation, rhythm, and accent), grammar and vocabulary (500 words), greetings and basic sentences (positive, negative, question and suggestion form), demonstrative pronoun, particles (10), time expressions, numbers (100, numerical notation above 100, time, date and month, money), verbs (13), adjectives (22)	6–8 lessons in the 'Nihongo Shoho', Japanese language overview (characters: 46 Hiragana, 46 Katakana, Yoon, 25 Dakuon and Han-dakuon, 33 Yoon, 33 special sounds, and 50 Kanji), grammar and vocabulary (500 words), greetings and basic sentences (positive, negative, question and suggestion form), demonstrative pronoun, particles (10), prepositions (9), colors (10), counter suffixes (10), time expressions, numbers (100, numerical notation above 100, time, date and month, money), verbs (13), adjectives (22)
Grade system	Attendance with mini-tests: 30%, group work speaking project/oral assessments: 35% final oral exam (listening and speaking) 35%	Attendance with mini-tests: 30%, individual assessments: 35% final exam (reading and writing) 35%

- **Spoken Japanese basics course**
  This course systematically and practically introduces spoken Japanese without using Japanese characters. The primary objective of the course is to facilitate the development of fundamental communication skills in Japanese and to share knowledge applicable to real-life situations. Additionally, students will be exposed to diverse cultural experiences and basic Japanese business manners: traditional attire, calligraphy, the exchange of business cards, etc.

Fig. 5 A large-scale Japanese language class at IITH

- **Japanese reading and writing basics course**
  This course introduces the basic knowledge and skills of Japanese reading and writing systematically and practically. Its overarching goal is to facilitate students in acquiring fundamental reading and writing skills in Japanese and practical knowledge applicable to real-life situations. Students will have acquired a comprehensive set of Japanese characters, comprising 46 Hiragana, 46 Katakana, and 50 Kanji as the learning outcome through this course.

2.3 Issues in a Large-Scale CLT-Based Language Class

Figure 5 depicts a classroom wherein a teacher is engaging students in CLT-based activities. Generally, oral repetition practice (ORP) serves as a crucial activity for novice students to assimilate appropriate utterances and enhance speaking accuracy. However, the opportunities for individual oral production are considerably constrained in classes characterized by large numbers of students (Bahanshal 2013). Consequently, in such class settings, various strategies such as simultaneous ORP, the form of which includes 'repeat after me' practice all together with students immediately following the teacher's utterance, pair-work, group-work, and pattern practice involving selected students, are frequently employed.

2.4 Focused Problems and Our ICT-Based Approaches for a Large-Scale CLT-Based Language Class

To make our classroom activities interactive and meaningful even in a large-scale class and improve learning effects in speaking accuracy, the Japanese courses proactively utilized ICT. As Table 2 shows, we focused on three specific problems during classroom hours based on our careful classroom observation and introduced three ICT-based approaches using (1) a web-based Learning Management System (LMS),

Table 2 Our ICT-based approaches to complement problems during classroom hours

	Problem during classroom hours	Our ICT-based approach
1	Lack of individual monitoring and **in-person** feedback to support students' speaking accuracy	A web-based Learning Management System (LMS) with individual monitoring and **e-feedback** mechanism
2	Lack visual aids for teachers to examine students' speaking and generate quantitative **in-person** feedback	Quantitative **e-feedback** by employing Automatic Speech Recognition (ASR)
3	Lack of opportunities for students to practice speaking and receive individual **in-person** feedback	Online Oral Repetition Practice (ORP) Gym as an online platform to practice speaking online and receive individual quantitative **e-feedback**

(2) Automatic Speech Recognition (ASR), and (3) ORP Gym (an online oral repetition practice support system which we designed and developed). Since classroom hours are limited, our ICT-based approaches aim to provide "electronic feedback (e-feedback)" to complement the tightly limited teachers' "in-person" feedback, which refers to feedback provided (orally in most cases) to students in a physical classroom (not a virtual classroom).

2.4.1 A Web-Based LMS with Individual Monitoring and Feedback Mechanism

To address the identified issues of insufficient monitoring and feedback, a learning management system (LMS) was implemented. Figure 6 illustrates the system design, outlining interactions between the teacher and each student within the LMS. The system operates through three main databases: (1) Question bank, (2) Mini-test and exam results, and (3) Student group list. Backend programs facilitate data retrieval based on user requests. The system randomizes questions from the question bank when students engage, and for viewing learning history and exam results, it references the mini-tests and exam results database. Consequently, the LMS successfully enhanced homework quality, expanded class topics, established effective monitoring and feedback mechanisms, allowed flexible class operation modes, and ensured transparency in speaking exam grading.

2.4.2 Quantitative Feedback by Employing ASR

Evaluating speaking performance is (1) labor-intensive for teachers, (2) qualitative depending on teachers' auditory perception, and (3) causing delayed feedback to students. To mitigate the workload and time for teachers to examine audio answers and a perceptional gap between students and teachers, ASR is employed as a visual aid for teachers and online speaking evaluators to examine submitted audio answers

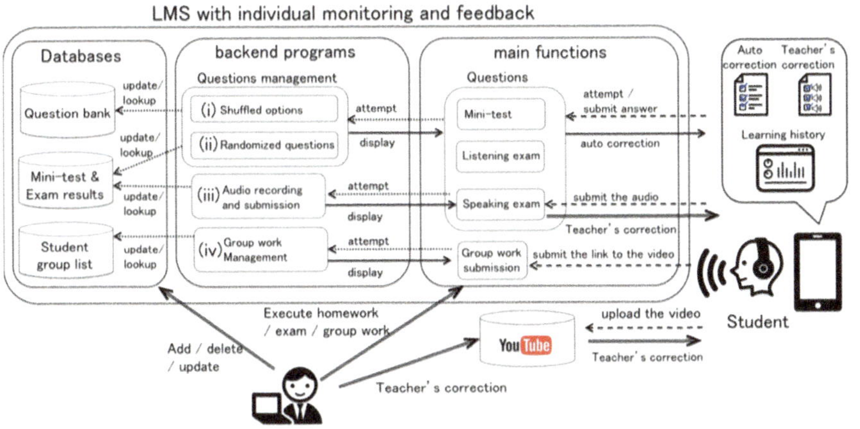

Fig. 6 Transaction between the teacher and each student on the LMS with individual monitoring and feedback mechanism (Kataoka et al. 2018)

quantitatively and objectively. Figure 7 outlines Google Cloud Speech-to-Text integrated into oral assessment for Quantitative e-feedback. After students' audio submissions, the process involves converting the audio, generating a transcript with a confidence rate, allowing teachers to compare and provide feedback, and finally displaying the feedback to students.

ASR aids teachers by identifying students' error tendencies with evidence before reviewing the audio data, bridging the perceptual divide between teacher and students, and (3) analyzing students' audio submissions with less effort in generating e-feedback. However, ASR transcripts are not displayed to basic students because

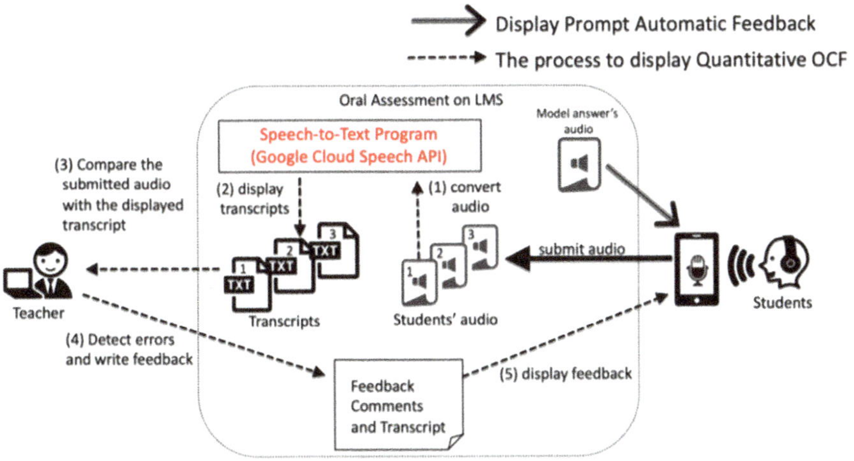

Fig. 7 Proposed oral assessments with "Speech-to-Text" as a visual aid on our LMS (Kataoka et al. 2019)

they are not comprehensible for basic students who cannot read Japanese characters. Although teachers can change the display setting whether the ASR is displayed or not according to the level of learners, there is a concern that ASR transcripts sometimes show irrelevant sentences and negative words that are different from the context because ASR is not designed especially for language learning. It might cause confusion and unnecessary negative psychological effects on students. In our preliminary experiment among a total of 214 audio datasets by Indian basic students, ASR exhibited high accuracy (82.4%) in word-based errors but low precision (28.1%) due to factors such as students' erroneous utterances, intonation, or background noise (Kataoka et al. 2019). Our proposed online assessment utilizing ASR holds promise as a complementary system. It provides teachers with evidence and priority orders, reducing the workload and time required for scoring audio data from students.

2.4.3 Online Oral Repetition Practice (ORP) Gym as an Online Support System for Students' Speaking Practice

We expanded our LMS by introducing the Online Oral Repetition Practice (ORP) Gym in 2019. It aims to offer online speaking practice and individual quantitative e-feedback (Kataoka et al. 2023a). As Fig. 8 shows, ORP Gym provides two types of e-feedback: Hybrid Recast (HR), a combination of audio and text-based automatic feedback, and Explicit Error Correction (EEC), which involves teachers' manual evaluations, offering scores and written feedback on students' problematic utterances. To aid EEC generation, Online External Assistants (OEAs) were trained as speaking evaluators using a standardized handbook (Kataoka et al. 2023a).

3 Results and Findings

We conducted 11 courses and a total of 1687 students completed the course during the span of the past 7 years (Table 3). Using three ICT-based approaches since 2018, we enabled individual monitoring and quantitative e-feedback outside classroom hours on LMS, and students' speaking performance was quantified through online oral assessments and exams. The two types of HR and EEC were actively utilized by students for their utterance improvement. Automated HR that was provided immediately after their attempt encouraged their multiple speaking practice. Although EEC was delayed feedback by teachers but points out the most problematic parts in their utterances, the two features of ORP Gym, ASR and OEAs, contributed to distributing the workload and time to evaluate students' audio answers and generate EEC. Since ORP Gym was enabled in the Spoken Japanese Basics Course and Japanese Reading and Writing Basics Course in 2019, the total number of students who utilized ORP Gym reached 1457 at present. Although the focus of the Japanese Reading and Writing Basics Course is introducing Japanese written systems using Japanese characters along with basic sentences in actual scenes, the read-aloud tasks are actively

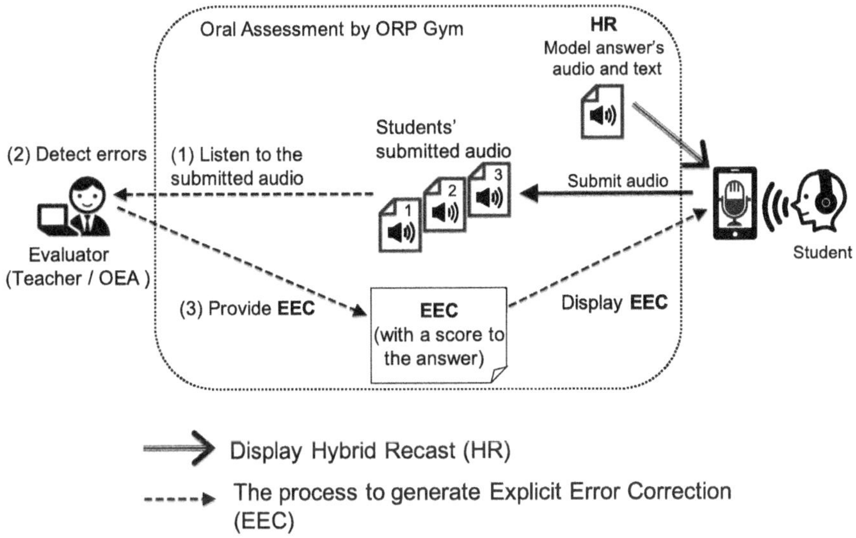

Fig. 8 Oral assessments by ORP Gym (Kataoka et al. 2023a)

utilized by students in the course. In addition, due to the COVID-19 pandemic issues in 2020, the form of most language classes was suddenly shifted to online for a certain period all over the world. The courses in 2021 were also conducted fully online using our web-based system relatively smoothly because this research has already had more than five years of technical and operational experience with online speaking assessments/exams using ORP Gym.

Furthermore, students' practical experience in using Japanese language was encouraged by group work project. Many video contents based on the theater art have been created by students (e.g. 26, 2, 41 video skits produced in 2018, 2019, 2022 respectively) after enabling our ICT-based approaches (LMS in 2018, ASR and ORP Gym in 2019).

Table 3 The number of students who completed the Japanese courses in IIT Hyderabad during the past 7 years

	2016	2017	2018	2019	2020	2021	2022	Total
Spoken Japanese basics	126	104	146	45	248	258	230	1157
Japanese reading and writing basics				23	106	215	186	530
Total	126	104	146	68	354	473	416	1687

4 Related Work

4.1 Large-Scale Online Language Courses in IITK

IIT Kanpur has provided Japanese language courses since 1995. They provide two courses 'Introduction to Japanese Language and Culture' annually (NPTEL 2023a, b). The courses I and II have been integrated into an online learning platform initiated by IITs and the Indian Institute of Science (IISc), NPTEL, in 2019 and 2022 respectively. The total number of certified students is 703 in 2023 (654 for the course I and 49 in the course II). The courses consist of around 40 digitally archived lectures, and a one-time live session was conducted in 2019. This is also one of the efficient ICT-based approaches in terms of the scalability of language education.

However, we have tackled how to support existing communicative language activities even in large-scale language classes. To the best of the authors' knowledge, digitally archived lectures or live sessions for large-scale students tend to be one-way in most cases, and such archived streaming or synchronous large-scale language sessions also persist in insufficient online repetition practice and individual feedback they need. Using ORP Gym as their asynchronous speaking practice, students can reflect and understand their speaking performance and improvement better and teachers can also provide more focused guidance and instructions for their speaking misuses in various error categories. In addition, the exams are administered manually to ensure identity verification and rigorous evaluation. We believe that ORP Gym can be utilized as oral assessments and exams for online Japanese language courses as long as identity verification is ensured.

4.2 Japan Hour by Suzuki Innovation Center (SIC)

Suzuki Innovation Center (SIC) was established by Suzuki Motor Corporation (SMC) and IITH under a vision of 'Creating Innovations for India and Japan' (Suzuki Innovation Center 2023). SIC aims to provide an open platform to promote and develop innovation between the two nations. Their activities cover three areas: Engagement (including education and networking), Exchange, and Innovation. To enhance the mutual understanding, they provide 'Japan Hour' program throughout the year. Since Japan Hour focuses on cultural exchange and human networking, it also encourages Indian students to make a presentation to introduce their own culture. Using video conferencing tools such as Zoom (Zoom 2023), Japan Hour sessions foster real-time interactions between Indian and Japanese people. For example, a student introduced his own culture in food, clothing, and housing, Japanese traditional textile company made a presentation of traditional Japanese dyeing and textile called 'Nishijinori' (Fig. 9), etc. These cultural exchange sessions give us inspiration for new compound materials and designs as well as promote opportunities for bilateral collaboration in R&D and new business. While ORP Gym provides asynchronous communication

Fig. 9 Cultural presentation during Japan Hour (left picture: learning culture of Arunachal Pradesh, right picture: learning Japanese traditional dyeing culture and Nishijin in Japan)

on the LMS, Japan Hour sessions help students enhance their presentation skills in Japanese through synchronous communication.

5 Discussion and Future Implication

Our ICT-based approaches including ORP Gym have the potential to facilitate further optimization of the CLT-based language course contents during classroom hours by aggregating individual corrective e-feedback outside classroom hours. 'Communication cannot take place in the absence of structure, or grammar, a set of shared assumptions about how language works, along with a willingness of participants to cooperate in the negotiation of meaning' (Savignon 2002). Therefore, practice and experience should be blended well in the communicative language course design (Savignon 2017, 2002). ORP Gym can reinforce 'practice' with input and output learning processes to enhance learners' speaking accuracy in their language use and delivery (linguistic/grammatical competence). On the other hand, the classroom can be utilized to foster their 'experience' to develop their communicative competence, which means 'expression, interpretation, and negotiation of meaning'. Through practice and experience, learners gradually expand their communicative competence which comprises grammatical competence, discourse competence, sociocultural competence, and strategic competence (Savignon 2002).

As course contents are more optimized or blended well in practice and experience, teachers' roles (e.g., facilitators, coaches, guides, or mentors) might be flexibly adapted to various communicative activities. As a major role of teachers is defined as 'facilitators' in CLT, teachers still need to support students using scaffolding strategies so that students can participate in conversation during classroom hours. Of course, teachers are not only the source of learning for students, students can also learn from learning tools and more competent students through peer work or group work task-based activities ('language arts') during classroom hours. In addition, one of the communicative activities, 'theater arts' can provide students with the opportunities to act. In such activities, the role of the teacher can be a coach who provides support, strategies, and encouragement for students (Savignon 2002). If language course-work is a problem-solving style or independent research for advanced students, the teachers' role will be a skillful 'mentor' to guide them toward their learning goal. As one potential form of cooperation, the opportunities for experience can be reinforced through various activities and real-world projects such as Japan Hour by the Suzuki Innovation Center.

Towards collecting further best practices, ORP Gym should be horizontally deployed and operated in multiple universities or educational institutes. To do that, distributed systems are required to accommodate further diversified demand and distribute simultaneous access from users, not centralized systems. Future work is sharing documentation on accumulating knowledge of how to operate ORP Gym with other teachers (including native and non-native JF teachers) so that they can learn how to operate ORP Gym in their local network (campus network). In addition, autonomous rating mechanisms against OEAs or feedback might be also required for OEAs to ensure the quality of their feedback. In this research, the quality of feedback was ensured manually by training native Japanese speakers who are not certified JF teachers using a handbook. However, a certain rating mechanism can encourage and train OEAs autonomously on ORP Gym (e.g., OEAs who are highly rated by students can receive incentives on the system).

6 Conclusion

At IITH, we have developed and operated a learning management system (LMS) with individual monitoring and personalized feedback in a large Japanese class for science and engineering students. Our ICT-based approaches aimed to improve the learning effect in Japanese language education by devising and implementing methods for effective use of ICT, quantifying and visualizing the effects of learning in speaking accuracy, and reflecting them in teaching methods and learning content during limited classroom hours. By building a platform to utilize the human resources of Japanese speakers around the world as "online external assistants (OEAs)", we can address diversified and increasing needs for Japanese language learning and the growing scale of the Japanese language education system.

The utilization of ICT can expand the existing theories and teaching methods that have been clarified in Japanese language education in a scalable manner without being bound by the constraints of classroom time and location. It also can be expected to have a leveraging effect on education by quantitatively improving the number of learners, amount of practice, frequency of assessment and examinations, etc. While speaking practice is reinforced in terms of both quality and quantity by ORP Gym outside of classroom hours, face-to-face in-class activities will allow students to apply the knowledge reinforced by practice through interaction with the instructor and peers, thus further enhancing their speaking skills. Furthermore, the field of Japanese language education will be further optimized with educational content specialized for each learner's field of study. Instructors engaged in Japanese language education will be able to provide more focused instruction and more personalized feedback by reflecting on learning data (e.g., misuse tendencies in pragmatics considering learners' cultural backgrounds) accumulated on the LMS.

We believe that our efforts to date have had a significant academic and social impact in the sense that they will also promote demonstration experiments and research on educational technology in actual educational settings. Future tasks include increasing the number of OEAs to realize a more scalable system, and horizontally expanding the system to other educational institutions in India and other countries. We will construct a system to increase the incentives of OEAs who contribute directly to error correction and to further deploy the system horizontally and internationally in collaboration with science and engineering universities in ASEAN countries, e.g., working with the SOI Asia project (SOI Asia Project 2023) and educational projects in the WIDE project (WIDE Project 2023). In addition, the non-native speaker's speaking data accumulated in this system can be analyzed for error tendencies in different countries and regions where students have different mother tongues. This research is expected to be applied to learning analytics and Artificial Intelligence (AI) in the next-generation Japanese language education, such as providing more focused instruction according to error tendencies. Our approaches will contribute to the construction of a platform that will serve as a fundamental platform for such applications.

According to the PwC report (PwC 2023) for the year 2050, India is a driving force of the economy among emerging countries, and its average economic growth rate is expected to exceed that of the U.S. by 2040, ranking second only to China, which is ranked first. India will be an important economic partner for Japan as well as other countries among emerging countries, and the demand for language education and culture for science and engineering personnel is expected to increase further according to diversified needs. It is essential to contribute to international society by building an educational platform that will serve as an important foundation to meet the expected future demand for Japanese language education and culture, and by expanding the platform in a wide range of fields.

References

ACTFL (2012) CTFL position statements most current as of July 30, 2012 summary document. https://nadsfl.org/wp-content/uploads/2017/04/ACTFL_Position_Statements.pdf

Bahanshal DA (2013) The effect of large classes on English teaching and learning in Saudi secondary schools. Engl Lang Teach 6(11):49–59

Butler YG (2011) The implementation of communicative and task-based language teaching in the Asia-Pacific region. Annu Rev Appl Linguist 31:36–57

Eguchi H, Dumas D (2020) Japan: then and now. The Japan Times, Tokyo

FRIENDSHIP Project (2023) Overview of FRIENDSHIP scholarship. https://www.iith.ac.in/japandesk/overview-scholarship.html

Indian Institute of Technology Hyderabad (n.d.) IITH. https://www.iith.ac.in/about/aboutiith/
Japan Foundation (2023) Survey on Japanese-language education abroad. https://www.jpf.go.jp/e/project/japanese/survey/result/index.html
Japan - IIT Hyderabad collaboration (2020) Japan - IIT Hyderabad collaboration (Phase1 closed). http://friendship.iith.ac.in/
JLPT (2023a) Past test data. https://www.jlpt.jp/e/statistics/archive.html
JLPT (2023b) Objectives and history. https://www.jlpt.jp/e/about/purpose.html
Kataoka Y, Thamrin AH, Murai J, Kataoka K (2018) Effective use of learning management system for large-scale Japanese language education. In: Proceedings of the 10th international conference on education technology and computers (ICETC'18)
Kataoka Y, Thamrin AH, Murai J, Kataoka K (2019) Employing automatic speech recognition for quantitative oral corrective feedback in Japanese second or foreign language education. In: Proceedings of the 11th international conference on education technology and computers (ICETC'19)
Kataoka Y, Thamrin AH, Van Meter R, Murai J, Kataoka K (2023a) Investigating the effect of computer-mediated feedback via an LMS integration in a large-scale Japanese speaking class. Educ Inf Technol 28:1957–1986
Kataoka Y, Thamrin AH, Van Meter R (2023b) ORP Gym: online oral repetition practice support system for large-scale Japanese-speaking courses abroad. Syst Pract Action Res
Kato F (2020) Strategies for growing and enhancing university-level Japanese programs. Routledge, London
Ministry of Foreign Affairs of Japan (2008) Report of the Japan-India Working Group on the cooperation toward establishment of a new IIT. https://www.mofa.go.jp/region/asia-paci/india/pmv0810/report.pdf
NPTEL (2023a) Introduction to Japanese language and culture, IIT Kanpur. https://nptel.ac.in/courses/121104005
NPTEL (2023b) Introduction to Japanese language and culture - II, IIT Kanpur. https://nptel.ac.in/courses/109104195
PwC (2023) The World in 2050: the long view: how will the global economic order change by 2050?. https://www.pwc.com/gx/en/research-insights/economy/the-world-in-2050.html
Richards JC (2006) Communicative language teaching today. Cambridge University Press, New York
Savignon SJ (2002) Communicative language teaching: linguistic theory and classroom practice. https://videa.ca/wp-content/uploads/2015/08/Communicative-language-teaching2.pdf
Savignon SJ (2017) Communicative competence. In: Liontas JI, International Association T, DelliCarpini M (eds) The TESOL encyclopedia of English language teaching, pp 1–7
SOI Asia Project (2023) SOI Asia Project. https://www.soi.asia/
Suzuki Innovation Center (2023) Suzuki Innovation Center. https://sicglobal.org/
The Japan Foundation Japanese Language Institute (1981) Nihongo Shoho. Bonjinsha
The World Japanese Language Center (2014) Spoken Japanese beginners 1. The World Japanese Language Center, Hornsby
WIDE Project (2023) WIDE Project. https://www.wide.ad.jp/index_e.html
Zoom (2023) Zoom. https://explore.zoom.us/ja/products/meetings/

Expansion of Learning Methods Using XR: India Japan Collaboration Potentials

Sachiko Takane and Rajib Shaw

Abstract XR (Cross Reality) is a generic term for virtual space technology that merges the real world and virtual world to enable users to perceive things that do not exist in reality. Currently, the field of XR (AR/VR/MR, etc.) is evolving rapidly, with many companies around the world competing to develop both the hardware and software. This chapter explores the utilization of XR (Cross Reality) in education, focusing on the development and implementation of immersive learning content in primary and secondary education. Collaborating with India Japan Lab (IJL) and Technology Innovation Hub of IIT Delhi (IHFC), we conducted a VR (Virtual Reality)-based experiential learning program titled "Let's create the energy stations of the future," centered around Sustainable Development Goal 7 (Affordable and Clean Energy). The program was conducted in September 2023 at Dr. Bhimrao Ambedkar School of Specialized Excellence in Delhi and Amity International School in Noida for students aged 11 to 14. We present the content used, methodology, learner responses, and prospects for future XR applications.

Keywords XR · VR · Education · Metaverse · SDGs · Clean energy

1 Introduction

XR (Cross Reality) is a generic term for virtual space technology that merges the real world and virtual world to enable users to perceive things that do not exist in reality. Through devices, users can perceive virtual spaces created using different technologies, as well as people and objects that exist in virtual spaces, as if they were in reality. As technology has developed, content can now be created by combining two or more of those technologies listed above, rather than using them alone. This

S. Takane
C-Sagyoba Corporation, Tokyo, Japan

R. Shaw (✉)
Graduate School of Media and Governance, Keio University, Fujisawa, Japan
e-mail: shaw@sfc.keio.ac.jp

R. Shaw and S. R. Choudhury (eds.), *India, Japan and Beyond*,
https://doi.org/10.1007/978-981-97-3282-1_15

has made it difficult to draw a line between technologies, such as "this is VR," and the term "XR" is said to have been created as a generic name for all of them. These technologies have the potential to significantly impact the field of education in various ways.

Japan has been using digital technologies in different levels of education under the umbrella of digital transformation (DX), and also with the new Digital Agency. This is also part of the Society 5.0. The Society 5.0 is "A human-centered society that balances economic advancement with the resolution of social problems by a system that highly integrates cyberspace and physical space" (CAO 2023). Society 5.0 was proposed in the 5th Science and Technology Basic Plan as a future society that Japan should aspire to. It follows the hunting society (Society 1.0), agricultural society (Society 2.0), industrial society (Society 3.0), and information society (Society 4.0). In the past information society (Society 4.0), there was a problem that knowledge and information were not shared and cross-disciplinary collaboration was insufficient. Because of the limitations of human capacity, the work of finding and analyzing necessary information from the overflow of information was burdensome, and there were restrictions on the scope of work and activities due to age, disability, and other factors. In addition, it was difficult to respond adequately to various constraints on issues such as the declining birthrate, aging population, and depopulation of rural areas. Thus, Society 5.0 aims at utilizing different technologies, and XR is considered as one of the key aspect for the Metaverse in Society 5.0 (NRI 2023).

In contrast, India is also progressing tremendously on new and disruptive technologies, and there have been several start-ups who are working on drones, virtual reality, internet of things (IOT), block chain etc. The new education policy of India (NEP 2020), flexibility, creativity, critical thinking are some of the key aspects of the principles of education, and there use of different types of technologies are encouraged, from elementary/early childhood to higher education level.

This chapter provides an overview of the implementation and research conducted on an experiential learning program utilizing VR in India for middle school students. It discusses the content used, methodology employed, the learning outcomes achieved, and explores new perspectives on incorporating XR into education.

2 XR and Education

XR is often considered as an innovative tool for immersive learning. Audrey Tang, Taiwan's Minister of Digital Affairs, once said that the difference between IT (Information Technology) and digital is that "*IT is what connects machines to machines, and digital is what connects people to people*". Yoshimitsu (2023) in his analysis has pointed out that among the OECD countries, Japan is relatively slow or low use of ICT (information and communication technology) in education. ICT is used as games but not learning. To reduce this gap, the GIGA (Global and Innovation Gateway for All) School program was adopted in 2019 by the Ministry of Ministry of Education, and special budget was provided for this (MEXT 2019). Currently, the field of XR (AR/VR/MR, etc.) is evolving rapidly, with many companies around the world competing to develop both the hardware and software. For example, until now, XR has mainly focused on "visually immersive" experiences. However, now, XR technology has been commercialized that allows users to experience the five senses (sight, hearing, smell, taste, and touch), including the sense of sight, such as "taste manipulation" in combination with AI. These advanced technologies thus will become essential for education and learning within a few years, and it is time for the student and Generation Z to be tuned with these technologies.

In a recent work, Yang et al. (2018) have proposed XR education framework, which incorporates designing instruction driven and learning centered XR system, which provide varied choices to the instructors as well as learners to customize to their needs. Personalized education becomes one of the key for XR education, where the learners can personalized the education and learning modules based on their interests and choices. Ryoo and Winkelmann (2021) developed blueprint of Innovative Learning Environment (ILE), which was found to be effective during the COVID-19. As a part of STEM education, XR education can provide a unique yet personalized and adaptive learning platform, which can be linked to other emerging technologies like AI (Artificial Intelligence), machine learning to make a holistic platform. Thus, there are increasing availability of learning and instruction platform for the teachers to use XR for different levels of learning (Kosko et al. 2021; Ferreira and Qureshi 2020; Wang et al. 2020).

XR education can be used for many different aspects as follow:

- **Immersive Learning Environments**: XR can create immersive simulations and environments that allow students to explore and interact with subjects in a more hands-on and engaging manner. This is particularly beneficial for subjects like science, history, and geography.
- **Virtual Laboratories**: In fields such as chemistry and biology, where physical experiments may be challenging or expensive, XR can provide virtual labs for students to conduct experiments in a safe and controlled environment.
- **Virtual Field Trips**: XR can take students on virtual field trips to historical sites, museums, or even outer space, providing an enriching experience that may not be feasible in the physical world.
- **Skill Training**: Industries such as healthcare and engineering can use XR for practical training. Medical students can practice surgeries in a virtual environment, and engineering students can simulate the construction and testing of structures.
- **Language Learning**: XR can create immersive language-learning experiences by placing students in virtual environments where they can interact with native speakers and practice language skills in real-life scenarios.
- **Accessibility**: XR can help address learning challenges by providing alternative ways for students with different learning styles or disabilities to engage with educational content.
- **Collaborative Learning**: XR can facilitate collaborative learning experiences, allowing students from different locations to interact and work together in virtual spaces.

- **Soft Skills Development**: XR can be used to develop and assess soft skills such as communication, teamwork, and problem-solving through simulated scenarios.

It is worth noting that the adoption of XR in education depends on factors like cost, accessibility, and the development of content that aligns with educational objectives. XR, encompassing the fusion of the real and virtual worlds to create novel experiences, is the focus of this study (DHW 2023). By leveraging technology aligned with learning themes, we anticipate an expansion of imagination leading to better retention of knowledge and information. Proven learning effects beyond mere repetition may open pathways for learning unbound by existing frameworks (MRI 2023). Providing opportunities for teenagers to familiarize themselves with technology at an early stage contributes to their knowledge enhancement. In our VR content, aimed at envisioning "future clean energy stations," we designed the program to stimulate imagination and foster the idea that anything is achievable. The program's framework revolves around "knowing, thinking, creating, experiencing, and sharing." Notably, the inclusion of "creating" allows learners to construct VR scenes, while we emphasize time management within the school's class schedule when implementing this method.

3 Target Audience, Theme and Implementation Environment

In collaboration with India Japan Lab (IJL: www.indiajapanlab.org) of Keio University and Technology Innovation Hub (IHFC) of Indian Institute of Technology (IIT) Delhi, the program targeting 12 to 15-year-old students in two middle schools, accommodating 25 students per class, in Delhi, India. India being considered an ICT (Information and Communication Technology) advanced nation, there was high interest in understanding the reactions of the teenagers who will shape the next generation. Simultaneously, conducting the program in both public and private schools aimed to provide insights into the current ICT infrastructure. Envisioning young minds sharing technological skills and ideas using the soon-to-be Metaverse across distances was considered a remarkable prospect.

The topic was titled "*Let's create the energy stations of the future.*" It was centered on Sustainable Development Goal (SDG) 7, "Affordable and clean energy," prompting students to visualize current issues in their living environment and consider future developments within VR scenes. While introducing basic concepts of SDGs from poverty, hunger, environmental issues to economic growth and gender equality, the initial proceeding highlighted the broad spectrum of sustainable development goals. Most participants had prior exposure to SDGs through school education, indicating a high awareness of environmental conservation on a global scale.

The purpose of expanding learning through XR was to utilize XR in primary and secondary education, anticipating the following effects: (1) Deepening interest in learning themes, (2) Generating interest in XR technology, (3) Enabling students to

Table 1 Time schedule (120 min)

Activity	Duration (min)
1. Introduction	10
2. Speech and think	10
3. VR experience	25
4. Basic operation of STYLY	10
5. Create scene	25
6. Experience and presentation	25
7. Comments and questionnaire survey	15

confront challenges and think more fundamentally about solutions, and (4) Anticipating an expansion of interest in subsequent subjects such as energy development, mathematics, and physics, among others. By employing VR goggles (Head-mounted displays), the entire field of vision of the user becomes a virtual space synchronized with their bodily senses. This experience, often the first for many teenagers, is believed to enhance their active participation in class. Incorporating the act of "creating" within the content is a proposal for a new learning method utilizing VR. However, relying solely on experiential learning through VR goggles presents challenges related to equipment weight, discomfort during prolonged use, and the potential for VR sickness (AIST 2023). Therefore, in this content, VR experiences are limited to around 5 min before and after normal class time, considering time allocation within regular class schedules (Table 1).

The necessary environment for implementing VR-based programs includes a Wi-Fi environment with secured bandwidth, which needs verification both domestically and internationally. In Japan, the GIGA School Project (MEXT 2019) has been completed in 99.9% of 1,018 local governments as of March 2023 (MEXT 2023). However, challenges persist, such as external network management and ensuring communication speed and volume during simultaneous access, although progress in infrastructure development is expected to accelerate (Takahashi 2023). It is due to these ICT advancements that previously unprecedented methods of learning expansion become feasible.

The two schools in India involved in this study also possessed adequate ICT infrastructure, facilitating the setup of VR equipment. Notably, schools address issues regarding wireless wifi connection, security measures, which sometimes require additional time for setting up externally brought VR goggles or connecting computers. The VR goggles used for students in this study were the Meta Quest 2.

4 Explanation of Implementation Methods and Exchange Between Japanese and Indian Educational Settings

Maintaining the same structure as before, we tailored the schedule to the preferences of the participating schools. For Japanese schools, we assumed a two-period frame, designing for a 120-min session, replicating the same structure in this instance. Here is the explanation of the overview according to the timetable in Table 1. Introduction involved self-introduction, acquainting each other, introducing the content, discussing the sequence of activities, and ensuring comprehension. The school welcomed the visitors in their unique cultural manner. In Japan, such acts are termed "*omotenashi*," symbolizing a school's hospitality towards visitors from overseas. Witnessing the school imparting their national culture to visiting foreigners was a remarkable and delightful experience. In the Speech and Think session, learners were given a lecture on fundamental aspects of VR and its current applications. Understanding this was crucial as we wanted them to imagine how they could utilize this technology. In the VR Experience session, students experienced the prepared VR scene titled "XRSDGs7_ The Future Energy Station in Delhi, India" (Takane 2023). This scene's backdrop aligned with their visualization of ideas. Each student spent around 2 min without using controllers, immersing themselves in the 360-degree space, enhancing their understanding. Most students were thrilled and excited to explore the world of VR. Moving on to "Think" and "Create," we used the XR platform "STYLY" (Style 2023).

STYLY serves as a VR/AR/MR creative platform providing space for artists to express spatial representations. Students used web browsers on their computers, creating scenes with various objects, widening their choices. Our focus was on how students could demonstrate their capabilities within the designated class time. Basic operations were straightforward for students accustomed to computer usage. Consequently, we projected the instructor's operations on slides, facilitating simultaneous instruction and operation on the computers. Subsequently, working in groups on assigned VR scene templates (Fig. 1) students brainstormed about the theme, envisioned an ideal future energy station, considered objects needed for their local environment, and positioned them in the scene. Since this task was group-based, discussing how to visualize witty statements from others' ideas became a collaborative learning experience for the students (Fig. 1). In "Experience and Share," students experienced the VR scenes they created. In a three-dimensional space, they confirmed their created scenes, similar to corporate VR-based development simulations. These experiences were shared on monitors with other students (Fig. 1).

During "Share," scenes were published on the XR platform, allowing sharing among distant students or those interested in their ideas. It's a significant feature and goal of utilizing VR. While the literal translation of "Virtual Reality" conveys something essentially "real," it's astonishing that unconventional ideas visualized through this medium garner admiration. When conveying this, one can observe heightened enthusiasm and motivation among students, evident from their excited expressions

Fig. 1 Glimpses of VR activities (top: VR scene template, bottom left: collaborative learning, and bottom right: presentation of ideas)

and remarks. Moreover, in this content, VR usage provides a new learning environment that transcends physical limitations, aiding those with mobility challenges. By initially placing and visualizing objects in VR space and exchanging opinions within the metaverse, essential thoughts can be stimulated. Surprisingly, among the VR scenes created by students (Fig. 2) some visualized ideas resembling wireless power transmission technology (Fig. 2, NTT 2023). While resembling existing technology, their novel perspectives highlighted their consideration of innovative methods. Witnessing students envisioning power supply methods without harming the environment, even within this short class, was a fantastic discovery. Several other remarkable ideas were presented (Fig. 3).

Fig. 2 Line-up of student-created and published VR scenes

Fig. 3 NTT automated power transfer image in Japan

5 Learner Responses and Feedback

Providing teenagers with an opportunity to encounter a fragment of new technology may contribute to the evolution of future digital technologies and DX (Digital Transformation). The driving force behind promoting such learning lies in students'

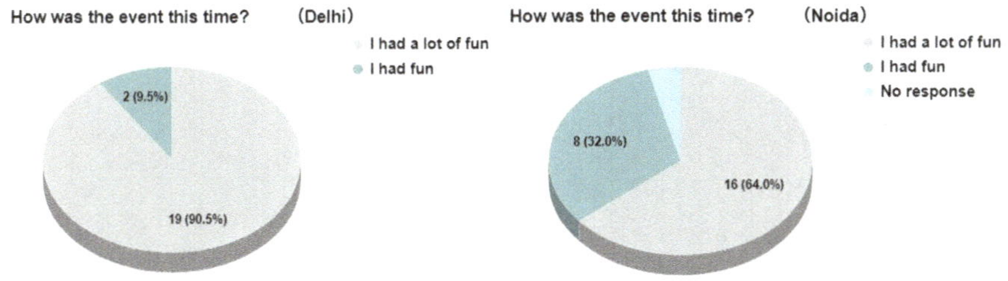

Fig. 4 Survey result on the satisfaction rate of the program (left public and right private school)

intrinsic motivation to learn. Setting learning themes relevant to one's surroundings is crucial to motivate learning. Encouraging a mindset of not just thinking but also doing is equally important. Therefore, the VR utilized in this instance successfully captured interest, fostered collaborative learning, and, most importantly, made learning enjoyable. These aspects were reflected in the free descriptions provided in the post-implementation survey. Understanding how students learn and continually exploring new learning methods is essential for those providing education.

Offering content to Indian middle school students allowed us to gauge their high interest in VR (Fig. 4). Moreover, their high satisfaction with VR-based content was an encouraging result. Nearly half of them had prior VR experience, showcasing India's advancement in this aspect compared to Japan. It was surprising but also insightful, considering India's recognition as an IT advanced nation. Simultaneously, the students' first experience of exploring VR scenes they created and their burgeoning interest in using it further were delightful observations.

6 Prospects for Future XR-Based Educational Content

The utilization of VR in educational methods is expected to further proliferate in the future. The prospect of next-generation youths utilizing metaverse spaces to share their learned skills and ideas is remarkable. We aim to develop content that introduces encounters with fragments of new technology, lowers technological barriers, and initiates the next step without overwhelming difficulties in utilizing technology, thus supporting the growth of youths worldwide.

This chapter summarizes the report on implementing a content-based experiential learning method utilizing VR among Indian middle school students. While conducting research and developing content to expand learning through XR, we believe that with equipment becoming more affordable and user experience improving, the introduction into educational settings can accelerate. We anticipate that learners themselves will devise innovative uses. Recognizing the importance of providing the process of "knowing new technology," we acknowledge the need for creating content, continual implementation, and comprehensive manual creation to facilitate easier adoption within educational settings. Reporting the significance of implementing educational content developed in Japan in an international setting.

References

AIST (2023) 'VIMS countermeasures' essential for the widespread use of VR. AIST Magazine. https://www.aist.go.jp/aist_j/magazine/20221214.html. Accessed 4 Jan 2024

CAO (2023) What is Society 5.0? https://www8.cao.go.jp/cstp/english/society5_0/. Accessed 4 Jan 2024

DHW (2023) What is XR (cross reality): differences between VR, AR, MR, etc., and examples of its use. https://www.dhw.ac.jp/now/list/howtobe/cross_reality/. Accessed 4 Jan 2024

Ferreira JMM, Qureshi ZI (2020) Use of XR technologies to bridge the gap between higher education and continuing education. https://ieeexplore.ieee.org/abstract/document/9125346?casa_token=pJ3es4-dh8QAAAAA:bCaQ6sPOhitCcY6_mLjMmeKmIOGpZv61H1hLkjrl-xEEq9kwNlri552XgscGSeGcafB6xpXPFg. Accessed 4 Jan 2024

Kosko KW, Ferdig RE, Roche L (2021) Conceptualizing a shared definition and future directions for extended reality (XR) in teacher education. J Technol Teach Educ 29(3):257–277. Society for Information Technology & Teacher Education, Waynesville, NC, USA. https://www.learntechlib.org/primary/p/219894/. Accessed 3 Jan 2024

MEXT (2019) Ministry of Education, Realization of the GIGA school concept. https://www.mext.go.jp/a_menu/other/index_00001.htm. Accessed 4 Jan 2024

MEXT (2023) Ministry of Education, Culture, Sports, Science and Technology, Status of one terminal per student at the compulsory education stage. https://www.mext.go.jp/content/20230711-mxt_shuukyo01-000009827_01.pdf. Accessed 4 Jan 2024

MRI (2023) Mitsubishi Research Institute, Game-changing education with VR. https://www.mri.co.jp/knowledge/mreview/201811-4.html. Accessed 4 Jan 2024

NEP (2020) National Education Policy, Government of India. https://www.education.gov.in/sites/upload_files/mhrd/files/NEP_Final_English_0.pdf. Accessed 4 Jan 2024

NRI (2023) Nomura Research Institute, What is the metaverse? https://www.nri.com/jp/knowledge/glossary/lst/ma/metaverse. Accessed 4 Jan 2024

NTT (2023) What is wireless power transfer? Explanation of the benefits, principles, types and efficiency of its use. https://www.rd.ntt/se/media/article/?tag=tag7. Accessed 4 Jan 2024

Ryoo J, Winkelmann K (2021) Innovative learning environments in STEM higher education: opportunities, challenges, and looking forward. https://library.oapen.org/handle/20.500.12657/47325. Accessed 4 Jan 2024

Style (2023) STYLY, The VR platform: official site. https://styly.cc/. Accessed 4 Jan 2024

Takahashi J (2023) Wi-Fi in school education. https://www.soumu.go.jp/main_content/000548778.pdf. Accessed 4 Jan 2024

Takane S (2023) XRSDGs7_ The Future Energy Station in Delhi, India. https://gallery.styly.cc/scene/d512f2e5-2680-4480-bc5f-a915c04d56ba. Accessed 4 Jan 2024

Yang K, Zhou X, Radu I (2018) XR-Ed framework: designing instruction-driven and learner-centered extended reality systems for education. https://doi.org/10.48550/arXiv.2010.13779.pdf. Accessed 4 Jan 2024

Yoshimitsu T (2023) Use of XR in education and future prospect. https://www.jstage.jst.go.jp/article/bplus/16/3/16_185/_pdf. Accessed 4 Jan 2024

Wang M, Ryoo J, Winkelmann K (2020) Preface to the special issue on Cross Reality (XR) and Immersive Learning Environments (ILE) in education. https://www.tandfonline.com/doi/full/10.1080/10494820.2019.1696845. Accessed 4 Jan 2024

Youth Voices and India Japan Relation

Tomo Kawane and Rajib Shaw

Abstract India Japan Laboratory of Keio University in Japan was initiated in November 2019. The year 2023, the fourth year for conducting India Class, saw the registered number of the students swelled to 180 from 115 in 2020. Every year since 2020, India Class has offered learning opportunities about India and Japan-India bilateral relationship via online mode during spring semester through 14 weekly lecture-based classes. Some students have had direct interactions with Indians or having lived in India before joining India Class. Majority of the students' experience with India has been principally coming from the media. After the entire course was over, the students were requested to choose one particular topic of their liking out of 14 classes and write a report of 1000 words. Most of the students have read further on their subject of interest. While political and economic analysis have been mostly based on what they have learnt in the class and additional readings, many interesting interpretations and observations were expressed by those who reflected on what they have learnt through their own experiential immersive learning through yoga, music of both Indian and Japanese instruments, and virtual cooking class of an Indian meal. The reports by the students in 2023 carried varied aromas with elaborate details on food culture and music tradition of India while some constructive suggestions were offered on Indian polity, economy environment, and society. This chapter provides some of the key analysis from these reports.

Keywords Youth voice · India Japan relation · Diversity · Technology · Collaborative platforms

T. Kawane (✉) · R. Shaw
Graduate School of Media and Governance, Keio University, Fujisawa, Kanagawa 252-0882, Japan
e-mail: kawane@sfc.keio.ac.jp

R. Shaw and S. R. Choudhury (eds.), *India, Japan and Beyond*,
https://doi.org/10.1007/978-981-97-3282-1_16

1 Introduction: Culture, Technology, and Environment of India

1.1 India Japan Laboratory Initiates "India Class"

India Japan Laboratory of Keio University in Japan was initiated in November 2019. With the scope of offering opportunities to the students at Keio University, the oldest private university in Japan established in 1858, to learn about various facets of India, the class titled "Culture, Environment, and Technology of India" was initiated in April 2020. It coincided with the time when the world was forced to acknowledge the enormity of the impact of the COVID-19 pandemic. Health hazard to human lives due to the pandemic pushed education to transform drastically. The educational institutions needed a measure to prioritize minimizing human contacts. Classrooms were compelled to go online. Educational institutions had to improvise their own teaching method to adapt the 'new normal' as well as maintain their educational standard while conducting regular classes online. Face-to-face interactions, vital to establish the means for evaluating students as well as their participation, got halted. Ensuring regular attendance and interactions in person came to a temporally standstill. Educational pedagogy demanded educators to be inclusive as well as creative, while allowing the students to choose either being seen or not on live screen during the class hours. Conducting online introductory class on India has demanded us to pay attention to include themes that characterizes India—about its people, its culture, its society, and its polity as well as its economy. The momentum of the start of online class on India was, thus, timely—without anticipating that the beginning of our "India Class"—fondly so called by the students—had faced a pedagogical challenge to make it inclusive and diverse. The early stage of the class coincided with the onset of COVID-19 pandemic that impacted globally and changed the world forever. Thus, since the COVID-19 pandemic situation restricted human contacts, India Class started only in virtual mode to maintain social distance to be in line with the guidelines and regulations of the university. This chapter focuses on the analysis of the youth voices emerged from the class.

1.2 A New Chapter for "India Class"

The year 2023, the fourth year for conducting India Class, saw the registered number of the students swelled to 180 from 115 in 2020. This marked 56.5% increase in registration in just 4 years. The students joined from diverse backgrounds, majoring economics, environmental studies, policy studies, medicine, law, as well as literature studies. Because it was completely online, India Class could invite students from all the six campuses of Keio University. After the initial three years of the COVID-19 pandemic when the classes had to be conducted online, in 2023 we could open a new chapter for India Class. While the principal mode of imparting education for this class remained online, only one class on Indian classical music by Tim Hoffman was conducted in hybrid-mode where 140 students out of 180 registered students attended their first in-person class.

The drastic increase in the number of the students who opt for India Class could be because of several reasons.

(1) The most prominent development about India was India had been anticipated to achieve the status of the most populated country in the world in 2023 surpassing the population of China. This has been most prominently discussed among the students.
(2) As far as the bilateral relationship between Japan and India is concerned, its geo-political significance for Japan was more prominently emphasized through the Indo-Pacific.
(3) It is also significant to mark that, regardless of the COVID-19 pandemic, India recorded an exponential growth in the number of start-ups and unicorns. This made the students consider that Japan has lots to learn from India.
(4) India also could show the potential of growing further in IT field and becoming a formidable ally for Japan in collaboration through R&D in India.
(5) Aging Japanese society requires some good impetus to rejuvenate the country. Bringing Indian brilliance and youth in Japanese society could be considered as a good opening for Japan to plan for sustainable society with cooperation from India regarding its manpower.
(6) Students are more keen to understand the background of bilateral relationship between Japan and India and possible developments in coming years especially when the world is facing military conflicts that will put further pressure on energy and security issues.

1.3 "India Class" and Its Students at Keio University

Every year since 2020, India Class has offered learning opportunities about India and Japan-India bilateral relationship via online mode during spring semester through 14 weekly lecture-based classes. Both Japanese and international students from different countries who are studying at Keio University participated in the class. English has been the medium of instruction for India Class. For the students with the lineage of more than one nationality, India Class has given a point of reflection to see where they stand in the society.

Some students have had direct interactions with Indians or having lived in India before joining India Class. Majority of the students' experience with India has been principally coming from the media. Several students know Indian food, Indian music, and yoga, but they remain as a minority throughout 4 years since 2020 when we initiated the course. Some students have had direct interactions with Indians, but such cases were rare. Most of those students did so because one of their parent's professional engagements with India. Some students have studied in the third country where they used to have classmates and neighbors of Indian origin. Some students lived in India mainly because of some professional engagements of their parents. Yet some other students claimed that the hobbies of their close family members or themselves have had deep connection with India: improvising yoga postures while pursuing their favorite sports or playing some fusion of Indian music through their cherished musical instruments got highlighted as their real-life experiences that made them feel very close to India.

Overall, the students found India embraces many competing elements in one country. India shows matured face in some aspects of its economic growth yet embracing vulnerable sections of people within its economy. Indian diplomacy has significant impact upon Japan today where the Indo-Pacific has been considered as a major playground for harnessing the bilateral relations.

1.4 Reference-Based Analysis Versus Immersive Learning

Barring a few students from India, majority of the registered students for India Class joined the course with no direct or deep interaction with India and Indian people. The course was conducted through lectures on diplomacy and ODA, Indian society, Japan's economic as well as diplomatic cooperation with India, digitalization, start-up ecosystem and innovation, infrastructural development corporation, exploring new energy and Japan India collaborations, interactive sessions with our consortium members of India Japan Laboratory, a music concert, a virtual yoga session, a lecture cum virtual sessions of astrology, and a lecture cum live cooking of Indian curry.

After the entire course was over, the students were requested to choose one particular topic of their liking out of 14 classes and write a report of 1000 words. Most of the students have read further on their subject of interest. While political and economic analysis have been mostly based on what they have learnt in the class and additional readings, many interesting interpretations and observation were expressed by those who reflected on what they have learnt through their own experiential immersive learning—through yoga, music of both Indian and Japanese instruments, and virtual cooking class of an Indian meal. The reports by the students in 2023 carried varied aromas with elaborate details on food culture and music tradition of India while some constructive suggestions were offered on Indian polity, economy environment, and society. Some curious observations were presented by the students in their reports on India Class, reflecting upon their own background by engaging their thought process with some interesting elements of India they found through the course. This article tries to elaborate on how India has been perceived by the students who attended India Class in 2023 and what could be envisioned for enhancing the bilateral relations between Japan and India.

2 Why India Matters

2.1 *Population*

Today, India is the most populous country in the world with 1.4 billion population, surpassing the number of the population of China that used to be the world's No.1 till 2023 (United Nations 2023). India's working-age population against total population will reach its highest level at 68.9% by 2023. By the year 2030, India's dependency ratio is projected to reach its lowest point at 31.2% (EY India 2023). While many developed countries have marked declining birth rate and growing aging population, reaching the limit of growth, India is a country of opportunity with significant possibility for growth. Thus, populous India is considered as an advantage in terms of providing labor and creating opportunities as growing consumer market.

Double-edged Sword: However, students also consider that India's population could be described as a "double-edged sword." They consider that Indian government needs to get engaged with the situation urgently to turn the risks into opportunities for growth. India's GDP is the fifth largest in the world (World Economic Forum 2022). Japan now considers India as the leading economic power. On the other hand, about 20% of India's population lives below the poverty line, which earns $1.9 per day. Further, when it turns to per capita GDP, it is about one-fifth of China which has similar number of populations (The World Bank 2022).

2.2 *Entrepreneurship and Social Business in India*

Government of India has created the catalyst "Startup India" initiative launched in 2016. The number of startups surged from over 471 in 2016 to more than 72,993 in 30th June 2022, making India the world's third-largest startup ecosystem after the United States and China (The Times of India 2022). Among the various industries, information technology (IT) and IT-enabled services are prominent. A three-year income tax exemption for qualifying startup companies, the establishment of a 100-billion-rupee (approximately 1.7 billion USD) fund for indirect investments in startups, and the provision of an online platform connecting startups with investors have helped its entrepreneurial ecosystem grow. Today India has more than 100 Unicorns and more than 88,000 recognized startups in the country (Press Bureau of India 2023).

Social Business has also attracted the imagination of the students as it could possibly tackle the issues such as (1) environmental challenges, (2) poverty, and (3) gender inequality. The number of unicorns in India far exceeds that of Japan. India has more unicorns compared to Japan's total of 10 unicorns (Kayo 2022). Globally ranked as the third largest startup ecosystem, India has witnessed a significant surge in startups, surpassing European countries and Israel. The country has experienced a large increase in the number of companies growing into unicorns in recent years. In 2021 alone could see the growth of unicorns surpass the total number of unicorns since 2014. Indian startups significantly contributed to India's growth and innovation in various sectors such as technology, e-commerce, fintech, and healthcare.

2.3 Growth of IT Industry in India

Growth in IT sector within a short span of time has given most impressive image of India to the students. Indian students overall are perceived as excellent and their proficiency in STEM subjects is highly appreciated. India's demographic dividend is anticipated to peak around 2041, when the share of working-age, i.e. 20–59 years, population is expected to hit 59 per cent (Government of India 2020). India allows own IT companies ensure a development opportunity to collaborate with international clients while they sleep, further enhancing their global competitiveness. In the case of the United States of America, the time difference with India is about 12 hours, which allows any task to be carried out round the clock. English, the language of former colonial power, became the driving force of the modern software development in India. English usage has helped Indians, especially IT professionals, to form international collaborations and expand their global reach by seamlessly interacting with clients from across the globe.

3 How India is Perceived by the Students

India has two extremes: growing unicorns on the one side and less empowered vulnerable people. Students have categorically pointed out that more empowering and sustainable approach is crucial to the large number of people to share the fruit of development and socio-economic opportunities so that Indian society will be truly inclusive. The current phase of "Make in India" policy and "Invest India" of the Government of India has placed much importance to turn India as the most competent manufacturing hub and the market attractive to foreign investors. The students consider the manufacturing sector will lead India to give employment opportunities for the vast population, though they acknowledge fulfilling educational needs and other requirements for the vast number of people in the country needs to be prioritized.

3.1 *Both Sides of One Coin: Indian Brilliance and Vulnerability*

Indian talents are reaching worldwide. On the other hand, the students have marked that there has been a significant number of people that has been remained vulnerable, not getting sufficient education to become self-sustainable. Thus, Indian brilliance and vulnerability of many communities in India coexist.

3.1.1 Indian Heritages Lead to Diversity

There are many rich heritages of India. Ayurveda, India's ancient system of natural medicine, is embraced for promoting overall well-being. Indian art and craft traditions, encompassing embroidery, pottery, painting, and weaving, demonstrate the nation's artistic brilliance. Religious rituals and practices vary across the numerous religions practiced in India, including Hinduism, Islam, Christianity, Sikhism, Buddhism, Jainism, and others. Folk dances like Bhangra and Garba, along with folk music, further testify India's cultural wealth and are cherished in different regions, honoring local heritage. Diversity allows harmonious coexistence of different religions and cultural practices, making the country thriving with its trails of rich heritage. Students consider religious pluralism and mutual respect promise harmony within Indian society as India's identity, appreciating its rich tradition and culture that consider diversity and pluralism as the core of the society.

3.1.2 Narrating on Philosophical Practice of the East

The late nineteenth century saw prominent Yogis and spiritual leaders visit Japan. Yoga was introduced to Japan at that time. Swami Vivekananda, with his admiration for Japan and its culture and people, became an important link for Japan and India. Together with Okakura Tenshin, Swami Vivekananda wrote extensively on Eastern culture and philosophy. Confluence of Indian and Japanese ideas could be studied further through some testimonies by such philosophers who interacted with people in both societies while being able to communicate with the West. This part of history, Japan and India exploring common footprints in view of western philosophy, needs to be documented more by those who have had direct access to their diaries and writings in original languages. Some serious efforts need to be done for this purpose in near future.

3.1.3 Yoga in Japan: Its History and Practices

It was only during the 1920s when scholars from Japan travelled to India and start practicing Yoga. Post the World War II there were more publications and workshops on yoga. Further, recognition of Yoga as alternative health practices have grown with the increasing number of Indian immigrants to western countries. In late twentieth century, popularity of Yoga mostly owed to its western style practices in urban Japan. Yoga as a brand in modern Japanese society has been practiced individually as a means of personal care and healing. Yoga in Japan used to have westernized approach where the practices are held in yoga studios, more as a sport and exercise, being taught by trained instructors while paying attention to mindful use of body. Through internationalization of yoga, its relevance in spiritual and meditative practice has come to be duly highlighted today.

3.1.4 Indian Music: Celebrating Diversity

Indian music manifests India's rich cultural heritage with regional flavours. Woven with profound and introspective classical tunes, vibrant and energetic folk melodies, soul-stirring devotional music, infectious beats of Bollywood, and innovative contemporary fusion, India's musical tapestry reflects the country's vibrant diversity and the indomitable spirit of its people. Celebrating unity in diversity, music plays the role of the universal language connecting hearts of people, transcending boundaries and resonating with the very essence of our existence. One student describes Indian music as follows:

> Like a skilled painter, India blends a vibrant palette of musical colors, weaving together ancient classical traditions, regional folk melodies, soul-stirring devotional chants, and the infectious rhythms of Bollywood, creating a symphony that transcends cultural boundaries and touches the depths of the human soul.

Indian classical music evolved through oral traditions, passing from *gurus* to disciples. Its reach has gone global, forming a valuable rich heritage of traditional music. Performances by Ravi Shankar's at Woodstock and the Beatles' visit to India in the late 1960s let the presence of Indian music felt throughout the world, including Japan.

3.1.5 Food Culture: Widening Culinary Horizons

One student maintained that "Indian curry has become a global culinary ambassador, representing India's rich cultural heritage and diversity," suggesting that "culinary horizons" of people from different cultures could allow us to learn how to appreciate other cultures. How to prepare meals by using spices, aromas, and flavors of Indian curry literally captivated students, as the way of cooking Indian home food is so different from the way Japanese curry is prepared. Indian curry exemplifies a vernacular narrative that carries its cultural heritage forward—to its own people through generations and beyond oceans. Preparing food and sharing meals are important part of social life in India, connecting families, friends, and communities. Food signifies hospitality and togetherness, through festivities, rituals, and religious gatherings. It is also very important to note that the lecture on food greatly inspired students to type their comments and questions in the chat box freely. The chat box was literally flooded with the comments of the students expressing their curiosity and astonishment by observing cooking procedures. This went on parallel to the proceeding of preparing a basic home meal. Students were also able to interact with the instructor on Indian food.

3.2 Technology

3.2.1 India's Digitization

India-Japan relationship in the field of IT technology could be considered first as supplying Indian talents to Japanese workforce: as described by one student, "desirable for Japan to adopt India's high technological capabilities and actively exchange limited human resources." Some vision to enhance practical collaborations are yet to be narrated fully. The anticipation for enhancing collaborations was abundant in the reports of the students. India's technological progress is fueled by its diversity and the quality of its human resources. India is one of the most populous countries in the world, and its growing educational standards and IT industry have nursed many talented engineers and programmers. In addition, India has become an attractive market for investment and collaboration from global technology companies, which in turn bring the latest technology and knowledge to India. The Indian government is also actively encouraging technological progress. Policies and initiatives such as the Digital India Campaign are driving the adoption of digital technologies and the development of digital infrastructure. This has increased the penetration of Internet connectivity and facilitated the growth of the digital economy. India's technological advancements are evident in a variety of areas. For example, innovative technologies such as artificial intelligence (AI), big data, cloud computing, the Internet of Things (IoT), and blockchain are being utilized. This has led to greater efficiency and innovation in areas such as education, agriculture, healthcare, energy, and finance, contributing to the development of society.

3.2.2 Learning Digitized Administrative Processes from India

India is more digitized than Japan. Enforcing digitalization of identification has created agencies that assist the processes for the residents/citizens of India to process formal interactions digitally. For instance, India government provides digital infrastructure for all citizens such as consolidation of high-speed internet, digitization of identification, mobile phones, bank accounts, through digital infrastructure. They also provide On-demand administrative services such as real-time administrative information utilizing online, mobile devices digitization and cashless financial services. Through biometric based ID system in India called Aadhaar, Open API, e-KYC, Digital Locker and UPI (Unified Payment Interface) and digital payment such as Google Pay, PayTM, PhonePe, Zomato and Mobikwik, India has been strengthening digital literacy to improve services. In India's digital transformation startups have played a crucial role. E-commerce has remarkably grown while coexisting with traditional trade. India's response to the COVID-19 pandemic has further accelerated the digitization of smaller businesses. On the other hand, digitalization in Japan has not progressed yet. In particular, it is notable that digitalization of administrative services has not progressed compared with India. Penetration of My Number Card is not complete, which is not solely because of the issue of digitization. ICT utilization and data utilization using appropriate digital infrastructure have not progressed sufficiently in Japan.

3.3 Environment: Air Pollution as Health and Economic Hazard

Concentrations of the air pollutant PM2.5 have in some places been recorded at 30 times the WHO standard, taken as a pressing hazard for India by the students. The health hazards of air pollution negatively impact overall population, especially children who are still growing up. Ameliorating this situation in a short time in India would not be easy since economic growth and industrialization are progressing in the county. But environmental degradation has been noted as one of the major concerns for India by the students.

4 Challenges

While the students are overwhelmed by the potentials of India to grow further, they are also very concerned about the enormity of the issues India faces due to having vulnerable groups of people.

4.1 Informal Employment in India

Though India has become the most populous country in the world, the prevalence of informal employment in the country has been impacting society with far-reaching social and economic implications. From 2005–2006 to 2019–2021, incidence of poverty fell from 55.1 to 16.4% in India (Oxford Poverty and Human Development Initiative and United Nations Development Programme 2023). Informal employment refers to a work by not regulated labor laws and lacks formal employment benefits,

Table 1 Informal Employment in India: Issues, Challenges, and Remedial Measures as Perceived by the Students (2023)

Issues	Challenges	Remedial measures
Unemployment Social disparity Need for improvement of human rights approaches Need for improvement of educational opportunities	• Lack of education • Limited skills • Limited legal protection • Lack of access to health care • Lack of formal employment benefits • Child labour • Low wages • Long working hours • Minimal job security	• Creating more formal job opportunities by promoting inclusive economic growth • Supporting entrepreneurship • Providing skill development programs • Strengthening labor laws and social protection measures

Source Authors (Compilation of extracted elements from the reports of the students who attended India Class of 2023 curated by India Japan Laboratory, Keio University, Japan)

such as job security, social protection, and access to basic rights. The informal sector encompasses a wide range of activities, including street vending, small-scale manufacturing, agriculture, and domestic work. The authors have summarized the issues, challenges and remedial measures suggested by the students regarding informal employment in India as follows (Table 1).

The matter of socio-economic development for minorities and the poor remains a great concern for the people. Women and the vulnerable tend to internalize the idea that they are good enough to do informal work and home-based work only by being 'trapped' in conventional ideas and social stigma. Awareness of marginalization should come both from those who are marginalized and from those who are not marginalized. The mindset of accepting marginalization comes from both sides, further reinforcing the idea and letting the idea dominate the society.

4.2 Vulnerability of Neglected People Deserve More Attention

What could be seen from the students' writing is the kind of despair and urgency to rectify the uneven development that leaves people to fend on their own. In this context, it is important to study more on various cases of what we would call "Small India"—a real face of India in day-to-day life of ordinary people. The situation of such uneven development does not go well with India 's global position as a leading economy and its geo-political impact. "Great India" co-exists with "Small India." Informal employment for survival of an entire family leaves some women exceptionally vulnerable. The loop of informal employment" persists in India. Be it urban or rural, there are women and vulnerable communities who would need urgent attention and appropriate policy measures and supports in their trade to make the growth and development of India sustainable. For addressing such vulnerability of Indian society, conditional cash transfer was recommended by some students to protect such vulnerable people from poverty and mitigate already existing poverty to be eradicated by raising the overall level of educational attainment. The long-term positive outcomes such as improved height and behavioral development of a child could be anticipated.

4.3 Centralization and Centre-State Politics

Indian political scene requires to have reflections upon the urban and rural lives in terms of local power structure and national politics. Post economic liberalization since 1990s, Centre-led Panchayat legislation to promote decentralization of power has taken turn towards centralization of power under Modi government. Centre-State relationship in politics in India has gone through a significant change. A large number of foreign enterprises developed industry through economic liberalization since 1990s. Three-tier panchayat legislation by the Central Government exacerbated the rift between states and local governments is considered as the source of difficulties and complexities to hinder reforms since regional forces would not take heed on Centre-led reforms. The Centre has created administrative connection through nearly 1200 federal government officials who have been appointed as "Nodal Officers" to coordinate implementation of and supervision over the state affairs across the country, where direct reporting of the outcome to the Center would not be required to go through state government (Aiyar 2018). Students' opinions vary on the matter of centralization and its effectiveness. Some students are in favor of centralization saying that would certainly accelerate further reforms that would give a push for economic development of the country. Yet other students are of the opinion that regional voices need to be valued more. The topic of Centre-State politics could be discussed through case examples.

4.4 Disparities Among the States in India

Regional disparities are manifested among the states in India as income levels in some states are as low as $600 while some are as high as $6,000. Major states in India have similar income levels compared to major cities in ASEAN countries like Jakarta, Manila, and Ho Chi Minh City, while 70% of India's population depends its livelihood in rural areas. A large number of people from such rural regions are often seeking opportunities to migrate to urban areas in India or even to other countries, seeking better economic opportunities.

4.5 *India and Its Higher Educational Institutions*

The students have reasoned the growth of IT industry owes a lot to Indian Institute of Technologies (IITs, the number reached to 23 in 2023) that helped India to nurture skilled graduates and researchers. Science and Technology has been promoted on computer science, information technology, and related disciplines. Around 10,000 applicants clear entrance, but over all, only one out of 53 applicants (506,000 applicants a seats in 2012) got through, "leaving 98% examinee chokes back the tears." Still, India is anticipated to remain as a global IT powerhouse primarily thanks to young and skilled workforce who are fully conversant in English with its time zone advantage that allows them to carter to the requirements of international clients. Learning about Tier-2, Tier-3 towns that have nursed competitive colleges and universities as well as private institutions to respond to the need of educating Indian youth could be an interesting way to see future possibilities of education in India. Further, it is also important to note that limited seats in domestic higher educational institutions and regular employment opportunities at both government agencies and private enterprises have become major reasons for Indian youth to migrate abroad in search of better life and higher earnings.

5 Enhancing Bilateral Relationship Between India and Japan

5.1 *Possible Areas of Further Collaborations*

There are variety of suggestions from the students to address the challenges India faces. The authors have collected their recommendations and organized below (Table 2).

In order to facilitate further bilateral collaborations, students consider that cultural and environmental factors, local needs, and financial constraints in addition to the implementation of technology need to be reflected upon.

5.2 *Importance of Shared Experiences*

India Class has shown what makes one bridges the boundaries: enjoying music, dining together, visiting other people's homes, and having friends from other countries have helped the students feel close to a country like India. Through their writing, it was revealed that understanding India starts from what we see, hear, smell, feel, and taste (Table 3).

The authors have several recommendations to further enhance the scope of university education regarding India and its relationship with Japan.

Table 2 Students' recommendations on possible areas of further collaborations between India and Japan

Field	Issues	What is missing	Solutions/remedies
Infrastructural development	Traffic congestion in urban areas	Efficient means of transportation Good public transformation High-speed railroads and subways Government initiatives	• Building roads and ports • Construction of dams • Sewage systems • Water supply systems Introducing Japanese rail technology and transportation systems
Urban development	Rapid growth of Indian cities Concentration of population in developed areas Uneven or Skewed development	Sustainable urban planning Energy efficient architecture	• Japan's urban planning and design • Japan's efficient use of public space • Paying attention to rural and underdeveloped areas
Renewable energy	India's rapidly growing energy demand	Effective support for India's renewable energy policies and projects	Japanese renewable technologies such as solar and wind power

Source Authors (Compilation of extracted elements from the reports of the students who attended India Class of 2023 curated by India Japan Laboratory, Keio University, Japan)

Table 3 Different means generating narratives for people-to-people connections

Elements	Means	Goals	Impact
Food Music Yoga Languages	Narratives Visuals Dining together Concerts Lectures	Interactions among people Understanding about other culture	Friendship Curiosity towards different cultures Warmth Hospitality

Source Authors (Compilation of extracted elements from the reports of the students who attended India Class of 2023 curated by India Japan Laboratory, Keio University, Japan)

1. To create opportunities for students from Japan to visit India.
2. To invite Indian students to interact with the students studying in Japan.
3. To learn from the Indian residents in Japan to understand how two cultures do meet on a foreign land.

5.3 Experience with Students Exchange Program

The students may still not be ready to step out of the country of residence to explore India on their own. India Japan Laboratory had planned the first student exchange in India with Indian Institute of Science Bengaluru for the month of February 2024. We found only 2 Japanese students applied for the program. For a student to make up mind to travel to India for an exchange program is still not easy.

5.4 *Immersive Program in Japan 2024*

Therefore, we thought to introduce immersive program in Japan where we can invite students from India. Interacting with Indian students while involving residents in Japan, sharing Indian and Japanese meals, and meeting community people from India in Japan will help us create opportunities for both students from India and those students studying in Japan to have co-learning experiences. In February 2024, Indian Institute of Management Kozhikode, Kerala, India, sent 39 students of Management. Interactions of people will generate narratives. Narratives will help the students as well as the educators to understand the most impactful effects of India Class upon the students. And those narratives can happen outside of the regular classrooms.

6 Next Steps: Creating Platforms for More Interactions

For the first time since India Japan Laboratory started our India Class in 2020, many reports of the students in 2023 elaborated upon very detailed accounts of Indian traditions of food, medicinal effects of main spices, and the usage of various music instruments that are key to compose and play traditional music. The authors have observed in the writing of the students that there are several interesting topics that could be elaborated further: the students are very keen to explore the richness of regional cultures of India to celebrate diversity in regional flavors and uniqueness.

1. **Interactions with Japanese Residents in India**: Learning from observations of Japanese residents in India who are proficient in Indian culture, arts, music, languages, and so on.
2. **Interaction with Indian Residents living in Japan**: Getting feedback from those Indians in Japan who have been working in various fields, not necessarily limited to IT sector, would give the chance to the students to reflect upon the opportunities of collaborations between India and Japan.
3. **Voices from Rural and Urban Areas of India and Japan**: Some region-specific and theme-based interactive sessions could be curated to interact with ordinary people in both rural and urban areas in both countries. This will help us to understand real life issues through the voices of people.

The following suggestions could be incorporated in coming lectures in India Class.

A. **Curry Culture Speaks for India Being Plural**

There is a huge potential in imparting education on other countries to discuss about their food culture. In the case of India, the variety of food according to the regions needs to be appreciated. At higher educational institutions, too, student exchange program can be further facilitated with dining together. Meeting and eating together allows us to share quality moments together. This could be a shortest possible way to get to know others. It also let us have an opportunity to reconsider our observations and opinions about other culture and people.

B. **Decoding Entrepreneurship in India**

The students are quite fascinated and mesmerized to find Indian start-ups and unicorns letting their impact known to the world. Innovation and entrepreneurship need to be taken as one of the main themes of the future sessions of India Class. It will help review the approaches taken in Japan to foster entrepreneurship. While India requires to create income-generating economic activities for the masses, Japan needs to address the lack of social and economic impetus to initiate and manage enterprises.

C. **Conducting Studies on Micro, Small and Medium Enterprises in India and Japan**

One common thread Japan and India share is the presence of micro enterprises as well as SMEs as the major economic activities to sustain livelihood of the major part of the population. India has regarded the MSMEs as the important crucial source of economic activities of the country and has formed an individual ministry in the Central Government. The fate of many SMEs in Japan is at stake due to the aging of the management and challenges for internationalization of their business when domestic market is dwindling. Effective use of government supports in private business at small scales and regional SME collaboration for sustainability and revitalizing regional economy are the two aspects Japan should address. The bilateral collaboration between Japan and India could be elaborated further by creating opportunities in sustaining existing businesses and exploring new business opportunities for both Japan and India.

D. **Learning About Indian Buddhism and Japanese Buddhism**

While personal enlightenment is emphasized in Indian Buddhism, compassion and devotion to society is valued in Japanese Buddhism. It is curious to see that in Japan Buddhism as a salvation—a sort of resilience—got evolved against natural disasters, famines, and epidemics that impacted local populace in Japan during Kamakura period. It will be interesting to have some lectures on the evolution of Buddhism in India and Japan as it has got different connotations and political implications in the history of both the countries.

E. **Experiential Learning of Yoga**

India Class found many students had very limited experience of Yoga but there are some students who have integrated Yoga as a part of their sport activities and/or daily practices. Students were surprised to find that Yoga could give solutions to anxiety, stress, enhancing concentration, apart from improving body movements. Some synergies with sports and music have been also pointed out. Spirituality of Japanese martial arts such as Karate, Aikido, and Sumo and Bushido practices find some synergies. It will be interesting to invite some students who would be interested to share their own experiences with Yoga among the peers during the class. A hybrid venue could be arranged for such interactive sessions.

7 Conclusion

While most of the analysis shown in the reports of the students were done through their further reading on the subjects, some other students have tried to express what they have felt and understood through experiential learning, especially through the lectures on Indian curry cooking and Indian music concert. Through past life experiences, daily or occasional interactions with their Indian friends and classmates in Japan, India, or yet some other third countries, for example, in Singapore, the U.S.A., or Europe, some students could elaborate more on their learning about various topics. One stark finding was how the students talked about 'poverty'. It suddenly becomes the data and quotes. That exactly could tell us that it was not enough to understand the issue of poverty.

7.1 Changing India and Sustainability

The importance of the presence of India in the world is of multi-fold. India does not remain static, rapidly changing, having evolved into a leading economy since its independence. The economic impact of India, however, is mired with the inherent issues of the society where certain sections of people remain vulnerable. The magnanimous population carries the issue of unemployment, while giving birth to small enterprises and budding entrepreneurs all over the country depending on the needs of such people to seek for the source of sustainable livelihood. When the country exists and depends on its economy to expand and grow further, those small business owners, regardless of the type of their enterprises being formal or informal, need to be taken care of. While India has been regarded as a promising country, it is also noted that the students are concerned with the issues in India that would hinder sustainable growth. The issue of sustainability in the context of environment could enable us to consider new ways of application of technology in view of functioning cultural values in Indian society.

7.2 *Innovation in Education*

Universities need to reexamine and rejuvenate what the educators offer. While reflecting ever-changing international relations—beyond the bilateral relationship between India and Japan, aspiration of the students could be taken into consideration to direct their course of leaning to understand that could be the collective values as a society. It is important to remind ourselves that we are living in the age when the relevance of traditional university education has gone through a complete overhaul due to the COVID-19 pandemic. In this context, exploring innovative approach of hands-on experience through online classes need to be embodied in our India Class. At the same time, it is important to remember that higher educational institutions in Japan need to be innovative and sensitive enough to engage diverse international students as well as Japanese students who may have origins of foreign nationalities.

References

Aiyar Y (2018) Are bureaucrats turning politicians? The Hindustan Times, 26 July 2018. https://www.hindustantimes.com/columns/are-bureaucrats-turning-politicians/story-7oKNAFIrmtrW6vIWGIN1LM.html. Accessed 18 Nov 2023

EY India (2023) Reaping the demographic dividend. https://www.ey.com/en_in/indiaat-100/reaping-the-demographic-dividend. Accessed 18 Nov 2023

Government of India (2020) India's demography at 2040: planning public good provision for the 21st Century, Chapter 7, pp 128–147. https://www.indiabudget.gov.in/budget2019-20/economicsurvey/doc/vol1chapter/echap07_vol1.pdf

Kayo T (2022) Uncovering Japan's 41 hidden unicorns. Coral Capital. https://coralcap.co/2022/05/uncovering-japans-41-hidden-unicorns/?lang=en. Accessed 18 Nov 2024

Oxford Poverty and Human Development Initiative and United Nations Development Programme (2023) Global multidimensional poverty index 2023, unstacking global poverty: data for high impact action, pp 1–26. https://www.undp.org/sites/g/files/zskgke326/files/2023-07/2023mpireportenpdf.pdf

Press Information Bureau (2023) India: world's 3rd largest startup ecosystem: 100+ unicorns; 88,000+ recognized Startups. Department for Promotion of Industry and Internal Trade, Government of India, 16 January 2022, updated on 16 January 2023. https://static.pib.gov.in/WriteReadData/specificdocs/documents/2023/jan/doc2023116152401.pdf. Accessed 15 Jan 2024

The Time of India (2022) Number of startups in India grows to 72,993 in 2022 from 471 in 2016, 23 July 2022 https://timesofindia.indiatimes.com/india/number-of-startups-in-india-grows-to-72993-in-2022-from-471-in-2016/articleshow/93077040.cms. Accessed 25 Nov 2023

The World Bank (2022) GDP per capita (current US$)—India, China. https://data.worldbank.org/indicator/NY.GDP.PCAP.CD?locations=IN-CN. Accessed 14 Dec 2023

United Nations (2023) UN DESA Policy Brief No. 153: India overtakes China as the world's most populous country. United Nations Department of Economic and Social Affairs. https://www.un.org/development/desa/dpad/publication/un-desa-policy-brief-no-153india-overtakes-china-as-the-worlds-most-populous-country/. Accessed 08 Jan 2024

World Economic Forum (2022) India overtakes UK as the world's fifth-largest economy. weforum.org. https://www.weforum.org/agenda/2022/09/india-uk-fifth-largest-economy-world/. Accessed 22 Dec 2023

Higher Education: Weaving the Future of "Indispensable Partnership"

Srabani Roy Choudhury and Rajib Shaw

Abstract As India and Japan ink their relationship from a global partnership to an indispensable partnership, the people-to-people exchange gains momentum. Youth, a significant segment, play a vital role as they build and enhance the capacity of a nation and contribute to the future trajectory. Education can act as a catalyst for diplomatic relations by creating newer avenues for cooperation. The higher education demographic profile is that of a mature youth with a distinct focus and purpose, and encouraging student exchange of this group allows for upgrading skill, capacity, and competence. Further, it allows for deepening interest in each other's countries as it fosters cultural understanding. There is a plethora of complementarities between India and Japan in STEM and liberal arts. This chapter will focus on identifying and delineating the complementarities in higher education of these two nations and discuss the scope of this vertical in creating a long-lasting bond between India and Japan.

Keywords Indispensable partnership · Education · Capacity building · Technology · NEP 2020

1 Introduction

The world is undergoing dynamic geopolitical, technological, societal, and environmental changes. The World Economic Forum Global Risk Outlook has underlined environmental risks (like disasters, extreme events, climate action failure, biodiversity losses, etc.) as critical risks regarding likelihood and impacts (WEF 2020). The risk landscape changed significantly over the last 4–5 years. We have seen new risks like infectious disease (WEF 2021), digital power concentration and digital divide

S. R. Choudhury (✉)
School of International Studies, Jawaharlal Nehru University, New Delhi, India
e-mail: sroyc2@gmail.com

R. Shaw
Graduate School of Media and Governance, Keio University, Fujisawa, Japan

R. Shaw and S. R. Choudhury (eds.), *India, Japan and Beyond*,
https://doi.org/10.1007/978-981-97-3282-1_17

(WEF 2021), cybersecurity (WEF 2022), energy crisis and cost of living (WEF 2023) etc. In the recent report 2024 (WEF 2024), new risks of dis- and misinformation are identified as the most prominent emerging risks in the next two years.

Higher education is indispensable in the global risk landscape as it equips individuals and societies with the knowledge, skills, and resources to effectively understand, respond to, and mitigate risks. It fosters resilience, innovation, collaboration, and ethical leadership in the face of uncertainty and complexity. Thus, while it is essential to have in-depth education and research in conventional subjects, higher education also needs to evolve to address new challenges and opportunities. Thus, an essential stream of future course higher education needs to address the following:

Complex problem solving: Many challenges facing the world today, such as climate change, global health crises, economic instability, and political conflicts, require sophisticated problem-solving skills. Higher education equips individuals with the knowledge, analytical abilities, and critical thinking skills to understand complex issues and develop practical solutions.

Research and innovation: Universities are hubs of research and innovation, driving progress and advancement across various fields. Higher education institutions contribute to developing new technologies, strategies, and methodologies that can help address emerging risks and challenges on a global scale.

Global collaboration: Higher education fosters collaboration and knowledge-sharing among scholars, researchers, and students from diverse backgrounds and regions. This collaboration facilitates the exchange of ideas, perspectives, and best practices, enabling more effective responses to global risks through collective action and cooperation.

Capacity building: Higher education plays a crucial role in building the capacity of individuals and communities to anticipate, adapt to, and mitigate risks. By providing specialized training, education, and skill development, universities prepare future leaders, professionals, and policymakers to tackle complex challenges and make informed decisions in an increasingly interconnected world.

Promotion of resilience: Education empowers individuals and societies to become more resilient in the face of adversity. Higher education helps individuals and communities develop strategies to enhance their resilience and withstand shocks and disruptions more effectively by fostering a deeper understanding of risk factors, vulnerabilities, and potential impacts.

Ethical leadership: Higher education promotes ethical leadership qualities, including integrity, accountability, and social responsibility. Ethical leadership is essential for addressing global risks responsibly and sustainably, ensuring that decision-making processes prioritize the well-being of people and the planet.

Policy Development and Advocacy: Universities are crucial in informing policy development and advocacy efforts related to global risks. Through research, analysis, and engagement with policymakers and stakeholders, higher education institutions contribute to formulating evidence-based policies and initiatives to address pressing global challenges and minimize potential risks.

Overall, higher education institutions have a broad range of social responsibilities that extend beyond traditional academic pursuits. By actively engaging with their communities, promoting diversity and inclusion, conducting ethical research, and fostering global citizenship, universities can make meaningful contributions to society and help address some of the most pressing challenges facing the world today. Higher education institutions should strive to provide equal opportunities for all individuals regardless of their socio-economic background, race, gender, or other characteristics. This includes initiatives such as scholarships, financial aid, and outreach programs to ensure that education is accessible to everyone. Higher education institutions are responsible for fostering diverse and inclusive environments where individuals from different backgrounds feel welcome and respected. Universities and colleges should actively engage with local communities and address societal challenges through research, service-learning programs, and community partnerships. This can involve projects focused on poverty alleviation, healthcare access, environmental sustainability, and more. Higher education institutions are responsible for minimizing their environmental footprint and promoting sustainable practices. This can include initiatives such as reducing energy consumption, promoting recycling and waste reduction, implementing sustainable transportation options, and integrating sustainability into the curriculum.

Identifying the above needs is critical to enabling the orientation of higher education. Japan has always given prime importance to education. Its economic miracle is often attributed to its education system, which has its roots in the Meiji period, followed by compulsory education as the country recovered from the war. India, too, has a long history of giving importance to education. India's knowledge system of the past is accoladed for its foresight and sagacity. Today, India stands at the juncture where its economic growth and leadership in the region are gaining dominance. To enable it to sustain and grow, the education of the children and youth needs to meet the expectations of industry and global bodies. Both nations have thus embarked upon aligning education with the new needs that have originated. This chapter draws upon understanding the education orientation of both nations. Further, it looks at the soft power element of people-to-people exchange through propelling exchanges among youth through the vector of higher education.

2 Higher Education Policy of Japan

Japan has three higher educational institutes: universities, technology colleges, and vocational schools. Universities in Japan can be broadly categorized into national, public, and private universities. The government runs national universities, and local governments and private entities fund private universities and operate public universities. Around 77% of the universities are private, 11% national, and 12% are with local authorities. Consequently, 73% of the students study in private universities, 22%

in national universities, and 5% in local universities. Admissions to universities in Japan are highly competitive, particularly for prestigious institutions. Students typically take standardized exams like the National Center Test for University Admissions (commonly known as "Center Test" earlier and "Common Test" currently) along with subject-specific exams conducted by individual universities.

Japan's modern higher education system goes back to 1858 when visionary Yukuichi Fukuzawa established Keio University. This was followed by establishing the University of Tokyo in 1877, Kyoto University in 1897, and other imperial universities in early 1900. Keio University was the first university in Japan to accept foreign students (two Korean students) back in 1881. Japan saw a strong population growth, resulting in an increasing percentage of students in higher education in the 1960s and 1970s. This prompted the establishment of more private universities, which the Ministry of Education approved through loosened regulatory reform. National universities remained under the control of the Ministry of Education until 2004, after which several became independent and autonomous agencies. Private universities, too, despite their autonomous status, were under the strict supervision and control of the Ministry at the initial stage and became fully autonomous in due course.

The Japanese government strongly emphasizes maintaining the quality of higher education institutions. The Ministry of Education, Culture, Sports, Science and Technology (MEXT) oversees quality assurance mechanisms and accreditation processes. Japan is known for its strong research culture and investments in science and technology. Many universities in Japan actively engage in research across various fields and contribute significantly to global scientific advancements. Japan particularly emphasizes science, technology, engineering, and mathematics (STEM) education to support its innovation-driven economy. There are numerous initiatives aimed at promoting STEM education at all levels, including higher education. Higher education institutions in Japan often provide career support services to help students transition into the workforce. This includes internship opportunities, job placement assistance, and career counseling.

There has been a push towards internationalization in Japanese higher education in recent years. Many universities offer programs taught in English to attract international students, and various scholarship programs are available to support international students studying in Japan. Despite its strengths, Japan's higher education system faces challenges such as an aging population, declining birth rates, and the need to adapt to rapidly changing economic and technological landscapes. Efforts are being made to address these challenges through various reforms and initiatives.

Universities, as centers of academia, are by their very nature to deeply explore the truth and to teach and research specialized arts and sciences, and in order to fully guarantee their activities, their fundamental characteristic is that they have traditionally been granted a certain degree of autonomy and self-governance. The twenty-first century is said to be the era of a "knowledge-based society" in which new knowledge, information, and technology will dramatically increase in importance as the basis for activities in all areas of society, including politics, economics, and culture. In the coming "knowledge-based society," higher education will be critical in forming individual character, as well as in national strategies for social, economic, and cultural development and promotion and for securing international competitiveness. In the future, as international competition intensifies, the overall strength of the nation's higher education system and higher education policy itself will be questioned. Japan's national government recognizes the importance of higher education in a knowledge-based society (MEXT 2005). In particular, as Japan's greatest resource is its people's intellectual activity and creativity, it is indispensable to foster excellent human resources and promote science and technology. A crisis in higher education is also a crisis in society. For Japan's society to continue to develop, it is essential to transform higher education into something that can fully respond to society's trust as the driving force of the times and to build a two-way relationship whereby society actively supports this transformation.

Universities have education and research as their essential missions. However, at the same time, the roles expected of universities are changing. Today, the importance of social contribution (contribution to the development of society in a broad sense, including local communities, economic society, and international society) is becoming increasingly emphasized. Naturally, education and research are social contributions from a long-term perspective. However, more direct contributions through international cooperation, open lectures, and industry-academia-government collaboration have also come to be demanded in recent years. This is an era in which such social contribution roles should be regarded as universities' "third mission" (MEXT 2005).

For higher education to continue to fully fulfill its role in the midst of increasingly sophisticated academic research, diversifying learning demands, changing societal values, and advancing internationalization and information technology, it is necessary for each institution of higher education to clarify its individuality and characteristics in a competitive environment, and to achieve diverse development as a whole. MEXT has introduced different innovative programs to enhance social responsibilities as well as internationalization of the universities, like: "Framework for 300,000 international student plan", "Developing 20 universities as centers of excellence for internationalization (Global 30), "supporting internationally competitive projects and COE (center of excellence)/GP (good practice projects)" etc. (MEXT 2011). In 2014, the MEXT issued specific guidelines for double degree and joint degree programs with overseas universities (MEXT 2014). The Guidelines have been formulated in light of "University Education and Global Human Resource Development for the Future" (third proposal) published on May 28, 2013, by the Education Rebuilding Implementation Council.

However, it is highly problematic whether higher education responds to recent social changes and maintains sufficiently high quality. It is no exaggeration to say that Japan's higher education is on the verge of a crisis. Even during the 1980s, when Japan had the highest economic growth, Japan's universities were not at the top of the world. Yonezawa (2023), in his recent analysis, pointed out that: "*Over the past 20 years, Japan's higher education policy has been strongly influenced by national policies for achieving world-class excellence as a tool for economic and social development. However, Japan's universities, especially the top universities, have faced difficulties maintaining an international presence, both in terms of academic excellence and developing globally competitive human resources*". In order to overcome the current situation, Japan's higher education must be transformed into one that can fully meet society's expectations as a driving force of the times through the enrichment of liberal arts education and graduate schools at universities, diversification of short-term higher education, and proactive response to internationalization.

The new generation, or Generation Z, prefers more action-based, problem-solving ways of learning, which provides them with Higher education and immersive learning, which are increasingly interconnected concepts in today's educational landscape. Immersive learning refers to educational experiences that deeply engage learners in the subject matter through various sensory stimuli, technologies, and interactive activities. In Japan, currently, this approach aims to create environments where learners are actively involved in the learning process, often blurring the lines between physical and virtual spaces. In higher education, immersive learning takes on various forms, including.

Virtual Reality (VR) and Augmented Reality (AR): These technologies offer simulated environments where students can explore concepts, scenarios, or historical events in a highly interactive and engaging manner. For instance, medical students can practice surgical procedures in a virtual environment, or history students can virtually visit ancient civilizations.

Simulation-Based Learning: Simulations allow students to apply theoretical knowledge in realistic scenarios. This is particularly useful in fields like engineering, business management, and healthcare, where hands-on experience is crucial for skill development.

Game-Based Learning: Educational games incorporate learning objectives into gameplay, making learning more enjoyable and engaging. Games can teach a wide range of subjects, from mathematics and science to language and history.

Experiential Learning: This approach emphasizes learning through experience, often involving internships, fieldwork, research projects, or service learning. By actively participating in real-world activities, students gain practical skills and insights that are difficult to acquire through traditional classroom instruction alone.

Immersive Laboratories: These are physical spaces equipped with advanced technology, such as interactive displays, 3D printers, and robotics kits, where students can engage in hands-on experimentation and exploration.

Online Immersive Courses: With the rise of online education platforms, immersive learning experiences are increasingly accessible to students worldwide. These courses often incorporate multimedia content, virtual labs, and interactive assignments to create engaging learning experiences.

The integration of immersive learning techniques in higher education offers several benefits, like: increased student engagement and motivation, improved retention and understanding of complex concepts, development of critical thinking, problem-solving, and decision-making skills, preparation for real-world challenges and career readiness, and accessibility and flexibility for diverse learning styles and preferences. However, there are also challenges associated with immersive learning in higher education, such as the need for adequate technological infrastructure, faculty training, and assessment methods that accurately measure learning outcomes. Despite these challenges, the potential of immersive learning to transform higher education and enhance the learning experience for students is significant, making it an area of ongoing innovation and exploration in academia.

3 India's New National Education Policy (NEP 2020)

India, a sizeable populated country, needs to find a solution for educating its nation to gain from population dividend. Broadly, the education system is categorized as school education and higher education. In India, higher education begins with an undergraduate program and ends with a doctoral program. Further, higher education in India is funded by the national and state (prefectural) governments. Since 1993, private sectors have also functioned within the system. The "Central Universities" and "Institutes of National Importance" are funded by the national government and are at the forefront of the education system.

India, too, welcomed the UN's SDG initiative. India adopted the Global Education Development Agenda (SDG4) in 2015. It seeks to ensure inclusive and equitable quality education and promote lifelong learning opportunities for all" by 2030.[1] This is a tall order for a large, diverse, populated country like India. While the literacy rate is 77.7 percent, which is limited to an individual able to sign his/ her name, the education percentage drops beyond primary school and reaches about 60 percent. Thus, a committed focus on education to build worthy human resources is critical for India as it continues to ascend in the economic growth path and looks towards global leadership.

Due to the quantum of technological leaps, the fast-changing global ecosystem requires a skill orientation significantly different from the past. The education system, which followed the principle of learning through understanding and reproduction of facts, is archaic. Today, children must hone their skills and "learn how to learn". Creativity, critical thinking, quick absorption of technological change, and orientation towards adapting to new ways and functions were identified as necessities, and immersion across the education system was vital. The aim for India was to have an education system by 2040 that is second to none, with equitable access to the highest-quality education for all learners regardless of social or economic background.[2]

In an attempt to overhaul the education system, the Government of India set about a consultation process to devise a policy that would address concerns that, over a period of time, had consumed the education sector from primary to higher education. The areas of concern were diverse. They included aligning the years to major international education systems, addressing curriculum lacuna, technology upgradation, making it relevant to global demands, and promoting Indian roots, to name the essential focus.

The process followed was all-encompassing. The government had a dedicated window available for all citizens on www.MyGov.in for a period of little less than ten months (26.01.2015–31.10.2015). This information was disseminated among schools, colleges, universities, and other relevant bodies. Many suggestions were received, then segregated according to the border areas and put forward to the T.S.R. Subramanian Committee and the Ministry of Human Resource Development (Currently known as the Ministry of Education). According to National Education Policy (NEP) Chairman Dr. Kasturirangan, "Given the reactions and responses of the public, our committee revisited the entire policy with additional consultations with over 300 institutions, individuals, and groups. These extensive consultations and their systematic incorporation have ensured that our Constitution's federal and secular nature is reflected in the NEP 2020".[3] In so doing, the new educational policy was able to find acceptance among various stakeholders and address the social, physical, emotional, and moral dynamics that help build a holistic education for children and youth.

3.1 Vision

India's vision for education is to amalgamate India's roots with global needs. Thus, an education system rooted in Indian ethos which contributes directly to transforming India sustainably into an equitable and vibrant knowledge society by providing high-quality education to all, thereby making India a global knowledge superpower. The curriculum and pedagogy of our institutions must develop a deep sense of respect towards the fundamental duties and constitutional values, bonding with one's country, and a conscious awareness of one's roles and responsibilities in a changing world. To instill a deep-rooted pride in being Indian, not only in thought but also in spirit, intellect, and deeds, as well as to develop knowledge, skills, values, and dispositions that support responsible commitment to human rights, sustainable development, and living, and global well-being, thereby reflecting a truly global citizen[4] Further, the emphasis was on developing creative individuals by addressing beyond the cognitive capabilities of literacy and numeracy to critical thinking and problem-solving.

In a nutshell, moving towards multidisciplinary universities and colleges, with more higher education institutions across India that offer a medium of instruction in local/Indian languages and internationalization of education. The NEP formulators worked on lifting higher education to global standards by advancing a more multidisciplinary undergraduate education and embedding a research culture in higher education. Increased access, equity, and inclusion were also incorporated in the vision. Independent boards with academic and administrative autonomy are also incorporated to create fair governance of higher education institutes. With a single regulator for higher education[5] ere excluded as their curriculum is different.

The policy of "light but tight" was infused. The idea was to ensure that a quality higher education must enable personal accomplishment and enlightenment, constructive public engagement, and productive contribution to society.

3.2 Structure

The new structure's goal was to create uniformity, thus addressing the fragmented system that had emerged in higher education. The attempt was to structure the system that encourages higher education and maintains its rigor yet has the scope to allow for the need for employment and income generation. A conscious decision was to make it flexible and job-relevant.

Therefore, the most significant development was in the structure formulated for higher education. The undergraduate program would be for three to four years. The fourth year would be dedicated to research orientation. The postgraduate program would be for one or two years, followed by a doctoral program of four years. An integrated five-year postgraduate program is planned to ensure certainty and compatibility with the global education system.

3.3 Higher Education Institutes[6]

A groundbreaking approach to the higher education system has been multiple entry and exit. This option allows the student to leave and return to continue studies. This addresses the discontinuation of higher studies among poor and socially disadvantaged students due to other personal commitments and societal pressure.

Thus, the undergraduate degree will be three- or 4-year, with multiple exit options within this period, with appropriate certifications. For example, a certificate after one year in a discipline or field, including vocational and professional areas. A diploma after two years of study. A Bachelor's degree after a 3-year program. The 4-year multidisciplinary Bachelor's program would lead to the same degree "with research" as deemed by the specified higher education institutes. This shall be the preferred option for those looking at postgraduate and doctoral programs.

This has led to a revolutionary digital approach to academic certification. An Academic Bank of Credit (ABC) will be established, which would digitally store the academic credits earned from various recognized higher education institutes so that the degrees from a higher education institute can be awarded, taking into account credits earned in yet another higher education institute. To create uniformity, the higher Educational Institutes would, under consultation, decide all assessment systems, including those that lead to final certification. The Choice-Based Credit System (CBCS) is reformulated to instill innovation and flexibility.[7]

Focus innovation centers would be created for start-up incubation, technology development, and new frontier areas. Academic-industry linkages will be encouraged, along with interdisciplinary research between science, technology, and liberal and humanities studies.

A new model public university called "Multidisciplinary Education and Research Universities" (MERU) will be established to attain the highest global standards in providing quality education.

The gap between education and the needs of the industry will be addressed through an Internship program for students. This will not be limited to local industry but span India and, if possible, other countries.

In its aspiration to become an international destination for education, the NEP also looks at an International Students Office at each higher education institute hosting international students. This office will be set up to coordinate all matters relating to welcoming and supporting students arriving from abroad.

The aim is to achieve a strong research-orientated higher education at par with the international education system but also to allow flexibility, such as enabling students to take care of commitments that led to discontinuation of education. The multidisciplinary approach will go a long way in integrating science, technology, and humanities, bridging a much desired element in research.

3.4 Vocational Learning

Vocational learning has also been revamped under NEP 2020. While the Bachelor's in Vocational Studies was introduced in 2013, it has always been perceived by society at large as a course that was less than that of the regular graduate program. This has been addressed in NEP 2020 by integrating it into all schools and higher education institutions. Such a change will require building infrastructure and curriculum and thus will be done in a phased manner over the next decade. MHRD will constitute a National Committee for the Integration of Vocational Education (NCIVE), consisting of experts in vocational education and representatives from across Ministries, in collaboration with industry, to oversee this effort.[8]

3.5 Internationalization

While formulating the NEP 2020, it was realized that India has a solid potential to become an education hub. The advantage of English education being at a world standard at a relatively low cost should be attractive to neighboring states and states of the Global South. Thus, keeping an eye on this, the above structure with model public universities has been designed.

Deconstructing the multiple entry and exit option and flexibility of credit transfer, from the viewpoint of attracting international students, are initiatives one can easily see that give greater mobility for both international and Indian students to study, transfer credits, and carry out research. Attractive courses and programs in subjects such as Indology, Indian languages, AYUSH systems of medicine, yoga, arts, music, history, culture, and modern India, internationally relevant curricula in the sciences and social sciences would provide opportunities for integrating and fostering global quality standards.

Indian Prime Minister Modi's aspiration to see India as Vishwa Guru is reflected in the policy. NEP 2020 singularly aspires to see India as a global study destination, providing premium education at affordable costs. A gamut of opportunities for institutes of higher education like Research/teaching collaborations and faculty/student exchanges with high-quality foreign institutions; relevant mutually beneficial "Memorandum of Understandings" with foreign countries; Leading higher education institutes would be given opportunities to set up campuses abroad. A legislative framework facilitating entries of the top 100 global universities through regulatory, governance, and content norms on par with other autonomous institutions of India will be expedited.

In the area of vocational courses, the desire is to align Indian standards with the International Standard Classification of Occupations maintained by the International Labour Organization so as to allow for the exchange of skilled labor with countries that desire such exchange.

Attentive to the SDGs and concern for global commons, as part of postgraduate and doctoral programs, internationalization will focus on environmental education will include areas such as climate change, pollution, waste management, sanitation, conservation of biological diversity, management of biological resources and biodiversity, forest and wildlife conservation, and sustainable development and living. Encouraging ethics and value development of humanistic, ethical, constitutional, and universal human values of truth (*Satya*), righteous conduct (*dharma*), peace (*shanti*), love (*prem*), nonviolence (*ahimsa*), scientific temper, citizenship values, and also life-skills will be promoted.

Globalization and the interconnected world have necessitated the formulation of Global Citizenship Education (GCED), a response to contemporary global challenges. Factoring this into the NEP 2020, it directs institutes to create curriculums that empower learners to become aware of and understand global issues and become active promoters of more peaceful, tolerant, inclusive, secure, and sustainable societies.

3.6 Capacity Building

Capacity building rests on three pillars—hard infrastructure, technology, and intellectual capacity building. Only when these are amalgamated into the capacity-building policy will the required changes envisaged in the vision come to fruition.

As far as hard infrastructure is concerned, it has to not only look at minimum facility buildings like drinking water, toilets, boards, libraries, labs, and updated classrooms but also integrate smart technologies to enable learning to happen at the global level.

Technology appropriate to the needs of the institutes would be facilitated. The core focus would be virtual Labs, the National Educational Technology Forum (NETF), and Digitally Equipping institutes, teachers, and students. NEP 2020 has only given a broad direction towards the use of technology. It is desired that each institute establish a center that will identify worthy technology and implement it in a phased manner. It is also vital to build continuous technology upgradation.

Faculty is the fulcrum of intellectual capacity. Giving due importance to this, NEP 2020 looks at establishing the National Research Foundation, the overarching goal being to grow the culture of research in all universities. National Research Foundation's major areas of operation would be funding grants for research proposals, competitive, peer-reviewed grant proposals of all types and across all disciplines, seed money to grow and facilitate research at academic institutions, and recognizing outstanding research and academic excellence in teaching. It would also act as a nodal point between researchers and relevant government agencies to allow for integrating research into applied research. Though a government agency, it would be independently governed by a Board of Governors on a rotational basis and consist of lead researchers from various higher education institutes.

MERU would be the institute that would lead the research and development. These institutes will exercise training and resource development for faculty by identifying cutting-edge research areas aligned to global research. They will also work today to identify collaborative work with international academia. Working in tandem with the National Research Foundation, these universities will direct, administer, and manage the pool of bright minds to build on India's growing reputation of quality, knowledgeable, and technologically skilled human resources.

4 Anchoring the "Indispensable Partnership" in Higher Education

India has the largest demography in the world. Further, India has advanced in providing education in English, making it a globally viable alternative. This has two strong potentials in the education sector. NEP 2020 attempts to achieve a 50% Gross Enrolment Ratio by 2035, which would also mean that a large number of students would also look at higher education abroad. With the upgradation of the educational system infrastructure and capability, India will become an attractive destination for Asian countries as Indian education is relatively low cost.

Over the last decade, Japanese universities have delivered many courses in English. This potentially opens doors for students seeking higher education abroad. The online option of rendering courses is opening up further opportunities. STEM subjects, which require lesser Japanese language capability, have the potential to draw Indian students who are conditioned toward STEM subjects. Further, as explained in the section above, as Japan gears up to use technology to make it more interactive, it can be a source of learning for the Indian education system. While NEP 2020 talks about the use of technology, it has not been defined, leaving it to the higher education institutes to find applications at the ground level.

India and Japan are interested in strengthening their educational ties to foster academic exchanges, promote research collaborations, and enhance cultural understanding. These collaborations often involve partnerships between universities, research institutions, and government bodies. In recent years, India and Japan have signed agreements and memoranda of understanding (MoUs) to facilitate cooperation in education and related areas. These agreements may cover student exchange programs, joint research projects, faculty development initiatives, and sharing best practices in education policy. Given the importance both countries place on education and human resource development, it is likely that collaborations on education policy involve discussions on curriculum development, teacher training, educational technology, and strategies for improving access to quality education.

5 Conclusion

Soft power has immense strength in improving a country's image and making a country more attractive, working in diplomatic influence, and resulting in an entrenching robust and resilient relationship among nations. Today, soft power is an integral part of any bilateral relations. India and Japan have long drawn upon their civilizational linkages to deepen their relations. As the relationship has grown, both nations, during their celebrations of 70 years of their friendship in 2022, identified people-to-people exchange as a weak link in this relationship.

In the joint statement of 2022, Prime Minister Kishida and Prime Minister Modi acknowledged the role of people-to-people exchange in vitalizing a relationship. They urged each of their governments to pay attention to it. The education sector is the linchpin of people-to-people exchange. The stakeholders of the education sector are children and youth, who are in their formative years, and exposure to the right kind has a substantial impact on the future. Moreover, the faculty is the next stakeholder. The exposure of a faculty through research and collaborations has an exponential impact on the coming generations.

As one can witness from the direction that higher education has taken in both Japan and India, there is immense potential in exploring this vertical between universities and other institutes of learning. Both governments have created the necessary environment for exploring collaboration possibilities in various forms. The significant hesitancy among universities has been due to language orientation. Students of both nations must be exposed to each other to release that English has percolated well into Japanese universities. In making of an "indispensable" partnership, appealing youth is vital. Image building among youth will go a long way in anchoring this relationship for the future.

Notes

1. National Education Policy-2020 https://www.education.gov.in/sites/upload_files/mhrd/files/NEP_Final_English_0.pdf.
2. National Education Policy-2020 https://www.education.gov.in/sites/upload_files/mhrd/files/NEP_Final_English_0.pdf.
3. The Hindu, " National Education Policy interconnects education, says Kasturirangan" https://www.thehindu.com/news/cities/Tiruchirapalli/national-education-policy-interconnects-education-says-kasturirangan/article65854092.ece.
4. New Education Policy 2020 https://www.education.gov.in/sites/upload_files/mhrd/files/NEP_Final_English_0.pdf.
5. The medical and law institutes w.
6. This refers to a whole university structure, be it central universities, state universities, private universities, or deemed universities.
7. The Salient Features of NEP 2020: Higher Education https://www.ugc.gov.in/pdfnews/5294663_Salient-Featuresofnep-Eng-merged.pdf.
8. Ibid.

References

MEXT (2005) The future of national higher education, by Ministry of Education, Sports Culture, Science and Technology (MEXT), Government of Japan. https://www.mext.go.jp/b_menu/shingi/chukyo/chukyo0/toushin/05013101.htm. Accessed 17 Feb 2024

MEXT (2011) Higher education in Japan, by Ministry of Education, Sports Culture, Science and Technology (MEXT), Government of Japan. https://www.mext.go.jp/en/policy/education/highered/title03/detail03/1374111.htm. Accessed 18 Feb 2024

MEXT (2014) Guidelines for building international joint diploma programs including double and joint degree programs, by Central Council for Education, Working Group on the Internationalization of Universities, Ministry of Education, Sports Culture, Science and Technology (MEXT), Government of Japan. https://www.mext.go.jp/component/b_menu/shingi/toushin/__icsFiles/afieldfile/2015/04/17/1356863_1.pdf. Accessed 17 Feb 2024

WEF (2020) The global risks report 2020, Published by World Economic Forum. https://www.weforum.org/reports/the-global-risks-report-2020/. Accessed 23 Dec 2022

WEF (2021) The global risks report 2021, Published by World Economic Forum. https://www.weforum.org/reports/the-global-risks-report-2021/. Accessed 23 Aug 2023

WEF (2022) The global risks report 2022, Published by World Economic Forum. https://www.weforum.org/reports/the-global-risks-report-2022/. Accessed 23 Aug 2023

WEF (2023) The global risks report 2023, Published by World Economic Forum. https://www.weforum.org/reports/the-global-risks-report-2023/. Accessed 23 Aug 2023

WEF (2024) The global risks report 2024, Published by World Economic Forum. https://www.weforum.org/reports/the-global-risks-report-2024/. Accessed 17 Feb 2024

Yonezawa A (2023) Japan's higher education policies under global challenges. Asian Econ Policy Rev 18:220–237

GPSR Compliance

The European Union's (EU) General Product Safety Regulation (GPSR) is a set of rules that requires consumer products to be safe and our obligations to ensure this.

If you have any concerns about our products, you can contact us on ProductSafety@springernature.com

In case Publisher is established outside the EU, the EU authorized representative is:

Springer Nature Customer Service Center GmbH
Europaplatz 3
69115 Heidelberg, Germany

Batch number: 10725971

Printed by Printforce, the Netherlands